Lecture Notes in Economics and Mathematical Systems

Editorial Board:

H. Albach, M. Beckmann (Managing Editor)
P. Dhrymes, G. Fandel, G. Feichinger, W. Hildenbrand
W. Krelle (Managing Editor)
H. P. Künzi, K. Ritter, U. Schittko, P. Schönfeld, R. Selten, W. Trockel

Managing Editors:

Prof. Dr. M. Beckmann
Brown University
Providence, RI 02912, USA

Prof. Dr. W. Krelle
Institut für Gesellschafts- und Wirtschaftswissenschaften
der Universität Bonn
Adenauerallee 24-42, W-5300 Bonn, FRG

M. J. Beckmann M. N. Gopalan
R. Subramanian (Eds.)

Stochastic Processes and their Applications

Proceedings of the Symposium held
in honour of Professor S. K. Srinivasan
at the Indian Institute of Technology
Bombay, India, December 27-30, 1990

Springer-Verlag
Berlin Heidelberg New York
London Paris Tokyo
Hong Kong Barcelona
Budapest

Editors

M. J. Beckmann
Brown University
Providence, RI 02912, USA
and
Institut für Angewandte Mathematik
Technische Universität München
München, FRG

M. N. Gopalan
Department of Mathematics
Indian Institute of Technology
Bombay, India

R. Subramanian
Department of Mathematics
Indian Institute of Technology
Madras, India

This Symposium was sponsored by the Indian Society for Probability and Statistics

ISBN 3-540-54635-9 Springer-Verlag Berlin Heidelberg New York
ISBN 0-387-54635-9 Springer-Verlag New York Berlin Heidelberg

Printed in Germany

Typesetting: Camera ready by author
Printing and binding: Druckhaus Beltz, Hemsbach/Bergstr.
42/3140-543210 - Printed on acid-free paper

PROFESSOR S.K.SRINIVASAN

CONTENTS

4. Operations Research

PREFACE

A volume of this nature containing a collection of papers has been brought out to honour a gentleman - a friend and a colleague - whose work has, to a large extent, advanced and popularized the use of stochastic point processes.

Professor Srinivasan celebrated his sixty first birth day on December 16,1990 and will be retiring as Professor of Applied Mathematics from the Indian Institute of Technology,Madras on June 30,1991. In view of his outstanding contributions to the theory and applications of stochastic processes over a time span of thirty years, it seemed appropriate not to let his birth day and retirement pass unnoticed. A symposium in his honour and the publication of the proceedings appeared to us to be the most natural and suitable way to mark the occasion. The Indian Society for Probability and Statistics volunteered to organize the Symposium as part of their XII Annual conference in Bombay. We requested a number of long-time friends, colleagues and former students of Professor Srinivasan to contribute a paper preferably in the area of stochastic processes and their applications. The positive response and the enthusiastic cooperation of these distinguished scientists have resulted in the present collection.

The contributions to this volume are divided into four parts: Stochastic Theory (2 articles), Physics (6 articles), Biology (4 articles) and Operations Research (12 articles). In addition the keynote address delivered by Professor Srinivasan in the Symposium is also included.

A large number of individuals have helped to make this volume a worthy scientific monument for Professor Srinivasan. First among these are the authors and the referees. Several authors have added personal comments expressing their high esteem and respect for Professor Srinivasan as a researcher,colleague and friend.

We gratefully acknowledge the financial assistance and other facilities provided by the Indian Institute of Technology, Madras for the preparation of the manuscript. We thank the Springer-Verlag for the production of this volume.

M.J.Beckmann

April, 1991. *M.N.Gopalan*

R.Subramanian

BIOGRAPHICAL SKETCH OF PROFESSOR S.K.SRINIVASAN

The following brief account of the life of Professor S.K.Srinivasan might help the reader to put him and his work in proper perspective.

Professor Srinivasan was born in Kanchipuram in Tamil Nadu, India on December 16, 1930. After completing his high school education in Cheyyar, he joined the Vivekananda college to study Mathematics and obtained his B.A. (Hons) degree of the University of Madras in 1953. Then he had to decide what to do next. This was a turning point for him and his inclination towards an academic career took shape then. The University of Madras had just started a new Department of Physics and attracted by the opportunity to work with Professor Alladi Ramakrishnan, he did his Master's programme in Theoretical Physics there, receiving his M.Sc.degree (by research) in 1955. Subsequently he continued for the Ph.D. programme with Professor Alladi Ramakrishnan as his guide and submitted his thesis entitled "Theory of random integrals and their applications to physical problems" in 1957 and was awarded the Ph.D. degree of the University of Madras in 1958. Just as his later contributions, his thesis work had a fundamental character and was directed towards physical applications.

After submitting his doctoral thesis, he went to the University of Sydney as a postdoctoral fellow, where he had the opportunity to work with Professor Messel and to familiarize himself with the first computer in Australia — SILLIAC. After a year he returned to the University of Madras, accepting a senior research fellowship of the National Institute of Sciences of India. Subsequently he joined the Indian Institute of Technology, Madras in 1959 at its very inception. Here he went on to become a professor in 1967 and a senior professor in 1974. We can only hint here at the magnitude of Professor Srinivasan's contribution to the Indian Institute of Technology, Madras in general and to the Department of Mathematics in particular. He spent a great deal of his time to improve and maintain the quality of the Department. This is true not only when he was the Head but also before and after.

Teaching ability should be measured by one's effectiveness as a teacher as well as one's contribution to course design, curriculum development and continued improvement of teaching methods. Judged from these aspects Professor Srinivasan is a very good teacher. He was particularly creative with regard to the introduction of modern subjects.

At the Indian Institute of Technology, Madras, Professor Srinivasan has served as the Head of the Department of Mathematics, as the Dean of Research Programmes and as a Member of the Board of Governors. Each position demanded a different kind of management and administrative skill and he was quite successful in each of these positions. In addition he has served in a host of other important committees.

Research competence should be measured on the basis of distinguished and continuing contribution to scholarship; contribution to scholarship is expressed mainly in the form of publications as well as evidence of ability to provide academic leadership. Judged by these norms Professor Srinivasan is an excellent researcher. In fact, the phenomenal growth of advanced research in the field of stochastic processes and their applications at the Indian Institute of Technology, Madras during the last three decades can be justifiably attributed to the dynamic and purposeful stewardship of Professor Srinivasan. He has research interests in a large number of allied areas. To his credit he has 10 books and 170 research papers in established journals. He has successfully guided 20 students in their Ph.D. programme. One of them got his degree in Electrical Engineering, one in Civil Engineering, two in Physics and the rest in Mathematics.

No wonder several honours and distinctions came to Professor Srinivasan in the course of his brilliant and illustrious career as a scientist. Many of his publications have had wide recognition and have been quoted by several well-known authors. Professor Srinivasan has had visiting assignments at various Universities abroad: University of Sydney, State University of New York at Buffalo, University of Waterloo, National University of Singapore, Technical University of Munich and University of Texas at Austin. Professor Srinivasan is a life member of the Indian Mathematical Society, Operational Research Society of India (President, 1975-1976) and Bernoulli Society of Probability and Mathematical Statistics, an elected Fellow of the Indian Academy of Sciences and a founder Fellow of the Tamil Nadu Academy of Sciences. He was on the editorial board of "OPSEARCH" (1973-1977) and is a founder editor of the "Journal of Mathematical and Physical Sciences" now in its 24th year of publication. He is on the editorial board of "Solid Mechanics Archives" and "Annals of the Institute of Statistical Mathematics" and is a member of the Research Board of Advisers of the American Biographical Institute. Professor Srinivasan is listed in several directories of men of achievement and distinction.

His background in Physics combined with mathematical insight enables him to understand and relate to people interested in the applications of stochastic processes — probabilists, engineers, physicists, operations researchers etc. — and to bridge the gap among them. Professor Srinivasan feels completely at home in this diverse community.

Professor Srinivasan is a original thinker and his work is always of the highest quality. His papers are marked by their clarity and originality. He is one of the early workers in the field of stochastic processes and to him are due some important results. A characteristic of Professor Srinivasan is the strong application bias, maintaining at the same time strong theoretical support and mathematical rigour for his work. His scientific curiosity and creativity seem to increase rather than decline over the years. In addition he was and continues to be tremendously helpful in advising and influencing many graduate students. He is selfless and extraordinarily generous in giving his time and concern to his students and colleagues.

Professor Srinivasan is one of the most intellectual persons I have come across. I do not know anyone else who has as many interests as he and pursues them so deeply. He reads books like others read news papers. He is a great lover of music and likes gardening. He is a devoted walker. He is very pleasant and amiable. Any one who came into contact with him cannot but recall his acts of kindness and love. His warmth and affection reach out to any one who needs them. I am touched by the feelings expressed by a number of people. Discussions with them revealed a person who sometimes differed from the one I know. It is probably true that no one ever thoroughly knows another person in all his aspects and it made me feel good that there are other aspects to Professor Srinivasan's personality that others had seen. He is truly a person worth knowing.

It is very difficult to express how I admire him. The qualities of devotion, generosity, honesty and kindness are actualized in him, not from time to time but consistently in his daily life. He always sets the highest standards for himself, his collaborators and his students. He is an outstanding scholar and teacher. His career is a shining example of what a sincere and dedicated academician can achieve in this country in spite of the unsurmountable obstacles in the path of progress.

R. Subramanian.

SCIENTIFIC CONTRIBUTIONS OF PROFESSOR S.K.SRINIVASAN

All of Professor Srinivasan's work is in the area of Applied Probability. Even so, it is rather difficult to summarize his contributions, as they are scattered in different areas like the theory of stochastic processes and their applications to physics, operation research, biology and engineering. His work in the fifties is a consequence of his intimate contact with Professor Alladi Ramakrishnan and resulted in his doctoral dissertation on the theory and applications of stochastic processes in physics and allied fields which in turn laid strong foundations for his work in cascade theory, stochastic integrals and differential equations dealing with noise phenomena. His familiarity with cascade shower theory naturally provided the motivation and desire for his investigations in high energy physics, pion physics and scattering theory. While he had adequate reasons for pursuing theoretical physics, with the appointment to a full professorship in applied mathematics at the Indian Institute of Technology, Madras in 1967, he had shifted his interests to other typical applied probability areas like operations research and biology, besides noise and fluctuating phenomena where there was abundant scope for the study of specific non-Markov processes.

Theory of Point Processes:

Professor Srinivasan's contributions in the area of point processes are manifold - extension of the theory of product densities to the multidimensional and non-Euclidean spaces on the one hand and irregular point processes on the other. His doctoral work on cascade theory was developed further in the early sixties in collaboration with Professor Iyer, Dr.Koteswara Rao and Professor Vasudevan and this led to the evolutionary sequent correlation functions as a means for the study of the general theory of point processes. His early research work on twins and multiplets in the age dependent population growth culminated in 1961 in the formal theory of multiple point processes and multiple product densities. Apart from this he had introduced several innovations for the study of special point processes and in particular he characterized the space charge limited shot noise processes (now studied in the literature as inhibited point processes) and the resulting cumulative response processes. A significant outcome of his investigations is the emergence of the new formalism for the study of the kinetic theory, fluctuation phenomena and onset of turbulence apart from original ideas on second order point processes. The formulation of stochastic integrals and differential equations through the sample path analysis engaged his attention in the fifties and the encouragement of Professor Richard Bellman and his teacher Professor Alladi Ramakrishnan enabled him to take up a systematic study of the processes arising in physics and engineering (Srinivasan and Vasudevan, 1971). His early work in this relates to the characterization of processes arising from differential equations where the forcing terms are stochastic in nature (Srinivasan and Mathews , 1956, Srinivasan, 1960). With the advent of McShane-Ito calculus he was naturally led to many innovations culminating in an appropriate definition of stochastic differentials of jump processes, more specifically of point processes (Srinivasan, 1977, 1982).

In the late sixties he started investigating interacting point processes with Dr.Rajamannar and Dr.Rangan and made innovations in modelling firing sequence of single neurons. The interacting point processes or for that matter even renewal processes are a bit tricky to handle; his main contribution lies in the discovery of the appropriate regenerative structure which in turn enables one to obtain the characteristics of the process. These are codified in his work on single neurons.

Cosmic Rays and Elementary Particle Physics:

The cascade theory of cosmic ray showers as formulated by Bhabhaand Heitler deals with the distribution of particles over energy at a particular thickness. Motivated by the experimental findings in the mid-fifties, Professor Srinivasan in collaboration with Professor Ramakrishnan came up with a modified approach and expressed the expected value of the size of the shower in terms of production product densities defined over a two dimensional continuum. The estimation of the fluctuation of the size, which was normally considered to be formidably difficult to compute, was carried out in collaboration with Professor Iyer and Dr.Koteswara Rao (Srinivasan, 1961, Srinivasan and Iyer and Koteswara rao, 1964) through a set of compact equations obtained by the use of the invariant imbedding technique. This work on high energy cosmic rays has stimulated him further to take up investigations in high energy physics where he made significant contributions in the area of multiple production of particles. Some of his recent contributions in the area of multiplicity distribution and quark gluon cascade are further innovations.

Operations Research:

In the sixties he was attracted to the area of operations research due to the rich non-Markov structures present in the stochastic models of queues, inventories and reliability. Partly independently and partly in collaboration with Professor Subramanian and Professor Kalpakam, he developed a unified approach by discovering regenerative structures in the stochastic processes governing these models. We wish to make a special mention of the complete solution for the most general type of (s,S) inventory model (Srinivasan,1979) and the analysis of redundant systems (Srinivasan and Subramanian, 1980). All along Professor Srinivasan had the conviction that the study of point processes would lead to the characterization of non-Markov processes. Here he was able to satisfy himself and also convince fellow workers in ample measure.

Optics:

In the area of optics, he was drawn towards the statistical detection theory. His work in the sixties relates to the correlational structure of the point process of photo electron emissions of partially coherent light (Srinivasan and Vasudevan,1971). He then undertook detailed investigations leading to the determination of photo electron statistics of light beam with arbitrary spectral profile (Srinivasan and Sukavanam, 1971, Sudarshan *et al.*, 1973, Srinivasan, 1974). During the past seven years he is working on the population theoretic approach to the problem of light and its detection; as a result of his recent findings we now have a complete population point processes theory of light and its detection. An important outcome of this investigation is the identification of a class of doubly stochastic Poisson processes with the point process generated by the emigrations of a population process (Srinivasan, 1988). His latest work relating to intermittency in particle production and his projected work on quantum optics will introduce new vistas in the field.

R. Subramanian.

SCIENTIFIC PUBLICATIONS OF PROFESSOR S.K. SRINIVASAN : BOOKS

1969 1. **STOCHASTIC THEORY AND CASCADE PROCESSES.** American Elsevier.

1971 2. With R. Vasudevan. **INTRODUCTION TO RANDOM DIFFERENTIAL EQUATIONS AND THEIR APPLICATIONS.** American Elsevier.

1974 3. **STOCHASTIC POINT PROCESSES AND THEIR APPLICATIONS.** Griffin, London.

4. **STOCHASTIC POINT PROCESSES.** *Solid Mechanics Division*, University of Waterloo.

1977 5. With G. Sampath. **STOCHASTIC MODELS FOR SPIKE TRAINS OF SINGLE NEURONS.** Springer.

1980 6. With R. Subramanian. **PROBABILISTIC ANALYSIS OF REDUNDANT SYSTEMS.** *Lecture Notes on Economics and Mathematical Systems.* **175**, Springer-Verlag.

1981 7. With K.M. Mehata. **INTRODUCTION TO PROBABILITY AND RANDOM PROCESSES.** Tata McGraw-Hill, New Delhi. Second edition.

1982 8. **StOCHASTIC CALCULUS AND MODELS AND THEIR APPLICATION TO SOME PROBLEMS OF PHYSICS.** *Lecture Notes* **15**, Department of Mathematics, National University of Singapore, 203-263.

1987 9. With K.M. Mehata. **STOCHASTIC PROCESSES.** Tata McGraw-Hill, New Delhi. Second Edition.

10. **POINT PROCESS MODELS OF CAVITY RADIATION AND DETECTION**, Griffin, London.

SCIENTIFIC PUBLICATIONS OF PROFESSOR S.K.SRINIVASAN : RESEARCH PAPERS

STOCHASTIC PROCESSES AND APPLICATIONS

1956 1. With P.M.Mathews. Stochastic processes associated with integrals of a class of random functions. *Proc. Nat. Inst.Sci.* A **22**, 369-376.

2. With P.M.Mathews. Ordinary linear differential equations involving random functions. *Proc. Ind. Acad. Sci. A* **43**, 4-20.

3. With Alladi Ramakrishnan. On stochastic integrals associated with point processes. (in French) *Publ. Inst. Univ. Paris*, **5**, 95-106.

1962 4. Multiple stochastic point processes. *Zast. Mat.* **6**, 209-219.

1963 5. On a class of stochastic differential equations. *ZAMM* **43**, 259-265.

1964 6. With K.S.S.Iyer. Sequent correlations in evolutionary stochastic point processes and its application to cascade theories. *Nuovo Cimento* **33**, 273-285.

1966 7. With K.S.S.Iyer. Random processes associated with random points on a line. *Zast. Mat.* **8**, 221-230.

8. On a class of non-Markovian processes and its application to theory of shot noise and Barkhausen noise. in *Alladi Ramakrishnan (ed.) Symp. in Theo. Phy.* **3**, 107-120, Plenum Press, New York.

9. Sequent correlations in evolutionary stochastic point processes. *Symp. Theo. Phys.* **4**, 143-155, Plenum Press, New York.

10. Stochastic random integration and differential equations - a physical approach. *in Alladi Ramakrishnan (Ed.) Symp. Theo. Phys. and Maths.* **9**, 109-129, Plenum Press, New York.

11. With R.Vasudevan and N.V.Koteswara Rao. Sequent correlations in stochastic point processes-II. *Nuovo Cimento* **44**, 818-821.

1967 12. Stochastic point processes:A phenomenological approach. *J. Math. Phy. Sci.*. **1**, 1-39.

13. With Alladi Ramakrishnan and R.Vasudevan. Multiple product densities. *J. Math. and Phy. Sci.* **1**, 275-279.

1968 14. With N.V.Koteswara Rao and R.Vasudevan. Sequent correlations in stochastic point processes-III. *Nuovo Cimento.* **60 B**, 189-192.

15. With N.V.Koteswara Rao. Invariant imbedding technique and age-dependent birth and death processes. *J. Math. Analy. and Applns.*. **21**, 43-52.

1971 16. With S.Kumaraswamy. Delayed events and cluster processes. *J. Math. Phy. Sci.* **5**, 229-238.

1972 17. Correlation structure of a pair of interacting point processes. *J. Math. Phy. Sci.* **6**, 163-179.

1973 18. Stochastic point processes and their applications. *J. Math. Phy. Sci.* **7**, S107-S113.

1974 19. General theory of stochastic processes. *Stochastic problems in mechanics* **10**, 1-36.

1978 20. Stochastic integrals. *SM Archives* **3**, 325-379.

1980 21. With S.Udayabhaskaran. On a stochastic integral associated with a birth and death process. *J. Math. Phy. Sci.* **14**, 95-106.

1981 22. With G.Sampath. On a cumulative process resulting from interacting renewal processes and its applications. *J. Math. Phy. Sci.* **15**, 39-46.

23. With M.Gururajan. A two-valued process generated by a doubly stochastic Poisson process. *J. Math. Phy. Sci.* **15**, 297-313.

1982 24. With S.Udayabhaskaran. A generalized branching process and a stochastic integral associated with it. *J. Math. Phy. Sci.* **16**, 367-382.

25. With S.Udayabhaskaran. Modelling and analysis of dynamical systems subject to discontinuous noise processes. *J. Math. Phy. Sci.***16**, 415-430.

1984 26. Stochastic Calculii and models of physical phenomena. *J. Math. Phy. Sci.* **18**, S163-S168.

STATISTICAL MECHANICS AND FLUCTUATION PHENOMENA

1956 1. With Alladi Ramakrishnan. Correlation problems in the study of the brightness of the milky way. *Astrophys. J.* **123**, 479-485.

1963 2. On Katz's model for emulsion polymerization. *J. Soc. Indust. Appld. Math.* **11**, 355-359.

1965 3. A novel approach to the theory of shot noise. *Nuovo Cimento.* **38**, 979-992.

1966 4. With R.Vasudevan. On a class of non-Markovian processes associated with correlated pulse trains and their application to Barkhausen noise. *Nuovo Cimento.* **41**, 101-112.

5. A novel approach to the kinetic theory and hydrodynamic turbulence. *Zeit. Phys..* **193**, 394-399.

6. A novel approach to the kinetic theory of fluids-onset of turbulent motions. in *Alladi Ramakrishnan (ed.) Symposia in Theoretical Physics and Mathematics.* **7**, 163-186, Plenum Press, New York.

1966 7. A novel approach to the kinetic theory and hydrodynamic turbulence II. *Z. Phys.* **197**, 435-439.

1967 8. With K.M. Mehata. A Markov chain model for packed bed. *A.I.Ch.E. J.* **18**, 650-652.

9. Generalized Boltzman equation and kinetic theory of fluids. *in Alladi Ramakrishnan (ed.) Symp. Theo. Phys.* **7**, 163-186, Plenum Press, New York.

10. Fluctuating density fields and Fokker-Planck equations. *in Alladi Ramakrishnan (ed.) Symp. Theo. Phys.*, **5**, 163-166, Plenum Press, New York.

11. Theory of turbulence. *Zeit. Phys.*. **205**, 221-225

12. With R.Subramanian and S.Kumaraswamy. Response of linear vibratory systems to non-stationary stochastic impulses. *Jour. Sound Vib.* **6**, 169-179.

13. With R.Vasudevan. Fluctuating density fields. *Ann. Inst. H. Poincaré.* **7**, 303-318.

14. With Alladi Ramakrishnan and R.Vasudevan. Angular correlations in brightness of milky way. *J. Math. Phy. Sci.* **1**, 75-84.

1969 15. With R.Vasudevan. Stochastic kinetic equations and particle statistics. *Ann. Inst. H. Poincaré.* **A 10**, 419-429.

16. With R.Vasudevan. Response output from nonlinear switching elements with different types of finite dead times. *Kybernatik.* **6**, 121-124.

17. With S.Kumaraswamy and R.Subramanian. Unsteady motion of a viscous fluid subjected to random pulses. *Arch. Mech. Stos.* **21**, 191-198.

1970 18. With S.Kumaraswamy. Characteristic functional of a non-stationary shot noise. *Trans ASME, J. Appl. Mech.* **37**, 543-544.

19. With R.N.Sarkar. Vector-Meson contribution to the nucleon anomalous magnetic moments in Drell-Pagels' model. *J. Math. Phy. Sci.* **4**, 169-179.

1971 20. Stochastic point processes and statistical physics. *J. Math. Phy. Sci.* **5**, 291-316.

1972 21. With K.M. Mehata. A stochastic model for polymer degradation. *J. Appl. Prob.* **9**, 43-53.

1976 22. Stochastic models for fatigue failure of materials. *S.M. Archives.* **1**, 3-25.

1977 23. With S.Kumaraswamy. Two simple models for Gaussian noise. *J. Math. Phy. Sci.* **11**, 199-207.

1980 24. With S.Udayabhaskaran. Stochastic perturbation of a stable deterministic Plasma system. *Phys. Lett.* **A 80**, 387-389.

1987 25. With R. Subramanian. Stochastic models for fatigue failure of materials II: a discrete approach. *S M Archives*. **12**, 125-137.

PHYSICS OF FUNDAMENTAL PARTICLES

1957 1. With Alladi Ramakrishnan, N.R. Ranganathan and R. Vasudevan. Multiple processes in electron photon cascades. *Proc. Ind. Acad. Sci.*. **45 A**, 311-326.

1958 2. A note on charge independence and nucleon-antinucleon interactions. *Proc. Ind. Acad. Sci.*. **47 A**, 365-368.

1959 3. Applications of charge independence to annihilation of anti-nucleons. *Proc. Nat. Inst. Sci.*. **25 A**, 99-103.

4. With Alladi Ramakrishnan and N.R. Ranganathan . Meson production in nucleon nucleon collisions. *Nucl. Phys.*. **10**, 160-165.

5. With K. Venkatesan. Photo production of pion pairs by a nucleon. *Nucl. Phys.*. **12**, 418-525.

6. With Alladi Ramakrishnan, N.R. Ranganathan and K. Venkatesan. Photo-mesons from polarized nucleons. *Proc. Ind. Acad. Sci.*. **49 A**, 302-305.

7. With Alladi Ramakrishnan and N.R. Ranganathan. A note on the interaction between nucleon and anti-nucleon. *Proc. Ind. Acad.*. **50 A**, 91-94.

8. With K. Venkatesan. Photo production of pion pairs by polarized nucleons. *Proc. Ind. Acad. Sci.*. **50 A**, 392-397.

1960 9. With Alladi Ramakrishnan, N.R. Ranganathan and R. Vasudevan. A note on dispersion relations. *Nucl. Phys.*. **15**, 516-518.

1962 10 With K. Venkatesan. Multiple pion production in the sub-GeV region. *Nucl. Phys.*. **29**, 335-340.

1963 11. With K. Venkatesan. Angular momentum analysis of photo production of pion pairs by a nucleon. Nucl. Phys.. **48**, 337-344.

12. With K. Venkatesan. Photo production of pions from nucleons in the strip approximation. *Nuovo Cimento*. **30**, 151-162.

13. With K. Venkatesan. The strip approximation and the photo production of pions on pions *Nuovo Cimento*. **30**, 163-170.

1966 14. With P. Achuthan. Photo production of vector mesons. *Nucl. Phys.*. **76**, 638-656.

15. With P. Achuthan and V. Ramabhadran. Charge dependent corrections in photo production of positive pions. *Nuovo Cimento*. **42**, 997-999.

16. With Alladi Ramakrishnan and R. Vasudevan. Scattering phase shift in stochastic fields. *Zeit. Phys.* **196**, 112-122.

1967 17. With P. Achuthan. On the pionic production of eta meson. *Nuovo Cimento.* **48,** 1124-1136.

18. With P. Achuthan. Photo production of vector mesons-II. *Nucl. Phys..* **87,** 605-617.

1969 19. With P. Achuthan and R. N. Sarkar. On the photo production of eta - mesons. *Nuovo Cimento.* **59 A,** 171-180.

1970 20. With R. N. Sarkar. Vector meson contribution to the nucleon anomalous magnetic moments in Drell-Pagels' model. *J. Math. Phys. Sci.* **4,** 169-179.

1988 21. With R. Vasudevan. Multiplicity distributions and natural scaling. *Int. Math. Sci.* Preprint No. **88-011**.

1989 22. Multiplicity distribution, quark-gluon cascade and natural scaling. *Proc. of* **X** *Symp. High Energy Physics.*

23. With R. Vasudevan. A model of jet fragmentation and hadronization. (to be published.)

24. With R. Subramanian and V. Sridharan. Jet fragmentation and hadronization. (to be published).

THEORY OF COSMIC RAYS

1954 1. With Alladi Ramakrishnan. Two simple stochastic models of cascade multiplication. *Prog. Theo. Phys.* **11,** 595-603.

1955 2. With Alladi Ramakrishnan. Fluctuations in the number of photons in an electron photon cascade. *Prog. Theo. Phys..* **13**, 93-99.

1956 3. With Alladi Ramakrishnan. A new approach to cascade theory. *Proc. Ind. Acad. Sci..* **A 44**, 263-273.

1957 4. With N. R. Ranganathan. Numerical calculations on the new approach to cascade theory-I. *Proc. Ind. Acad. Sci..* **A 45,** 69-73.

5. With Alladi Ramakrishnan. A note on cascade theory with ionisation loss. *Proc. Ind. Acad. Sci..* **A 45**, 133-138.

6. With N. R. Ranganathan. Numerical calculations on the *new* approach to cascade theory-II. *Proc. Ind. Acad. Sci..* **A 45,** 268-272.

1958 7. With J. C. Butcher, B. A. Chartres and H. Messel. Numerical calculations on the new approach to cascade theory III. *Nuovo Cimento.* **9,** 77-84.

1961 8. Fluctuation problem in electromagnetic cascades. *Zeit. Phys..* **161,** 346-352.

9. A simple stochastic model of cascade multiplication with ionization loss. *Science and Engineering.* **2**, 20-26.

10. With Alladi Ramakrishnan. A note on electron photon showers. *Nucl. Phys..* **25**, 152-154.

1963 11. Electromagnetic cascades and polarization of the medium. *Nucl. Phys.*. **41**, 202-209.

12. With Alladi Ramakrishnan and R. Vasudevan. Some new mathematical features in cascade theory. *Proc. Int. Conf. on Cosmic Rays, Jaipur.* **5**, 498-501.

13. With K.S.S.Iyer and N.V.Koteswara Rao. Fluctuation problem in electromagnetic cascades. *Proc. Int. Conf. on Cosmic Rays, Jaipur.* **5**, 502-505.

1964 14. With K.S.S.Iyer and N.V.Koteswara Rao. Fluctuation problem in electromagnetic cascades II. *Zeit. Phys.*. **177**, 164-173.

15. With K.S.S.Iyer. On bursts produced by muons and electrons. *Nuovo Cimento.* **34**, 67-76.

1965 16. With K.S.S.Iyer. Fluctuation problem in electromagnetic cascades III. *Zeit. Phys.*. **182**, 243-256.

17. With Alladi Ramakrishnan and R.Vasudevan. Some new mathematical features in cascade theory. *Jour. Math. Anal. and Appl.*. **11**, 278-289.

1966 18. Recent mathematical developments in cascade theory. *in Alladi Ramakrishnan (ed.) Symp. Theo. Phys.* **2**, 195-208, Plenum Press, New York.

OPERATIONS RESEARCH

1967 1. With N.V.Koteswara Rao. Non-Markovian processes in dams and storage systems. *J. Math. Phy. Sci.* **1**, 180-189.

1969 2. With R.Subramanian. Queing theory and imbedded renewal processes. *J. Math. Phy. Sci.* **3**, 221-244.

1971 3. With S.Kalpakam. Imbedding equations in queueing theory. *J. Math. Anal. Applns.* **33**, 435-442.

4. With R.Subramanian and K.S.Ramesh. Mixing of two renewal processes and its applications to reliability theory. *IEEE Trans. on Reliability* **R 20**, 51-55.

5. With S.Ramani. A continuous storage model with alternating random input and output. *J.Hydrology.* **13**, 343-348.

6. Emptiness of a finite dam - time dependent theory. *Mathematical models in Hydrology-Proc. Warsaw Symp.*, **100**, 395-400.

1972 7. With R.Subramanian and R.Vasudevan. Correlation functions in queueing theory. *J.Appl. Prob.* **9**, 604-616.

1973 8. With M.N.Gopalan. Probabilistic analysis of a two-unit system with a warm standby and a single repair facility. *Operations Research* **21**, 748-754.

9. With M.N.Gopalan. Probabilistic analysis of a two-unit cold standby system with a single repair facility. *IEEE Trans. on Reliability.* **R 22**, 250-254.

1974 10. Analytic solution of a finite dam governed by a general input. *J. Appl. Prob.* **11**, 134-144.

1975 11. With S.Kalpakam. Finite dam with deterministic input. *J. Math. Phy. Sci.* **9**, 31-48.

1976 12. With N.Venugopalacharyulu. Analytic solutions of a finite dam with bounded input. *J. Math. Phy. Sci.* **10**, 141-164.

1977 13. With R.Subramanian. Availability of 2-unit systems with one repair facility. *J. Math. Phy. Sci.* **11**,331-349.

1979 14. With D.Bhaskar. Probabilistic analysis of intermittently used systems. *J. Math. Phy. Sci.* **13**,91-105.

15. General analysis of s-S inventory systems with random lead times and unit demands. *J. Math. Phy. Sci.* **13**, 107-129.

16. With D.Bhaskar. Analysis of intermittently used 2-unit redundant systems with a single repair facility. *J. Math. Phy. Sci.* **13**, 351-366.

17. With D.Bhaskar. Analysis of intermittently used 2-unit redundant systems with a single repair facility. *Microelectron Reliab.* **19**, 247-252.

1980 18. With C.Chudalaimuthupillai. Correlation functions in the G/M/1 system. *Adv. Appl. Prob.* **12**, 530-540.

1981 19. With D.Bhaskar. Intermittently used two unit system in warm standby. *J. Math. Phy. Sci.* **15**, 477-497.

1983 20. Reliability analysis of Redundant systems: A review of the state of art; Pacific-Asia Regional Conference on Operations Research, 1982.

21. Reliability Analysis of an intermittently used system *Microelectronics Reliab.* **23**, 295-308.

22. Analysis of a 2-unit parallel redundancy with an imperfect switch. *Microelectronics Reliab.* **23**, 309-318.

23. Analysis of (s,S) inventory systems with general lead time and demand distributions. *Research Report* **No.88**, Department of Mathematics, National University of Singapore.

1987 24. With Martin J.Beckmann. An (s,S) inventory system with Poisson demand and exponential lead time. *O.R. Spectrum.* **9**, 213-217.

1988 25. With N. Ravichandran. Analysis of (s,S) inventory systems of decaying items. *Engineering Cost and Production Economics.* **15**, 433-439.

26. Analysis of (s,S) inventory system with general lead time and demand distribution and adjustable reorder size. *Optimization* **19**, 557-576.

27. With N. Ravichandran. Multi-item (s,S) inventory model with Poisson demand, general lead time and adjustable reorder size. *Working paper* **No.762**, *IIM Ahmedabad.*

1990 28. With K.P. Kistner and R. Subramanian. Control of repair rates of cold standby systems. (to be published).

BIOLOGY

1958 1. With Alladi Ramakrishnan. On age distribution in population growth. *Bull. Math. Biophysics.* **20**, 289-303.

1969 2. With R. Vasudevan. On the response output from non-linear switching elements with different types of finite dead times. *Kybernetik.* **6**, 121-124.

1970 3. With G. Rajamannar. Renewal point processes and neuronal spike trains. *Math. Biosci.* **6**, 331-335.

4. With A. Rangan. Age dependent stochastic models for phage reproduction. *J. Appl. Prob.* **7**, 251-261.

5. With G. Rajamannar. Counter models and age-dependent renewal point processes related to neuronal firing. *Math. Biosci.* **7**, 27-39.

6. With G. Rajamannar. Selective interaction between two independent stationary recurrent point processes. *J. Appl. Prob.* **7**, 476-482.

7. With A. Rangan. Stochastic models for phage reproduction *Math. Biosci.* **8**, 295-305.

8. With A. Rangan. A stochastic model for the quantum theory in vision. *Math. Biosci.* **9**, 31-36.

1971 9. With A. Rangan and G. Rajamannar. Stochastic models for neuronal firing. *Kybernetik* **8**, 188-193.

1975 10. With G. Sampath. A neuron model with pre-synaptic deletion and post-synaptic accumulation, decay and threshold behaviour. *Bio. Cybernetics.* **19**, 69-74.

1976 11. With G. Sampath. On a stochastic model for the firing sequence of a neuron. *Math. Biosci.* **30**, 305-323.

1977 12. A stochastic model of neuronal firing. *Math. BiOsci.* **33**, 167-175.

1981 13. With C.R. Ranganathan. An age-dependent stochastic model for carcinogenesis. *Math. Biosci.* **57**, 155-174.

1982 14. With C.R.Ranganathan. On the parity of individuals in birth and death processes. *Adv. Appl. Prob.* **14**, 484-501.

1983 15. With C.R.Ranganathan. Parity dependent population growth models. *J. Math. Phy. Sci.* **17**, 279-292.

16. With C.R.Ranganathan. A Compartmental model with particles branching according to a general branching process. Report No.**91**, Department of Mathematics, National University of Singapore, 273-289.

1987 17. Analysis of age-specific population growth by phase approach. (to be published.)

18. With R.Subramanian. The phase approach to modelling of bacterial life cycle. (to be published)

OPTICS

1967 1. With R. Vasudevan. Fluctuation of photoelectrons and intensity correlation of light beams. *Nuovo Cimento* **B** 47, 185-193.

1968 2. With R.Vasudevan. Photo electron statistics due to mixing of different types of fields. *Nuovo Cimento.* **8 B**, 278-282.

1971 3. With S.Sukavanam. Photocount statistics of Gaussian light of arbitrary spectral profile. *Phys. Lett.* **35 A**, 81-82.

1972 4. With S. Sukavanam. Photocount statistics of Gaussian light with arbitrary spectral profile. *J. Phys.* **A5**, 682-694.

1973 5. With S.Sukavanam and E.C.G.Sudarshan. Many-time photocount distributions. *J. Phys.* **A6**, 1910-1918.

1974 6. Photocount statistics of Gaussian light of time limited spectral profile. *J. Phys. Lett.* **A47**, 151-152.

7. Dead time effects in photo counting statistics. *P*hys. Lett. **50 A**, 277-278.

1978 8. Dead time effects in photon counting statistics. *J. Phys.***A 11**, 2333-2340.

9. With S.Sukavanam. Approximate two-fold generating function of photocounts of Gaussian light. *Phys. Lett.* **A 66**, 164-166.

1979 10. With S.Udayabhaskaran. Canonical extension of a problem in parametric frequency up-conversion with stochastic pumping. *Optica Acta.* **26**, 1535-1540.

1981 11. With M.M.Singh. Counting statistics of stationary Gaussian field with Brilloui spectrum. *Optica Acta.* **28**, 1619-1635.

12. With M.M.Singh. Dead time corrected clipped correlations of Gaussian light. *Phys. Letters.* **A 86**, 409-413.

1982 13. With S.Sukavanam. Photocount statistics of Gaussian light with analytic spectral profile. *J. Math. Phy. Sci.* **16**, 75-86.

1983 14. Dead time corrections to photo counts and clipped correlations of Gaussian light. *Pramana J. Phys.* **20**, 547-558.

1984 15. A canonical model of optical up-conversion with stochastic pumping - exact results. *Optica Acta.* **31**, 785-793.

1985 16. With R.Vasudevan. A non-Markovian model of photo detection of cavity radiation. *Optica Acta.* **32**, 749-766.

1986 17. With R.Vasudevan. A non-Markov model of cavity radiation and its detection. *Optica Acta.* **33**, 191-205.

18. With R.Vasudevan. Approach to equillibrium of single-mode cavity radiation. *J. Math. Anal. Appln.* **119**, 249-258.

19. Analysis of characteristics of light generated by a space-charge-limited electron stream. *Optica Acta.* **33**, 207-211.

20. Generation of anti-bunched light by an inhibited Poisson stream. *Optica Acta.* **33**, 835-842.

21. A model of cavity radiation with anti-bunched and sub-Poissonian characteristics. *J. Phys. A.* **19**, L595-L598.

22. An age-dependent model of cavity radiation and **its** detection. *Pramana J. Phys.* **27**, 19-31.

23. A non-Markov model of cavity radiation with thermal characteristics. *J. Phys. A.* **19**, L513-L516.

1987 24. A non-Markov model of cavity radiation II. Antibunched and sub-Poissonian statistics. *J. Mod. opt.* **34**, 291-305.

25. With R.Vasudevan. The density matrix approach to cavity radiation and population point process models. *J. Mod. Opt.* **34**, 1545-1557.

26. An age-dependent model of cavity radiation and its detection II: anti-bunched and thermal characteristics. *Pramana J. Phys.* **30**, 59-70.

1989 27. Population point process model of antibunched light and its spectrum. *J. Phys. Lett.* **A 22**, L259-L263.

28. With V.Sridharan. A population model of thermal radiation. *J. Mod. Opt.* **36**, 711-724.

1990 29. With V.Sridharan. Multiphase evolution of population and its applications to optics and colliding-beam experiments. *J. Phys. A* **23**, 491-508.

30. Squeezing of light within the framework of the population : A theoretic approach. *J. Phys. Lett.* **A 23**, L369-L373.

FIFTY YEARS OF CASCADE THEORY : STOCHASTIC METHODS AND THEIR RAMIFICATIONS

S.K.SRINIVASAN
DEPARTMENT OF MATHEMATICS
INDIAN INSTITUTE OF TECHNOLOGY
MADRAS - 600 036
INDIA.

Keynote address delivered at the Symposium

1. Introduction

Cascade theory of cosmic ray showers has provided very good motivation for the development of the general theory of stochastic processes. Cosmic ray showers have been observed around 1930 when stochastic processes have not been studied in depth. The experimental data observed around that period needed a comprehensive stochastic model to comprise in itself the twin aspect of branching with respect to one parameter (energy) and evolution with respect to a different parameter (thickness of absorber). Unfortunately the study of branching processes was taken up by probabilists only much later; the celebrated work of Kolmogorov and Dmitriev leading to the limiting behaviour of population size emerged around 1947. The early work on point processes dates around 1943 when Palm published his work on streams of calls in a telephone exchange. Although population theory can be traced to the time of Galton and Watson, stochastic problems of population growth received attention only during forties. Thus when the experimental data on cosmic ray showers confronted the people around thirties, there was a search for a possible interpretation in terms of a statistical theory of population of points with specific labels. In 1937 Bhabha and Heitler and Carlson and Oppenheimer independently came out with their formulation in terms of what are now known as inclusive probability measures and thus laid a strong foundation for the theory of population point processes. Bhabha soon realized that his formulation of cascade showers was only a first step and was on the quest for further tools for the determination of the fluctuation in the size of the shower about its average value. His efforts culminated in a paper (jointly authored with Ramakrishnan) which provided a frame work for the calculation of the size fluctuation of showers. The work, from the probabilist's point of view, is highly significant in as much as it provided an explicit example of a bivariate population point process which is evolutionary rather than stationary in character. Both Bhabha and Ramakrishnan realized the importance and relevance of the work to the general theory of stochastic processes and soon independently published their theory of point distributions. By the time these results were published, there had been significant developments in the theory of population growth partly due to axiomatization by Kolmogorov and partly due to investigations in Galton-Watson processes. David Kendall had just then introduced his cumulant functions to deal with age structure in stochastic population growth. Quite unaware of Palm's contributions, Bellman and Harris developed the method of regeneration point by dealing with age-dependent population growth and were ultimately led to versions of limit theorems on population size of a nature more general than those attempted by Kolmogorov and Dmitriev (1947). After the publication of the results by Bhabha and Ramakrishnan, Bartlett and Kendall (1951) reformulated the problems in terms of differential equations satisfied by the characteristic functional of the population which has meanwhile been introduced by LeCam and Bochner. By an elegant application of the regeneration point technique , they were able to derive the differential equation satisfied by the characteristic functional which in turn yielded the Bhabha-Ramakrishnan equations in the context of shower theory on the one hand and the Bellman Harris age-dependent population growth equations on the other.

The story of cascade theory of showers is a long and interesting one to narrate; it cuts across many areas of the theory of stochastic processes and is highly suggestive of many other applications in other disciplines as well. The theory of population point processes which was the main offshoot remained to be an area of fruitful research until quite recently. Hence I feel it is worthwhile to present, in proper prospective, a brief account of the manner in which cascade theory of cosmic ray showers evolved and led to a rich theory of population point process.

2. Electron Photon Showers

To appreciate the impact of cascade theory on the development of the theory of stochastic processes, it is necessary to recall some of the salient features of the (experimental) cosmic ray research around 1912 when the extra terrestrial origin of radiation in the atmosphere was inferred from the baloon experiments of Victor Hess. It was soon demonstrated that cosmic rays at ground level can initiate cascades and are stopped in lead plates. With the advent of quantum mechanics, the cascade showers of particles generated by electrons or photons were sought to be explained by means of the two fundamental processes: Bremmstrahlung (radiation) by electrons and pair production by photons. Bhabha and Heitler (1937) analyzed the showers generated by a single primary by dealing with the expected value of the number of electrons (each with energy > E) that are detected at a definite depth of atmosphere (measured in an appropriate unit called cascade unit). By varying the parameter t representing the depth, they wrote down differential equations representing the differential mean (expected value) number of electrons and photons with energies in (E, E + dE). Carlson and Oppenheimer (1937) simultaneously and independently dealt with the problem using an identical approach. The main conclusion was that the mean size of the observed showers could be explained in a statistical frame work on the basis of the two fundamental processes: Bremmstrahlung and pair production. The problem also attracted the attention of the Russian Physicist Landau who in collaboration with Rumer came out with his own analysis providing some insight into energy degradation of individual electrons. In current day terminology, Landau's treatment is essentially a sample path analysis of continuous parameter Markov process. To appreciate this we note a typical shower can be represented as in Figure 1 .Solid lines represent electrons while wavy lines represent photons. The problem in the first instance was treated as one dimensional (with reference to depth of atmosphere) the lateral spread being neglected. If we persue the primary electron, its life history is easily narrated; it goes on emitting photons (due to interaction with matter) in a probabilistic sense until its energy is reduced considerably that it cannot emit photons of substantial energy. It should be appreciated that Landau and Rumer recognized the life history of an individual electron as a Markov process with respect to t, the state space being the one dimensional continuum representing energy. A rigorous mathematical treatment was later provided by Harris (1957).

3. Point Distributions and Product Densities

To get a proper perspective, we introduce the notation $N_i(E,t)$ representing the number of particles of type i (electrons correspond to i=1 while photons to i=2) with energy ≤ E at a definite thickness t^*. The function $N_i(E,t)$ is a random variable for a fixed t. At t = 0, we have a single electron of energy E_o. The quantum mechanical processes imply that the electron with energy E has a probability $R_1(E,E')dE'$ per unit thickness (of material traversed) of radiating a photon itself dropping to an energy in the interval (E',E'+dE'). Likewise a photon (if sufficiently energetic)of energy E has a probability $R_2(E,E')dE'$ per unit thickness of producing an electron-positron pair one of which has an energy in the interval (E',E'+dE').

Thus it is clear that the physical processes imply that an incident electron or photon gives rise to a shower as in Figure 1. The main problem was how to arrive at a frame work within which the statistical characteristics of the shower of particles can be described. For each t, we have particles (electrons/photons) distributed over a continuum (energy). Bhabha and Heitler dealt with the expected value of $N_i(E,t)$ and wrote down the conservation equation w.r.t. t.

The year 1937 appears to be a crucial one; in kinetic theory Yvon, a French Physicist, (see also de Boer) had to deal with a similar problem of distribution of a discrete population over a continuous parameter. Here the parameter represents the phase space and the population collection of molecules. Yvon not only dealt with the expected value of sum functions associated with the position momentum of the molecules but also its mean square value through an appropriate second order correlation function. In modern terminology of point processes, Yvon's formula corresponds to the mean square value of cumulative response of an arbitrary weighted point process. Apparently neither Bhabha nor any of the other proponents of shower theory was aware of Yvon's contribution. Furry and Arley (1943) studied the problem and they were able to propose simple multiplicative models starting from Poisson and population birth processes. Scott and Uhlenbeck (1942) circumvented the difficult problem (of describing the distribution of a discrete population over a continuum) by dividing the energy interval into a finite number of portions and dealing with the population in each of the portions; however they made some approximations which eliminated the correlation that otherwise would have persisted. Bhabha proceeded to refine his calculations by taking into account loss in energy by electrons due to ionization of the atoms in the media; in collaboration with Chakrabarti (1943, 48), he obtained the explicit Mellin transform solution and numerical estimates for the average number (first moment) of electrons. Since there was no awareness of Yvon's contribution on the part of shower theorists (and that too despite Bogliubov's reformulation of kinetic theory in 1946), the problem of estimation of the fluctuation of the size of the showers about its mean value remained open. Thus Bhabha continued his search for a new mathematical technique that would yield a formula for the mean square value of the number of electrons and ultimately succeeded in 1950 by building up a theory of point distribution. He essentially viewed the inclusive distribution which formed the basis for the differential mean numbers of electrons as the one arising from appropriate summation/integration over the points of the finite dimensional distribution. Bhabha dealt with a continuous parametric assembly of particles; he considered a possible eigenstate in which there are m systems (particles) present with the parametric values in the intervals (x_1, x_1+dx_1) (x_2, x_2+dx_2) ... (x_m, x_m+dx_m), $(x_1<x_2<x_3.....<x_m)$ and introduced the probability measure $\psi_m(x_1, x_2, \ldots, x_m)dx_1dx_2\ldots dx_m$ with the help of which he was able to obtain the expected value of any arbitrary function of the parameter. By introducing appropriate set functions he was able to obtain moment formulae. As an application of the formula, he in colloboration with Ramakrishnan, dealt with the problem of electron-photon cascade and wrote down the extension of the diffusion equations for the second order functions expressing the correlation in energy (for any fixed t). Since there are two types of particles, this gave rise to four functions leading to a linear simultaneous system of integro differential equations. By the combined use of matrix approach and Mellin transform, they were able to obtain an explicit solution and finally the second moment was expressed as double Laplace inversion integral. Although the expression was unwieldy, it was amenable to numerical inversion. In 1954, Ramakrishnan and Mathews prepared tables of variances of the size of the showers for various primary secondary energy ratios.

The probability measure induced by the density ψ_m can give rise to inclusive probability measures. As this is an important point, it is worth discussing the same in some detail. If we fix the parameter x_1 and integrate over the rest of the parametric values x_2 ... x_m taking into account the indistinguishability of the systems except by their parametric labels, we obtain

$$f_1^m(x_1) = \frac{1}{(m-1)!} \int\int\cdots\int \psi_m\,(x_1,\ x_2,\ \ldots\ x_m)\ dx_2\ldots dx_m$$

where the integration is over the entire range of the parametric values. Now the quantity $f_1^m(x_1)dx_1$ has a simple interpretation in as much it denotes the probability of finding m systems with a typical member having parametric value in $(x_1,\ x_1 + dx_1)$. If at this stage we sum over all values of m running from 1 to ∞ and denote the resulting function by $f_1(x_1)$, then $f_1(x_1)dx_1$ denotes the inclusive probability of finding a system in (x_1, x_1+dx_1). It is precisely this function for which differential equations were provided by Bhabha and Heitler and Carlson and Oppenhiemer On further integration of x_1 over any specific range, we obtain the expected value of the number of such systems in that range. An extension of this argument naturally leads to correlation functions of the type $f_n(x_1,\ x_2,\ \ldots\ x_n)$ where $f_n(x_1,\ x_2,\ldots x_n)\ dx_1\ dx_2\ \ldots\ dx_n$ denotes the probability of finding simultaneously systems in parametric ranges $(x_i,\ x_i + dx_i)$ $(i = 1, 2, \ldots n)$ irrespective of the number found elsewhere. Bhabha attempted to express the moment formulae in terms of these functions. Ramakrishnan, independently and around the same time, introduced the functions f_n directly through the random variable N(x) representing the number of particles in the interval [0, x] and obtained an explicit formula for the moments of the number of particles with parametric values in an arbitrary interval [a, b]. He called these functions as product densities; in fact such functions were introduced earlier by Rice (1945) in the context of zero crossings of a Gaussian process and were known under the name inclusive probability densities. Apparently such functions have been in vogue in kinetic theory of fluids (see Bogoliubov (1946)). However the merit of the work of Ramakrishnan lies in the moment formula and the formulation of evolution equations for the product densities.

The probability measures induced by the functions ψ_m were also studied by Janossy quite independently in the same year 1950. Janossy in fact studied the problem of showers and being aware of the inherent difficulty due to lack of general methods of handling such point distributions introduced in 1949 his famous G- equation by dealing with the mass function $\pi_i(n,E,E_o,t)$ representing the probability that there are n electrons at depth t each with energy > E in a shower generated by a primary of type i and energy E_o. Janossy used the fact that once a secondary particle is produced by the primary, the shower generated by the secondary is statistically independent of the residual shower. This in turn enabled him to obtain an equation, although integral in nature, for the π-functions:

$$\pi_i\,(n,E,E_o,t) \;=\; \int_0^{E_0}\int_0^t e^{-\rho_i(E_0)x} \sum_{n_1+n_2=n} \pi_1(n_1,E,E',t-x)\;\pi_{3-i}(n_2,E,E_o-E',t-x)\;\; R_i(E',E_o)\; dE\; dx \;+\; \delta_{i1}e^{-\rho_i(E_0)t}$$

where

$$\rho_i(E_o) = \int_0^{E_0} R_i(E'|E_o)dE'$$

If we now define

$$G_i(u,E,E_o,t) \;=\; \sum_n u^n\,\pi_i(n,E,E_o,t)$$

we obtain

$$G_i(u,E,E_o,t) = \int_0^{E_0}\int_0^t e^{-\rho_i(E_o)x}\; G_1(u,E,E',t-x)\; G_{3-i}(u,E,E_o-E',t-x)R_i(E',E_o)dE'\;\; dx + u\;\; \delta_{i1}\; e^{-\rho_i(E_o)t}$$

Janossy's idea stepped up the pace of research and one could calculate at least in principle the various moments of the electron distribution. In fact the equation itself is a generalization of the Bellman-Harris equation proposed in the context of age dependent population growth. In the same year Janossy came up with the next contribution wherein he dealt with the ψ_m functions and indicated how diffusion equations for ψ_m can be written down; he also indicated a method of obtaining the moments of the number of particles produced in a shower upto a depth t. Messel and his collaborators have always preferred to use the Janossy measure as the basis in as much as it is easy to justify the use of forward differential equation as a consequence of the Markov property being imposed on Janossy measure functions. The impact of the work of Janossy can be judged from the fact that more than fifty papers on the subject particularly by Messel and his collaborators appeared in the next few years.

It is rather difficult to accord priorities in these contributions. Point distributions are perhaps known to theoretical physicists well before 1950; in particular the inclusive probabilities have been used in kinetic theory as early as 1937. Bhabha used the ψ_m functions to arrive at the inclusive distribution; Ramakrishnan directly dealt with the inclusive distribution. Although diffusion equations were arrived at using physical arguments of number conservation in the first instance, a formal proof based on the Markov property enjoyed by the Janossy measures was provided a year later (Ramakrishnan 1951). Janossy introduced his measure quite independently and later on used Markov property to arrive at the equations for inclusive density functions. This contribution has been missed in view of his more important contribution leading to the celebrated G- equation.

4. Further Developments in Shower Theory

The second moment of the number of electrons in a typical electron photon shower as calculated by Bhabha - Ramakrishnan or Janossy and Messel was more or less in broad agreement with the experimental data and it was concluded that except for same highly energetic primary where there was a possibility of direct pair production of electron positron by electrons, the data more or less gave conclusive evidence of the showers being a consequence of the two fundamental processes of quantum mechanics. However around 1954 there were anamolous showers and a more extensive investigation particularly by Fay at Gottingen was taken up wherein showers were observed in emulsions; it was pointed out that it would be more convenient to count the particles with energy labels at the points of production rather than at any later depth. To facilitate comparison with experimental data, Ramakrishnan and Srinivasan came up with a new approach wherein the deal with electrons produced over a certain thickness say an interval [0,t] with energy of each particle (or pairs that are observed) is greater than E at the point of production. Thus in the product density function $f_n(x_1, x_2, \ldots x_n)$ the parametric space was identified to be the Cartesian product of E (energy) and t (depth). In this case, the main quantity of interest is the expected number of electrons produced in a small stack of emulsions. Tables of mean numbers were prepared for the various energy ranges using the then powerful computer SILLIAC and the analysis done by Srinivasan in collaboration with Messel and others (1958) showed that the anamalous showers could be explained within the framework of the basic quantum mechanical processes. The investigations gave a new lease of life for the cascade theory for two reasons. First the problem naturally led to a two dimensional point process. Second the calculation of higher moments had to be taken up so that the main conclusion that there was no anamaly could be confirmed beyond doubt. This time there was an added advantage. Janossy's G- equation could be adapted to suit the special requirements of the problem. Srinivasan (1961) came up with an idea that the unwieldy Bhabha - Ramakrishnan equations could be given up and in their place a new set of compact equations for the moments smaller in dimension arrived at by a judicious use of Janossy's regeneration point method. These equations were then solved by recourse to Mellin transform technique and by standard complex variable methods, inversion of the expressions were executed elegantly to yield simple formulae for the moments upto 3rd order. (Srinivasan, Iyer and Koteswara Rao (1964)).

There was also a parallel development in the theory of showers particularly by Dyson and Mc Voy who were motivated by some of the existing results in parity non-conservation and were led to the investigation of high energy showers initiated by electrons with a specific spin. There arose an interesting question whether the polarization properties were transferred to the secondary particles in the showers. Dyson and Mc Voy proceeded on the same lines as Carlson and Oppenhiemer and answered in the affirmative by dealing with the expected value of the number of electrons with polarization and energy labels. Koteswara Rao in his doctoral dissertation (1967) carried out the analysis and provided compact formulae for the various higher moments arising from such population point processes.

5. Population Point Processes and Further Outlook

The above developments in physics have naturally provided ample motivation for the construction of a formal mathematical theory of population point processes. The theory of product densities was generalized to include within its frame work multiple points or points which are themselves populations. For the cascade theory itself other questions like the non-finite nature of the quantum mechanical cross-sections were also resolved; the monograph of Harris (1963) contains an excellent summary of the salient features of electron-photon cascades as well as point distributions in general. With the publication of a monograph on analytic methods (Srinivasan 1969) leading to compact moment structures, it was generally believed that research in the area of cascade theory had reached a satisfactory state of culmination and that the techniques presented therein would be used in contexts other than cascade theory. However this is not to be; while there has been a wide appreciation of population point processes in areas like statistical physics, neuro-physiology, light amplification and detection (see for example Uhlenbeck (1971), Murthy (1974), Holden (1976), Sampath and Srinivasan (1977), Shephered (1981, 1987), Teich et al (1984) Srinivasan (1988)), there is also a marked revival of activity in the theory of cascade processes with special reference to characteristics like energy and rapidity in jets observed in collider experiments. Some of these activities are spearheaded by Carruthers and Shih (1987), Giovannini (1972, 79) and Van Hove (1986). The Bellman Harris regeneration equation and the improvized imbedding equation again due to Bellman, Kalaba and Wing (1960) have reappeared in the form of Altreilli-Parisi equation (See Hwa (1988)). Right through seventies, Feynman in collaboration with Field (1978) (See also Fukuda and Iso (1977)) studied the jets and being unaware of some of the powerful analytic tools advocated extensive use of Marte Carlo simulation techniques. In fact the Feymman-Field formulation of e^+e^- collisions resulting in big jets is essentially a reformulation of Janossy's nucleon cascade model in terms of the production product densities introduced earlier by Ramakrishnan and Srinivasan (1956) for the study of QED cascades in emulsions. In many workshops, there are tutorial sessions on branching models mainly to understand the particle multiplicity distributions in jets produced in colliders. Attempts are also being made to examine the applicability of some of the distributions used in light beam detection (Fowler et al 1988, Voordas and Weiner 1988); in this connection it is worth noting that light amplification and detection process is by itself an interesting population point process.

There are also important attempts to use the ideas that are currently in vogue in the modern theory of turbulence particularly like self-similarity, intermittency and fractal nature. (See for example Bialas et al (1988), Chiu and Hwa (1990)). Since the new particles and their interactions by themselves constitute a complex system, currently some investigations are on to verify whether the electron-photon cascade showers do enjoy some of these properties. These investigations are really opening up new vistas in our understanding. Let me illustrate by a simple problem which I have examined in the last few months in collaboration with Chiu of UT at Austin. In the cascade theory the main object of interest is the distribution of the number of electrons produced with energies > E_o. To examine the existence of fractal structure in the distribution or self-similarity in energy ranges, it is natural to look at the electrons with energy in a window $(\epsilon, \epsilon+\delta)$. There are two methods of analysis available : We can introduce an innovation in the Janossy G-equation by introducing another label to characterize the configuration in which the primary is found, since the primary can be to the left or right of the window or in the window itself. The G- equation can be written down and it in turn yields recursive integral equations for the moments. Right now we have found that although the equations have attractive features, quick inferences are not possible. Hence we examined the possibility of dealing with the product densities of the cascade process; it has turned out to our surprise that it is possible to obtain explicit analytic expressions for the second order product densities which in turn will enable us to obtain the moment structure of the number of particles in the window at least to the second order. The investigations are still continuing and the findings will enable us to conclude whether the branching nature will give rise to intermittency and fractal structure in multiplicity distribution.

In summary you will agree with me if I say that we have had an exciting period of over fifty years of research in cascade theory of showers that had dominated our activities of research in the area of stochastic processes and more particularly in population point processes.

REFERENCES

1. Arley, N., 1943, On the theory of stochastic processes and their application to the theory of cosmic radiation, Wiley, New York.
2. Bartlett, M.S. and Kendall, D.G., 1951, On the use of characteristic functional in the analysis of some stochastic processes in Physics and biology, Proc. Camb. Phil. Soc., 47, 65-76.
3. Bellman, R.E. and Harris, T.E., 1948, On the theory of age dependent stochastic branching processes, Proc. Nat. Acad. Sci. (U.S.A), 34, 601-604 .
4. Bellman, R.E. and Harris, T.E., 1952, On age dependent binary branching processes, Ann. Math., 55, 280-295 .

5. Bellman, R.E., Kalaba, R. and Wing, G.M., 1960, Invariant imbedding and Mathematical Physics I - Particle Processes, J. Math. Phys., 1, 280-308 .
6. Bhabha, H.J., 1950, On the stochastic theory of continuous parametric systems and its application to electron-photon cascades, Proc. Roy. Soc. (Lond.), A 202 , 301-32 .
7. Bhabha, H.J. and Chakrabarti, S.K., 1943, The Cascade theory will Collision loss, Proc. Roy. Soc. (Lond.), A 181 , 267-303 .
8. Bhabha, H.J. and Heitler, W. 1937. The passage of fast electrons and the theory of cosmic showers. Proc. Roy. Soc. (Lond.) A 159 , 432-58 .
9. Bhabha, H.J. and Ramakrishnan, A. 1950 The mean square deviation of the number of electrons and quanta in the Cascade theory, Proc. Ind. Acd. Sci. A 32 , 141-53.
10. Bialas, A. and Peschanski, R. 1986, Moments of rapidity distributions as a measure of short range fluctuations in high energy collisions Nucl. Phys. B 273, 703-718 .
11. Bochner, S. 1947, Stochastic processes Ann. Math 48, 1014-61 .
12. de Boer, J. 1949, Molecular distribution and equation of state of gases, Rep. Prog. Phys. (London) 12 , 305-340 .
13. Bogoliubov, N.N., 1946, Problems of Dynamical Theory in Statistical Physics, Translated by E.G. Gora in Studies in Statistical Mechanics, Vol. 1 edited by J de Boer and G.E. Uhlenbeck, North Holland publishing Co. Amsterdam, 1962.
14. Carlson, J.F., and Oppenheimer, J.R. 1937, On multiplicative showers, Phys. Rev. 51, 220-231 .
15. Carruthers, P. and Shih, C.C. 1987, The phenomenological analysis of hadronic multiplicity distributions, Int. J. Mod. Phys. 2 , 1447-1547 .
16. Chiu C.B., and Hwa R.C., 1990, Multifractal structure of multiparticle production in the branching models, Preprint OI TS 431, Institute of Theoretical Science, Orgon.
17. Crawford D.F. and Messel, H. 1965, The electron-photon Cascade in lead, emulsion and copper absorbers, 61 , 145-176 .
18. Dyson, F.J. and Mc Voy K.W. 1957, Longitudinal polarization of bremmstrahlung from polarized electrons in Born approximation Phys. Rev. 106 , 828-9 .
19. Fay, H. 1957, Electron- Photon Cascades of high energy in photographic emulsions, Nuovo Cim. 5 , 293-298.
20. Field, R.D. and Feynman, R.P. 1978, A parametrization of the properties of quark jets, Nucl. phys. B 136 , 1-76.
21. Fowler, G.N. Friedlander, E.M., Pottag, F.W., Weiner , R.M., Wheeler, J and Wilk, G 1988, Rapidity scaling of multiplicity distributions in a

quantum statistical approach, Phys. Rev. D37 3127-3135.

22. Fukuda, H and Iso, C. 1977, Unified analysis of inclusive spectra of meson and baryons by Quark Cascade model . Prog. Theo. Phys. 57 , 483-498.

23. Furry, W.H. 1937, On the fluctuation phenomena in the passage of electrons through lead Phys. Rev. 52 , 569-81.

24. Giovannini, A 1974, QCD Jets as Markov Branching Processes Nucl. Phys. B 161 , 429-448 .

25. Giovannini, A and Van Hove, L 1986, Negative Binomial Distributions in High Energy Hadron Collisions Zeit. Phys. C 30, 391-400 .

26. Harris, T.E. 1963, Theory of Branching processes, Springer-Verlag, Berlin .

27. Hwa, R.C. 1988, Branching Processes in Multi-particle production in P. Carruthers (Ed) Hadronic Multiparticle Production, World Scientific Singapore.

28. Janossy, L 1950, A note on the fluctuation problem of Cascades. Proc. Phys. Soc. (Lond.) A 63 , 241-49.

29. Janossy, L. 1950, On the absorption of nuclear Cascade, Proc. Roy , Irish. Acad. Sci. A 53 , 181-8 .

30. Janossy, L . and Messel, H. 1950, Fluctuations of the electron-photon cascade - moments of the distribution Proc. Phys. Soc. (Lond.), A 63 , 1101-15 .

31. Kendall, D.G.1949, Stochastic Processes and population growth, J. Roy. Statist. B 11 , 230-64 .

32. Kolmogorov, A.N. and Dmitriev, N.A. 1947, Branching Stochastic Processes, Doklady, 56 5-8 (in Russian) .

33. Koteswara Rao, N.V. 1967, Studies in stochastic point processes and their applications, Ph.D. thesis, Indian Institute of Technology, Madras .

34. Le Cam, L 1947 Un instrument d' etude des functions aleatories la functionale characteristique, CR Acad. Sci. Paris 224 , 710-11 .

35. Landau, L.D and Rumer, G 1938, The Cascade theory of electronic showers, Proc. Roy. Soc. (Lond.) A 166 , 213-26 .

36. Messel, H., Smirnov, A.D., Varlofomeav, A.A., Crawford, D.F. and Butcher, J.C. 1962, Radial and angular distributions of electrons in electron-photon showers in lead and in emulsion absorbers, Nuclear Physics, 39 , 1-88 .

37. Palm, C 1943, Intensitatchwankungen in Fernsprech verkeher, Ericsson Technics 44, 1-189.

38. Ramakrishnan, A 1950, Stochastic processes relating to particles distributed, Proc. Camb. Phil. Soc. 46 , 595-602 .

39. Ramakrishnan, A 1952, A note on Janossy's mathematical model of a nucleon

cascade , Proc. Comb. Phil. Soc. 48 , 451-56 .

40. Ramakrishnan, A and Mathews, P.M 1954, Studies on the stochastic problem of Electron-photon Cascades, Prog. Theo. Phys. 11 , 95-117 .

41. Ramakrishnan, A. and Srinivasan, S.K. 1956, A new approach to Cascade theory, Proc. Ind. Acad. Sci. A 44 , 263-73 .

42. Shepherd, T.J. 1981, A model for photodetection of single-mode cavity radiation Opt. Acta. 28 , 567-83 .

43. Shepherd, T.J. and Jakeman, E., 1987, Statistical Analysis of an incoherently coupled, steady-state optical amplifier, J. Opt. Soc. Am. 4 , 1860-69 .

44. Srinivasan, S.K. 1961, Fluctuation problem in electro-magnetic cascades, Zeit, fur Physik 161 , 346-52 .

45. Srinivasan, S.K., 1969, Stochastic Theory and Cascade processes, American Elsevier, New York .

46. Srinivasan, S.K., 1974, Theory of Point Processes and Their Applications , Griffin, London .

47. Srinivasan, S.K. 1988, Point Process Models of Cavity Radiation and Detection , Edward Arnold, 1988 .

48. Srinivasan, S.K., Butcher J.C., Chartres, B.A., and Messel, H., 1958, Numerical Calculations on the new approach to Cascade theory III, Nuovo, Cim. 9 , 77-84 .

49. Srinivasan, S.K., Iyer K.S.S., and Koteswara Rao, N.V. 1964 , Fluctuation Problem in Electromagnetic Cascade II - Zeit fur Physik , 177 , 164-73 .

50. Teich, M.C., Saleh, B.E.A. and Perina, J. 1984 Role of primary excitations statistics in the generation of antibunched and sub-Poisson light J. Opt. Soc. Am. B 1 , 366-89 .

51. Uhlenbeck, G.E., 1971, An outline of statistical mechanics in Cohen EGD (ed.) Fundamental Problems in Statistical Mechanics II p. 1-30, North Holland Publishing Co., Amsterdam .

52. Vordas, A. and Weiner, R.M., 1988, Multiplicity distributions and Bose Einstein Correlations in high energy multiparticle production in the presence of squeezed coherent states, Phys. Rev. D 38 , 2209-2217 .

53. Yvon, J. 1935, La Theorie statistique des fluides et 1' equation d' Etat, Hermann, Paris .

Figure 1: A typical shower of electrons and photons produced by a primary electron.

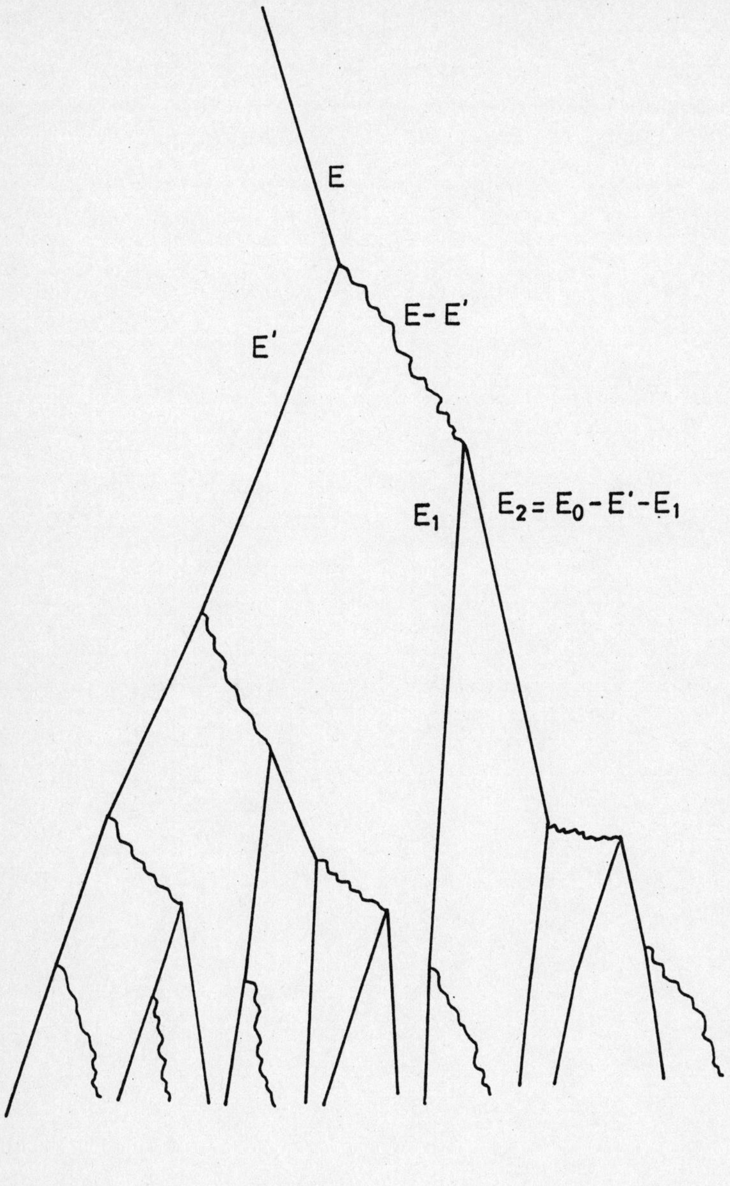

THE SQUARE WAVE SPECTRUM OF A MARKOV RENEWAL PROCESS

Marcel F. NEUTS
Department of Systems and Industrial Engineering
University of Arizona
Tucson, Arizona 85721
U.S.A.

Paper in Honor of Professor S.K. Srinivasan on the Occasion of his Sixtieth Birthday

ABSTRACT

The power spectral density of a stationary square wave process with sign changes at the transitions of a finite state Markov renewal process is derived. Particularly explicit formulas are obtained for the Markovian Arrival Process, a generalization of the Poisson process with a natural matrix formalism which commonly leads to useful explicit results. An application to a procedure for the quantification of burstiness is discussed.

1. INTRODUCTION

This paper deals with a descriptor of the random point process generated by the transitions of a finite-state Markov renewal process. We begin by defining some essential notation. We consider an irreducible, positive recurrent, m-state Markov renewal process with transition probability matrix $H(.)$. The matrix Laplace-Stieltjes transform of $H(.)$ is denoted by $h(s)$ and the finite positive vector of row sum means $-h'(0+)\mathbf{e}$ by $\underline{\beta}$. The column vector $\mathbf{e}$ has all its components equal to one. The stochastic matrix $H = H(\infty)$ has the invariant probability vector $\underline{\pi}$. The matrix $\Delta(\underline{\beta})$ is a diagonal matrix with the components of $\underline{\beta}$ as its diagonal elements. The inner product $E = \underline{\pi}\underline{\beta}$ is called the *fundamental mean* of the Markov renewal process. It plays an important role in many asymptotic results. The quantity $\lambda^* = E^{-1}$, the *fundamental rate,* is the rate at which transitions occur in the stationary version of the Markov renewal process.

It is well-known (see Pyke [13] or Cinlar [4]) that the stationary version of the Markov renewal process is obtained by choosing the initial state according to the probability vector $E^{-1}\underline{\pi}\ \Delta(\underline{\beta})$ and by making the first transition according to the semi-Markov matrix $H^*(x)$, defined by

$$H^*(x) = \Delta^{-1}(\underline{\beta})\int_0^x [H - H(u)]\,du, \text{ for } x \geq 0 .$$

In this paper we shall only deal with the stationary Markov renewal process so constructed. In order to avoid uninteresting issues, we shall also assume that, with probability one, there are only single transitions. That assumption is entirely "technical" and is satisfied in most applications.

For a stationary point process with single transitions, we may define an associated square wave process. We consider the *counting process* which is the random counting measure **N** induced by the stationary process on the Borel setsof R. In what follows, it is sufficient to consider the random variables

$$\mathbf{N}\{[0,t]\} = \mathbf{N}(t) = \inf\{n : S_n \leq t\}, \quad \text{for } t \geq 0 .$$

The *random square wave* corresponding to the stationary point process is the stochastic process defined by

$$Y_1(0) = 1 \text{ or } 0, \text{ each with probability } 1/2,$$

and

$$Y_1(t) = \max\{0, (-1)^{Y_1(o)+\mathbf{N}\{[0,t]\}+1}\}, \text{ for } t \text{ real} .$$

The process $\{Y_1(.)\}$ is clearly stationary and $E[Y_1(t)] = 1/2$ for all t. The *centered square wave* is simply the process $Y(t) = Y_1(t) - 1/2$.

The *square wave power spectral density* (SQSD) of the stationary point process is defined by

$$S(f) = \int_{-\infty}^{\infty} e^{-i\omega t}\, R(t)dt \tag{1}$$

where $\omega = 2\pi f$ and $R(t)$ is the autocovariance function

$$R(t) = E[Y(0)\ Y(t)]$$

of the process $Y(.)$. The function $R(.)$ is clearly even, and for $t \geq 0$, we have

$$R(t) = E[Y_1(0)Y_1(t)] - \frac{1}{4} = \frac{1}{2}P\{\mathbf{N}(t) \text{ is even }\} - \frac{1}{4} \tag{2}$$

The choice of the variable f, representing the "frequency" in cycles rather than in radians, is more prevalent in the engineering literature. Whether ω or f is used is entirely a matter of convention.

In terms of the probability generating function

$$P^*(z,t) = E\,\{z^{N(t)}\}, \text{ for } t \geq 0 \tag{3}$$

we may rewrite (2) as

$$R(t) = \frac{1}{4}\,P^*(-1,t) \tag{4}$$

The preceding considerations are well-known, at least for the Poisson process [5]. For a Poisson process of rate λ, one easily obtains that

$$R(t) = \frac{1}{4}\exp\{-2\lambda|t|\}, \text{ for all real } t \tag{5}$$

so that the SQSD is given by the Cauchy "density"

$$S(f) = \frac{\lambda}{4}\,[\lambda^2+\pi^2 f^2]^{-1} \tag{6}$$

For other instances of uses or discussions of the SQSD, see Castro, Kemperman and Trabka [3], Lamond [6,7], Neuts and Sitaraman [14] and Neuts [12]. In this paper, we shall give a derivation of a matrix-analytic expression for the SQSD of the point process generated by the transition epochs of a finite state Markov renewal process. We shall discuss an important particular case for which that density is highly tractable and conclude by describing a potential practical use of the SQSD .

2. THE DISTRIBUTION OF THE COUNTS

The derivation of the probability distribution of the counting variable $N(t)$ proceeds in classical fashion as in Pyke [13]. The following expressions are given for completeness. For the stationary Markov renewal process, we define

$$P(n;t) = P\{N(t) = n \mid N(0) = 0\}, \text{ and } P^*(n;s) = \int_0^\infty e^{-st}\,P(n;t)dt, \quad \text{for } n \geq 0 .$$

By direct probability arguments, we obtain that

$$P(0;t) = E^{-1}\underline{\pi}\int_0^t [H - H(u)]du, \text{ and } P^*(0;s) = s^{-1}-s^{-2}E^{-1}\underline{\pi}\,[I-h(s)]\mathbf{e} \tag{7}$$

and for $n \geq 1$,

$$P(n;t) = E^{-1}\underline{\pi}\int_0^t [H-H(u)]du\int_0^{t-u} dH^{(n-1)}(v)[H-H(t-u-v)]\mathbf{e} \tag{8}$$

so that

$$P^*(n;s) = s^{-2}E^{-1}\underline{\pi}\,[I - h(s)]^2\, h^{n-1}(s)\mathbf{e} \tag{9}$$

Routine calculations lead to the generating function

$$P^*(z;s) = \sum_{n=0}^{\infty} P^*(n;s)z^n \; :$$

$$P^*(z;s) = s^{-1} - (1-z)s^{-2}E^{-1}\,\underline{\pi}\,[I - h(s)]\,[I - zh(s)]^{-1}\mathbf{e} \tag{10}$$

The Markovian Arrival Process : A particularly tractable class of Markov renewal processes are those whose transition probability matrix is given by

$$H(x) = \int_0^x \exp(Cu)du\; D, \text{ for } x \geq 0 \tag{11}$$

where C and D are square matrices of order m whose sum Q = C + D is an *irreducible* infinitesimal generator. The matrix C has negative diagonal elements, nonnegative off-diagonal elements and its inverse exists. The matrix D is nonnegative.

Such a Markov renewal process (and the extension allowing for group arrivals) is known as a *Markovian arrival process,* and because of its appealing matrix-analytic properties, has already been extensively, used in queueing theory. We refer to Asmussen and Ramaswami [1], Blondia [2], Lucantoni [8], Lucantoni, Meier-Hellstern and Neuts [9], Neuts [11], for detailed discussions. Particular Markovian arrival processes are the PH-renewal process and the Markov-modulated Poisson process [11] and (independent) *superpositions* of these .

If we denote by $\underline{\Theta}$ the stationary probability vector of the generator Q, then it is readily verified that $\lambda^* = \underline{\Theta}\, D\, \mathbf{e}$, and $\pi = (\lambda^*)^{-1}\, \underline{\Theta}\, D$. The transform matrix h(s) is given by

$$h(s) = (sI-C)^{-1}D \tag{12}$$

By substitution in (10) or by a direct argument as in Neuts [11], we verify that

$$P^*(z;t) = \underline{\Theta}(sI-C-zD)^{-1}\mathbf{e} \tag{13}$$

For the sake of caution, we note that for some important cases, the stochastic matrix $H = (-C)^{-1}D$, is reducible (because some columns of D vanish,) but that rarely affects the validity of the calculations based on the useful matrix formalism for the Markovian arrival process. It is more convenient to conserve the matrix formalism than to work with the reduced transition probability matrices.

3. THE SQUARE WAVE SPECTRAL DENSITY

Theorem 1 : For a non-lattice, finite-state Markov renewal process, the SQSD is given by

$$S(f) = \omega^{-2}E^{-1}\underline{\pi}\,[I + h(i\omega)]^{-1}\,[I - h(i\omega)h(-i\omega)]\,[I + h(-i\omega)]^{-1}\mathbf{e} \qquad (14)$$

for $f \geq 0$, and where $\omega = 2\underline{\pi}f$.

Proof: By virtue o. (4), the Laplace transform $R^*(s)$ of $R(t)$ is given by

$$R^*(s) = \frac{1}{4}P^*(-1,s) = \frac{1}{4s} - \frac{1}{2s^2E}\underline{\pi}\,[I-h(s)]\,[I+h(s)]^{-1}\mathbf{e} \qquad (15)$$

Now observing that the Fourier integral in (1) may be written as $R^*(i\omega) + R^*(-i\omega)$, the expression for $S(f)$ is obtained by routine matrix manipulations.

Remarks: By rather belabored but straightforward calculations, we may express $S(f)$ in (14) in terms of the real and imaginary parts of the characteristic matrix $h(i\omega)$ (whose complex conjugate is the matrix $h(-i\omega)$). Since a complex matrix and its conjugate do not commute in general, the "real" expression for $S(f)$ is more involved than that given in (14).

The requirement that $H(.)$ is not a semi-Markov matrix of lattice type is essential. For that case, the square wave spectrum is discrete and a separate analysis is required. When $H(.)$ is non-lattice, the inverses in (14) exist by virtue of a corollary to the Perron-Frobenius theorem.

In what follows, we shall obtain an expression for the value of $S(0+)$ by series expansions of the matrices involved in formula (14). In that analysis, the case where the stochastic matrix H has -1 as an eigenvalue, or equivalently where the embedded Markov chain of the Markov renewal process is periodic with an even period, requires special treatment. The calculations for that case are, in general, quite belabored but follow the same lines as in Lamond [7].

Henceforth, we assume that the transition probability matrix $H(.)$ is non-lattice and that all sojourn time distributions have finite second moments. We denote the first and second order moment matrices respectively by

$$H_1 = -h'(0+) \quad \text{and} \quad H_2 = h''(0+)$$

Theorem 2 : Provided that the matrix I+H is nonsingular and under the stated assumptions on the moments of H(.), the value S(0) of the SQSD at zero is given by

$$S(0) = \frac{\lambda^*}{2} \{\underline{\pi} H_2 (I + H)^{-1}\mathbf{e} - 2\underline{\pi} H_1 (I + H)^{-1} H_1 (I+H)^{-1}\mathbf{e}\} \tag{16}$$

Proof : From the expansion

$$h(i\omega) = H - i\omega H_1 - \frac{\omega^2}{2} H_2 + o(\omega^2)$$

is follows that

$$\underline{\pi}\,[I - h(i\omega)] = i\omega\underline{\pi} H_1 + \frac{\omega^2}{2}\underline{\pi} H_2^2 + o(\omega^2)$$

Writing

$$[I + h(i\omega)]^{-1}\mathbf{e} = \mathbf{v}(0) + i\omega\mathbf{v}(1) \frac{\omega^2}{2} \mathbf{v}(2) + o(\omega^2)$$

we obtain, provided that I+H is nonsingular, the following expressions for the coefficient vectors $\mathbf{v}(0)$, $\mathbf{v}(1)$ and $\mathbf{v}(2)$:

$$\mathbf{v}(0) = (I + H)^{-1}\mathbf{e}, \text{ and } \mathbf{v}(1) = (I + H)^{-1}H_1(I + H)^{-1}\mathbf{e}$$

$$\mathbf{v}(2) = 2(I + H)^{-1}H_1(I + H)^{-1}H_1(I + H)^{-1}\,\mathbf{e} - (I + H)^{-1}H_2(I + H)^{-1}\mathbf{e}$$

Using these formulas, we compute the expansion of

$$\underline{\pi}\,[I - h(i\omega)]\,[I + h(i\omega)]^{-1}\mathbf{e}$$

and add its complex conjugate to it. The terms in ω and ω^3 cancel. Upon substitution into expression for $R^*(i\omega) + R^*(-i\omega)$, the stated result follows.

For the *Markovian Arrival Process*, most of the preceding expressions greatly simplify. By virtue of formula (4) ,

$$R^*(s) = \frac{1}{4}\,\underline{\theta}\,(sI - C + D)^{-1}\mathbf{e}$$

and since $\underline{\theta}(C - D) = \underline{\theta}(Q - 2D) = -2\underline{\theta}\,D$, we obtain by direct matrix calculations that

$$S(f) = \underline{\theta}\,D[4\pi^2 f^2 I + (C - D)^2]^{-1}\mathbf{e} \tag{17}$$

AN APPLICATION ; THE CLUSTERING OF A MARKOV RENEWAL PROCESS

The qualitative features of the point process generated by the transitions of a finite Markov renewal process can be very involved. We have been exploring analytic tools to quantify, for example, the time scales at which various degrees of clustering of the transition epochs manifest themselves. The SQSD is one such tool and the following discussion deals with its potential use in the qualitative analysis of such point processes. In that discussion, it should be borne in mind that a prevalence of "long gaps" in the point process will result in relatively higher spectral density values in the "low frequencies", whereas "dense clusters" tend to increase the spectral density at the higher frequencies.

We can consider various forms of thinning which progressively "remove" clusters of points from the point process. By studying the **SQSD** for various values of a "thinning parameter" we can expect to gain insight in the levels of clustering in the original process. Specifically, a simple form of thinning consists of imagining a Poisson process of rate α , independent of the given point process. We agree that a point of the given process is "preserved" if and only if there is at least one Poisson event between it and its predecessor in the given process.

The thinning corresponds to considering the arrival epochs of customers who receive service in a queue of capacity one, with a single exponential server and having the given Markov renewal process as its input process. At the expense of additional notation, we may readily replace the Poisson process by a renewal process, but for the sake of our example we shall not do so.

Let us write $a = 1/\alpha$, so that a is the mean interarrival time in the "thinning Poisson process." The original process may be viewed as the limiting case $a = 0+$, and as we increase the value of a, the procedure will tend to delete clusters selectively, starting with the densest ones first. We can therefore expect that the decrease in the tails of the spectral density and eventually also for smaller values of ω , will reflect (possibly) several levels of clustering in the given process.

For a finite Markov renewal process, it is readily seen that the thinned process is itself a Markov renewal process with the same state space. For a given positive value of a, the Laplace-Stieltjes transform h(s;a) of the transition probability matrix H(x;a) of the thinned Markov renewal process is given by

$$h(s;a) = [I - h(s + \alpha)]^{-1} [h(s) - h(s + \alpha)] \qquad (18)$$

The formula (14) can be numerically implemented for the tranform matrix h(s;a) and this for various values of the parameter a. However, the substantial analytic simplifications for the Markovian arrival process make it worthwhile to carry out this task for that particular class. This is not only because of its prevalence in applications, but also because the physical behavior of some MAPs is well understood and they can therefore serve as benchmark examples in assessing the utility of the SQSD as a statistical tool.

For the Markovian arrival process, the matrix h(s;a) is given by

$$h(s;a) = \alpha(sI + \alpha I - Q)^{-1}(sI - C)^{-1}D \tag{19}$$

A moment's reflection shows that h(s;a) is also the transform of the transition probability matrix of the MAP with the coefficient matrices C_1 and D_1 of order 2m, given by z:

$$C_1 = \begin{vmatrix} Q-\alpha I & \alpha I \\ 0 & C \end{vmatrix}, \qquad D_1 = \begin{vmatrix} 0 & 0 \\ D & 0 \end{vmatrix}$$

By elementary calculations, we find that the stationary probability vector $[\underline{\theta}_1(a), \underline{\theta}_2(a)]$ of the generator $Q_1 = C_1 + D_1$, is given by

$$\underline{\theta}_1(a) = \underline{\theta}D(\alpha I - C)^{-1}, \quad \text{and} \quad \underline{\theta}_2(a) = \alpha\underline{\theta}(\alpha I - C)^{-1} \tag{20}$$

The fundamental rate $\lambda^*(a)$ of the thinned process is therefore given by

$$\lambda^*(a) = \underline{\theta}_2(a)De = \underline{\theta}(I - aC)^{-1}De \tag{21}$$

We note that a graph of the function $\lambda^*(a)$ is already quite informative, particularly if there are intervals over which that function decreases rapidly. The ratio $(\lambda^*)^{-1}\lambda^*(a)$ is the fraction of transitions of the original process that survive in the thinned process. We have found a graph of that function useful as an easily computable means of quantifying the "burstiness" of an MAP.

In evaluating the SQSD S(f;a) for increasing values of a, no appreciable matrix-analytic simplifications arise from the particular form of the matrices C_1 and D_1. Items such as the vector $\underline{\theta}(2)D$ and the matrix $(C_1-D_1)^2$ are computed once and for all and the further numercial computation consists in evaluating the inverses $[\omega^2 I + (C_1-D_1)^2]^{-1}$ for given values of ω and a. We start by computing the SQSD for the given MAP by using formula (17). The spectral density should be computed for a sufficiently large value f_{max} of f, so that for larger values of f the function S(f) is negligibly small. To determine f_{max} is not entirely a hit-or-miss proposition. By using formula

(6) to determine the value of f for which the SQSD for a Poisson process of rate λ^* decays to a small value, say 10^{-5}, an adequate starting value of f_{max} is found. In an initial computation, we can divide the interval $[0,f_{max}]$ in, say 100 equal parts. Since the spectral density S(f;a) "decays" to zero with increasing values of a, that should be adequate to obtain plots for selected values of a . These will then suggest other values of a or specific ranges of f-values to be explored in further numerical computations.

REFERENCES

[1] Asmussen, S., and Ramaswami, V. *"Probabilistic interpretation of some duality results for the matrix paradigms in queueing theory".* Stoch., Mod. 6,715-733,1990.

[2] Blondia, C., *"The N/G/1 finite capacity queue".* Stoch Mod., 5, 2 73-294, 1989.

[3] Castro,P.E., Kemperman, J.H.B. and Trabka, E.A., *"Alternating renewal model of photographic granularity." J.Optical Soc. America,* 63,820-825., 1973.

[4] Cinlar, E., *"Markov renewal theory." Adv. Appl. Prob.*, 1, 123-87, 1969.

[5] Davenport Jr., W.B. and Root, W.L., *An Introduction to the Theory of Random Signals and Noise.* New York: McGraw Hill Book Company, 1958.

[6] Lamond, B.F., *"On the square wave spectrum of Markovian arrival processes."* University of Arizona, Department of Systems and Industrial Engineering, Working Paper Nr.89-009, March 1989.

[7] Lamond, B.F., "Approximate spectral analysis of Markovian point processes" University of Arizona, Department of Systems and Industrial Engineering, Working Paper No.89-030, December 1989 .

[8] Lucantoni, D.M., Meier-Hellstern, K.S., and Neuts, M.F., "A single server queue with server vacations and a class of non-renewal arrival processes". *Adv. Appl. Prob.*, 22, 676-705, 1990.

[9] Lucantoni, D.M., *"New results on the single server queue with a batch Markovian arrival process."* Stoch. Mod., 7, 1991 (forthcoming).

[10] Neuts M.F. and Sitaraman, H., "The square-wave spectral density of a stationary renewal process." *Journ. Appl. Math. Simul.*, 2, 117-130, 1989.

[11] Neuts, M.F., *"Structured Stochastic Matrices of M/G/I Type and Their Applications."* Marcel Dekker Inc., New York, New York, July 1989.

[12] Neuts, M.F., "On the packet stream generated by a random flow of messages of random durations." *Stoch. Mod.*, 6, 445-470,1990.

[13] Pyke, R., "Marrkov renewal processes with finitely many states." *Ann. Math. Statist.*, 322, 1243-59, 1961.

[14] Sitaraman, H., *"Approximation of a Class of Markov-Modulated Poisson Processes with a Large State Space"*, Ph.D. dissertation, Dept. of systems and Industrial Engineering, University of Arizona, 1989.

SIMULATION AND ESTIMATION PROCEDURES FOR STRESS RELEASE MODELS

Ann-Lee Wang
University of Malaya, Kuala Lumpur

David Vere-Jones and Xiao-gu Zheng
Victoria University of Wellington
Wellington, New Zealand

ABSTRACT

The stress release model is a piecewise linear Markov Process similar to the collective risk model, in which the observations consist of the times and sizes of the jumps, which are taken to represent the times and sizes of large earthquakes. Earlier studies have shown that the asymptotic behaviour of the likelyhood-based tests for this process against a Poisson null hypothesis show anomalous behaviour. The present paper develops simulation methods for the process and uses them to investigate quantitatively some of the qualitative predictions of theoretical studies. In particular it is confirmed that the distribution of the likelyhood ratio statistic is non-standard.

1. Introduction

This paper describes two methods for simulating a point process of "stress release" or "self-correcting" type, and uses these in an investigation of the distributions of parameter estimates and test statistics for this process. The results are of importance in fitting the model to data on large-scale historical earthquakes (see, for example, Zheng and Vere-Jones (1990)).

The data to be fitted consists of pairs (t_i, S_i) representing the "occurrence times" and "stress releases" from a list of large historical earthquakes. In other words, they can be regarded as observations on a marked point process with the times $\{t_i\}$ as points and the stress releases $\{S_i\}$ as marks.

The stress release model is a marked point process with conditional intensity function

(1) $$\lambda(t,S \mid \mathcal{H}_t) = \psi[X(t)]\, f(S \mid X(t))$$

where ψ is a monotonic increasing "risk function", $X(t)$ is a piecewise linear Markov process used to model the regional stress level, and the probability density $f(S \mid x)$ describes the distribution of the stress release S when the stress level has reached the value x at the time when the jump occurs. For the rest of the paper we shall assume that the distributions $f(S \mid x)$ are in fact independent of x, so that the successive values of the stress release form a sequence of i.i.d. random variables.

The process $X(t)$ will be taken to be of the form

(2) $$X(t) = X(0) + B[t - CS(t)]$$

where B and C are constants and

$$S(t) = \sum_{i:\, 0 \le t_i < t} S_i$$

is the cumulative sum of the stress releases up to time t. To be specific we shall assume that the intensity function can be written in the form

$$\psi(x) = \exp\{a + bx\}. \tag{3}$$

Then by combining (1), (2) and (3) we can write the intensity function for the times $\{t_i\}$ in the 3-parameter form

$$\lambda(t \mid \mathcal{H}_t) = \psi[X(t)] = \exp\{A + B[t - CS(t)]\}. \tag{4}$$

As for the stress releases, we shall consider three special cases:

(i) S_i constant;
(ii) S_i have a common exponential distribution with density $\lambda e^{-\lambda x}$ $(x \geq 0)$;
(iii) S_i have a common Pareto distribution with density $C_\theta x^{-\theta}$ $(x \geq 1)$.

As we shall see, the behaviour of the likelihood ratio statistic is sensitive to the tail behaviour of this distribution.

The case of constant jumps was studied by Isham and Westcott (1979), who called the process a "self-correcting" process because of the tendency of the number of events N(t) to closely track the target function t/C. Indeed, they established that as $t \to \infty$, Var N(t) remains bounded instead of increasing as O(t) as in the case of Poisson and Renewal processes. This is clear evidence of a high degree of dependence over time, and grounds for expecting that the asymptotic inference theory may not apply. This question was examined by Ogata and Vere-Jones (1984), who showed that while the asymptotic distribution of the Maximum Likelihood (M.L.) parameter estimates was normal, the likelihood ratio statistic for testing against the null hypothesis of a simple Poisson process was non-standard. Further results are given by Hayashi (1988) and Zheng (1990)

Inference for the case of variable jumps, while of greater importance in the earthquake context, has not been studied theoretically, although Vere-Jones (1988) did show that the bounded variance property for N(t) holds only for the case of constant jumps. Also, the data sets in practice are usually small (N = 30-100) and the applicability of the asymptotic theory is therefore open to doubt.

The present study was undertaken to supplement the theoretical investigations, and to give confidence to the application of the model to historical earthquakes.

The simulation methods are outlined in Section 2 below; Section 3 treats the distribution of the parameter estimates, and Section 4 the distribution of the likelihood ratio under the null hypothesis of a simple Poisson process.

2. Simulation Methods

We describe two methods, the first a special case of the "inverse transform" method, and the second a special case of the "thinning" method, both following the general discussion in Ogata (1981).

The applicability of the inverse transform method to simulating point processes rests on the observation that between events the conditional intensity function plays the role of the hazard function for the time to the next occurring event (see, for example, Chapter 13 of Daley and Vere-Jones (1989)). In particular, the time interval ξ from an arbitrary starting point t_0 to the next occurring event has a survivor function

$$S(\tau) = P(\xi > \tau \mid \mathcal{H}_{t_0}) = \exp\left[-\int_{t_0}^{t_0+\tau} \lambda(u \mid \mathcal{H}^*(u))du\right] \tag{5}$$

where $\mathcal{H}^*(u)$ denotes the history of the process up to t_0 augmented by the assumption that no further event occurs between t_0 and u, $u > t_0$.

Substituting the representation (4) for the conditional intensity of the stress release model we find

$$\int_{t_0}^{t_0+\tau} \lambda(u \mid \mathcal{H}^*(u))du = \int_0^{\tau} \exp\{A_0 + Bu\}du \tag{6}$$

$$= R_0(e^{B\tau} - 1)$$

where $A_0 = A + B[t_0 - CS(t_0)]$ and $R_0 = (\exp A_0)/B$. Note that A_0 and hence R_0 can be written down explicitly in terms of the parameters A, B, C and the events (t_i, S_i) with $0 \le t_i < t_0$.

The simulation can therefore proceed according to the following algorithm, supposing the simulation to terminate after generating a prescribed number of points N or after covering a prescribed time interval T. The parameter values A, B, C are supposed given.

Algorithm 1

1. Set $t_0 = 0$, $S(t_0) = 0$, $N(t_0) = 0$.

2. Calculate R_0 in (6).

3. Generate a uniform random variable U, and compute the time ξ to the next event from

$$\xi = B^{-1} \log_e[1 - (\log_e U)/R_0].$$

4. Generate a random variable S from the assumed distribution for the stress drops f(x).

5. Update t_0, $S(t_0)$, $N(t_0)$ according to

$$\begin{aligned} t_0 &= t_0 + \xi \\ S(t_0) &= S(t_0) + S \\ N(t_0) &= N(t_0) + 1 \end{aligned}$$

6. Stop if $N(t_0) > N$ or $t_0 > T$. Otherwise record (t_0,S) and return to step (2).

This provides a simple, effective and rapid algorithm for generating successive pairs (ξ_i,S_i) for interevent times and associated stress drops.

The thinning approach was first suggested by Lewis and Shedler (1979) in the context of non-homogeneous Poisson processes in arbitrary dimensions, where the inverse transforms method is inapplicable or involves excessive computations. This is not the case here, but nonetheless is may be of interest to examine an alternative method. The underlying idea is to first simulate points in a dominating (higher intensity) simple Poisson process, and then to successively accept or reject the points in this simulation, with a probability of acceptance equal to the ratio of the intensity of the required process to that of the dominating process at the point being examined.

One difficulty with this approach in the present example is that, although the conditional intensity (4) is everywhere finite, it has no uniform upper bound. This feature complicates the construction of a dominating Poisson process and therefore rather spoils the inherent simplicity of the thinning approach. The following version may be suggested. It requires an initial choice of the dominating intensity λ_0 which we discuss below.

Algorithm 2

1. Set $t_0 = 0$, $S(t_0) = 0$, $N(t_0) = 0$.

2. Generate a uniform random variable U and from it an exponential random variable ξ by setting

$$\xi = -(\log_e U)/\lambda_0.$$

3. Compute the ratio $\rho = \lambda(t_0 + \xi)/\lambda_0$, where

$$\lambda(t_0 + \xi) = \exp\{A + B[t_0 + \xi - CS(t_0)]\}.$$

(a) If $\rho \geq 1$ evaluate the time ξ^* for which $\lambda(t_0 + \xi^*) = \lambda_0$ from

$$\xi^* = \{\log \lambda_0 - [A + B(t_0 - CS(t_0))]\}/B.$$

Set $t_0 = t_0 + \xi^*$, $\lambda_0 = 3\lambda_0$.
Stop if $t_0 > T$. Otherwise return to step 2.

(b) If $\rho < 1$ generate a uniform random variable U.

(i) If $U \leq \rho$, accept ξ and generate a random variable S from the stress drop distribution.
If $\rho < 1/6$ set $\lambda_0 = \lambda_0/3$, otherwise retain λ_0.
Update t_0, $S(t_0)$, $N(t_0)$ according to

$$t_0 = t_0 + \xi$$
$$S(t_0) = S(t_0) + S$$
$$N(t_0) = N(t_0) + 1$$

Stop if $N(t_0) > N$ or $t_0 > T$. Otherwise record (t_0, S) and return to step 2.

(ii) If $U > \rho$ reject ξ. Set $t_0 = t_0 + \xi$.
Stop if $t_0 > T$. Otherwise return to step 2.

A suitable initial value of λ_0 is about three times the average stress level $\bar{\lambda}$, which may be found from the following consideration. For large t, assuming the process is in equilibrium

$$X(t) = \lambda(t - CS(t)) = 0(1)$$

and so as $t \to \infty$

$$S(t)/t \to C^{-1}.$$

But $S(t) \approx \overline{S}\, N(t)$ where $\overline{S}$ is the mean stress drop, and so

$$N(t)/t \to (\overline{S}C)^{-1}.$$

This means that after the process is started, X(t) will tend to drift upwards (or downwards as the case may be) until $\lambda(t)$ is fluctuating around a value in the vicinity of $(\overline{S}C)^{-1}$. In the case of constant jump, only the parameter B determines the local character of the realisations. When B is small the risk is close to constant and the process close to Poisson; when B is large, the intervals are close to regular, the risk starting at a low value just after an event and reaching larger and larger values as the interval increases.

In the case of variable jump size an important role is played also by the tails of the jump distribution. If the distribution is highly peaked around a mean value, the behaviour will be similar to the constant jump case. If occasional very large jumps can occur, the process will exhibit a quiescent period after the large jump, followed by more frequent smaller events until the next large jump occurs. Some examples of trajectories for different combinations of parameter values and distributions are shown in Figure 1(a)-(d).

3. Distribution of Parameter Estimates

In this section the simulations are used to examine the properties of the parameter estimates obtained by standard likelihood maximization procedures. The three main purposes are:

(i) to check that the algorithms being used are working correctly;

(ii) to confirm the theoretical results, in particular consistency and the asymptotic forms of variances, obtained for the maximum likelihood estimates by Ogata and Vere-Jones (1984);

(iii) to check the extent of small sample deviations from the asymptotic forms.

The log likelihood here has the characteristic point process form

$$\log L = \sum_{0 \le t_i < T} \log \lambda(t_i) - \int_0^T \lambda(u)du \tag{7}$$

where λ is given by (4) and contains the three unknown parameters A, B, C. The integral can be evaluated explicitly as a series of exponentials over the intervals between events. It can therefore be used directly as the target function in a standard optimization procedure. In our case the NAG routine (E04JAF) was used, which requires no derivatives to be given. Quadruple precision was used in all evaluations of the target function, to avoid rounding errors in these calculations causing instability in the iterations and hence producing error messages. Note that with this particular form of parameterization, a unique solution is guaranteed by convexity (cf. Proposition 2 of Ogata and Vere-Jones (1984)).

Since this paper treats only the constant jump case, this was chosen for the simulation, with unit jump size and parameter values A = 3, B = 2 and C = 1. The parameterization in (4) above differs slightly from that used in the earlier paper, viz

$$\lambda(t) = \exp\{\alpha + \beta[t - \rho_0 N(t)] + \gamma t/T\} \tag{8}$$

where $\rho_0 = C$ is the true value. The asymptotic relationships between the standard errors of the two sets of parameter estimates are given by

$$\begin{aligned} \sigma_{\hat{A}} &= \sigma_{\hat{\alpha}} \\ \sigma_{\hat{B}} &= \sigma_{\hat{\beta}} \\ \sigma_{\hat{C}} &= T^{-1},\ \hat{\sigma}_{\gamma} \end{aligned} \tag{9}$$

the standard errors on the right being given by

$$\mathrm{Var}[T^{1/2}(\hat{\alpha}, \hat{\beta}, \hat{\gamma})] \to J^{-1}$$

where

$$J = \begin{pmatrix} U & V & \frac{1}{2}U \\ V & W & \frac{1}{2}V \\ \frac{1}{2}U & \frac{1}{2}V & \frac{1}{3}U \end{pmatrix}$$

and

$$U = \lim_{T\to\infty} \frac{1}{T} \int_0^T \lambda(u)du$$

$$V = \lim_{T\to\infty} \frac{1}{T} \int_0^T X(u)\lambda(u)du$$

$$W = \lim_{T\to\infty} \frac{1}{T} \int_0^T X^2(u)\lambda(u)du$$

with $\lambda(u)$ given by (8) (or equivalently (4)) and $X(t) = t - \rho_0 N(t)$.

To use these results it is first necessary to carry out a preliminary simulation to evaluate U, V, W. This was done in two ways, firstly by directly tracking the trajectory of X(t) over a long period (T = 10,000) and substituting into the above limits, and secondly by using the expressions given in Vere-Jones and Ogata (1984) and Ogata and Vere-Jones (1984) for U, V and W as expectations with respect to a discrete skeleton of the process X(t). This required finding the stationary distribution of the discrete skeleton and evaluating the expectations of the necessary functionals. The two approaches gave values which agreed to within 1% which was considered adequate. Averaging the two sets of values gives

$$\begin{pmatrix} U \\ V \\ W \end{pmatrix} = \begin{pmatrix} 1.003 \\ -1.267 \\ 1.809 \end{pmatrix}; \quad J = \begin{pmatrix} 1.00 & -1.27 & .50 \\ -1.27 & 1.81 & -.635 \\ .50 & -.635 & .334 \end{pmatrix}; \quad J^{-1} = \begin{pmatrix} 11.5 & 6.0 & -5.8 \\ 6.0 & 4.8 & 0.1 \\ -5.8 & 0.1 & 11.9 \end{pmatrix}.$$

As a further check, we know that $N(t)/t \to \gamma^1 = 1$ from the bounded variance property, while $E(U) = E[\lambda(t)] = E[N(t)/t]$ in the stationary regime. Thus in theory U = 1. Also, from the form of J, the (2,3) position of J^{-1} must be 0.

Substituting finally from (9) we obtain for the limit standard errors and correlations, the predictions

$$T^{1/2}\hat{\sigma}_A \to 3.4 \qquad \hat{\sigma}_{AB} \to 0.8$$

$$T^{1/2}\hat{\sigma}_B \to 2.2 \qquad \hat{\sigma}_{AC} \to 0.5$$

$$T^{3/2}\hat{\sigma}_C \to 1.73 \qquad \hat{\sigma}_{BC} \to 0$$

The process was then simulated 100 times for T = 32, 64, 128, 256, 512, parameter values being estimated for each run. The means, standard errors and correlations are set out in the table below.

TABLE 1
Simulation results and theoretical limits for parameter estimates

	Length of Simulation					Limit
	32	64	128	256	512	
Means						
A	3.30	3.19	3.11	3.02	3.04	3.0
B	2.21	2.14	2.08	2.02	2.02	2.0
C	1.001	1.0002	1.0003	0.99997	1.00003	1.0
Standard Errors						
$T^{1/2} s_A$	2.60	3.57	3.41	3.78	3.89	3.4
$T^{1/2} s_B$	1.78	2.23	1.90	2.47	2.44	2.2
$T^{3/2} s_C$	1.59	1.84	1.98	1.91	1.62	1.73
Correlations						
r_{AB}	.67	.76	.68	.82	.88	0.8
r_{AC}	.49	.53	.67	.51	.58	0.5
r_{BC}	-.18	-.07	.01	-.00	.20	0.0

The results overall are reassuring. The procedures are producing sensible results, broadly consistent with the theory. There is evidence of positive bias in the estimates of A and B for small T, but it is at worst of the order of 10% and decreases with T. The results for the asymptotic variances are borne out well by the simulations, even for quite modest sample sizes.

4. Distribution of the Likelihood Ratio Statistic

The discussion in Section 3 of Ogata and Vere-Jones (1984) shows that in the case of constant jumps, the likelihood ratio statistic

$$\Lambda = 2\log(\hat{L}_1/\hat{L}_0) \tag{10}$$

where $\hat{L}_1$ is the likelihood (7) for the stress release model using the parameters estimated by maximum likelihood, and $\hat{L}_0$ is the likelihood for the simple Poisson model with estimated rate, has a non-standard asymptotic distribution. Instead of following the χ^2 distribution on 2-degrees of freedom, it has the distribution of a certain quadratic functional of standard Brownian motion on [0,1]. This phenomenon arises from the long-term memory property of the stress release model with constant jumps, which makes possible the "bounded variance" property of the trajectories N(t), and causes the Hessian matrix

(2nd derivative of the log likelihood) to converge to a random not a constant limit. The case of variable jump size has not been investigated theoretically, but it seems likely that similar phenomena will appear also.

Since the form of the distribution is crucial to the tests and selection procedures (AIC) used with the historical earthquake data, it is important to gain an idea of the extent of the departures from the standard χ^2 form. Since, moreover, the true distribution is either unknown (variable jump case) or hard to evaluate (constant jump case), and is in any case unknown if the sample sizes are small, simulation offers the only practicable route towards ascertaining the size of these departures. In this section we outline the results of a preliminary investigation of this kind, for various combinations of sample size and jump distribution.

On general grounds, we know that the distribution of the statistic (10) should be independent of the scales of measurement used for the time and stress dimensions. It must, therefore, be independent of the size of the jump in the case of constant jumps, and of the parameter λ in the case the jumps follow exponential distribution $1 - e^{-\lambda x}$. The case of most practical interest, however, is that of a Pareto distribution with power law tails, say

(11) $$1 - F(x) = \begin{cases} C_\theta x^{-\theta} & x \geq x_0 \\ 0 & x < x_0 \end{cases}$$

since a Pareto distribution for stresses corresponds to the exponential distribution ("Gutenberg-Richter relation") generally used for earthquake magnitudes. Here the ratio may be influenced by the shape parameter θ, but not by the threshold value x_0.

In all a total of 8 simulation runs were performed for constant jumps, exponential jumps, Pareto jumps with $\theta = 2$ (corresponding to $b = 1.5$), $\theta = 1$ (corresponding to $b = 0.75$ approximately in the Gutenberg Richter relation), each run for $T = 32$ and $T = 64$ time units. Note that in the second case the sufficient conditions for ergodicity in Zheng (1990) are not satisfied since the jump distribution has infinite mean.

Approximate values of some summary statistics for the distributions of Λ are shown in Table 2, and some sample density functions shown in Figure 2. They are based on 500 simulations each for $T = 64$, $T = 32$. Note that in the estimation part of the program, the parameters B and C of the stress release model were constrained to be non-negative.

TABLE 2
Simulation results for the distribution of the likelihood ratio statistic Λ

	Constant Jump		Exponential Jump		Pareto Jump (1)		Pareto Jump (2)		χ^2_2
	T = 32	T = 64	T = 32	T = 64	T = 32	T = 64	T = 32	T = 64	
Mean	3.7210	3.7862	2.5139	2.4854	2.1669	2.2539	1.1385	1.4079	2.000
Median	3.5633	3.4666	2.3644	2.0579	1.7980	1.8566	0.7686	0.9583	1.386
LQ	2.3446	2.1500	1.2299	0.9783	0.8023	0.7163	0.2315	0.3443	0.575
UQ	4.9671	5.0424	3.4754	3.3832	3.0003	3.1736	1.7059	2.0545	2.770
90%ile	6.2100	6.7771	4.5638	5.1747	4.7719	4.8911	2.7216	3.4282	4.605

From these results a number of surprising if tentative conclusions can be drawn.

(a) In all cases the use of the χ^2 results would be misleading. The exceedance probabilities (p-values) of a given observed value are higher than those for the χ^2 form for constant and exponential jumps and for the first Pareto form, but lower than the χ^2 value for the second Pareto form

(b) The discrepancy between the real and χ^2 form is at its most acute for the case of constant jumps.

(c) The discrepancies relate both to the shape and the scale of the distribution, and are suggestive of an "equivalent degrees of freedom" which decreases as the tail of the jump distribution becomes thicker.

It would clearly be highly desirable to have some theoretical underpinning of these results, and an attempt to provide such underpinning is the next stage in our project.

Acknowledgements

The authors are grateful for the opportunity to contribute to this celebration volume, and in so doing to recognize the outstanding contributions Professor S K Srinivasan has made to stochastic modelling in many fields, above all in point processes and their applications.

The work was commenced while the first author was a Visiting Fellow at Victoria University on sabbatical leave from the University of Malaya. The support of both institutions is gratefully acknowledged.

List of Figures

All graphs show values of Λ along the x-axis and frequency up the y-axis.

REFERENCES

Daley, D and Vere-Jones, D (1988) *An Introduction to the Theory of Point Processes.* Springer, New York.

Hayashi, T (1988) Local asymptotic Normality in self-correcting point processes. *Statistical Theory and Data Analysis II* (Ed K. Matusita), 551-559; North-Holland, Amsterdam.

Isham, V and Westcott, M (1979) A self-correcting point process. *Stoch. Proc. Appl.* **8**, 335-347.

Lewis, P A W and Shedler, J (1976) Simulation of non-homogeneous Poisson processes with log linear rate function. *Biometrika* **63**, 501-506.

Ogata, Y (1981) On Lewis' simulation method for point processes. *IEEE* Trans. Inf. Theory **IT-27**, 23-31.

Ogata, Y and Vere-Jones, D (1984) inference for earthquake models : a self-correcting model. *Stoch. Proc. Appl.* **17**, 337-347.

Vere-Jones, D (1988) On the variance properties of stress release models. *Austral. Jl. Statist.* 30A, 123-135.

Vere-Jones, D and Ogata, Y (1984) On the moments of a self-correcting process. *J.Appl. Prob.* **21**, 335-342.

Zheng, X (1991) Ergodic theorems for stress release processes. *Stoch. proc. Appl.* (to appear)

Zheng, X and Vere-Jones, D (1990) Applications of stress release models to historical earthquakes from North China. *Pure and Applied Geophys.* (to appear)

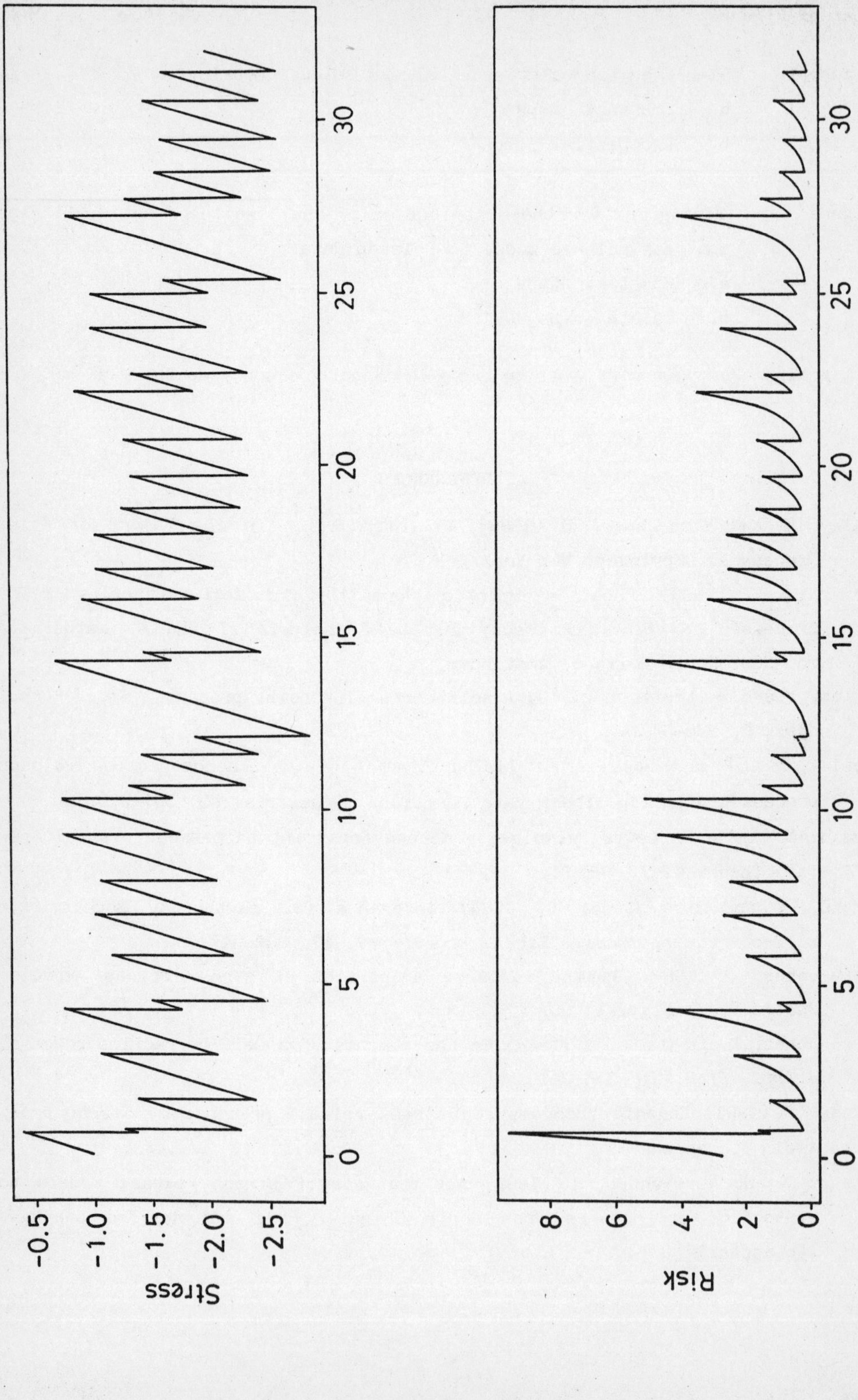

Figure 1a

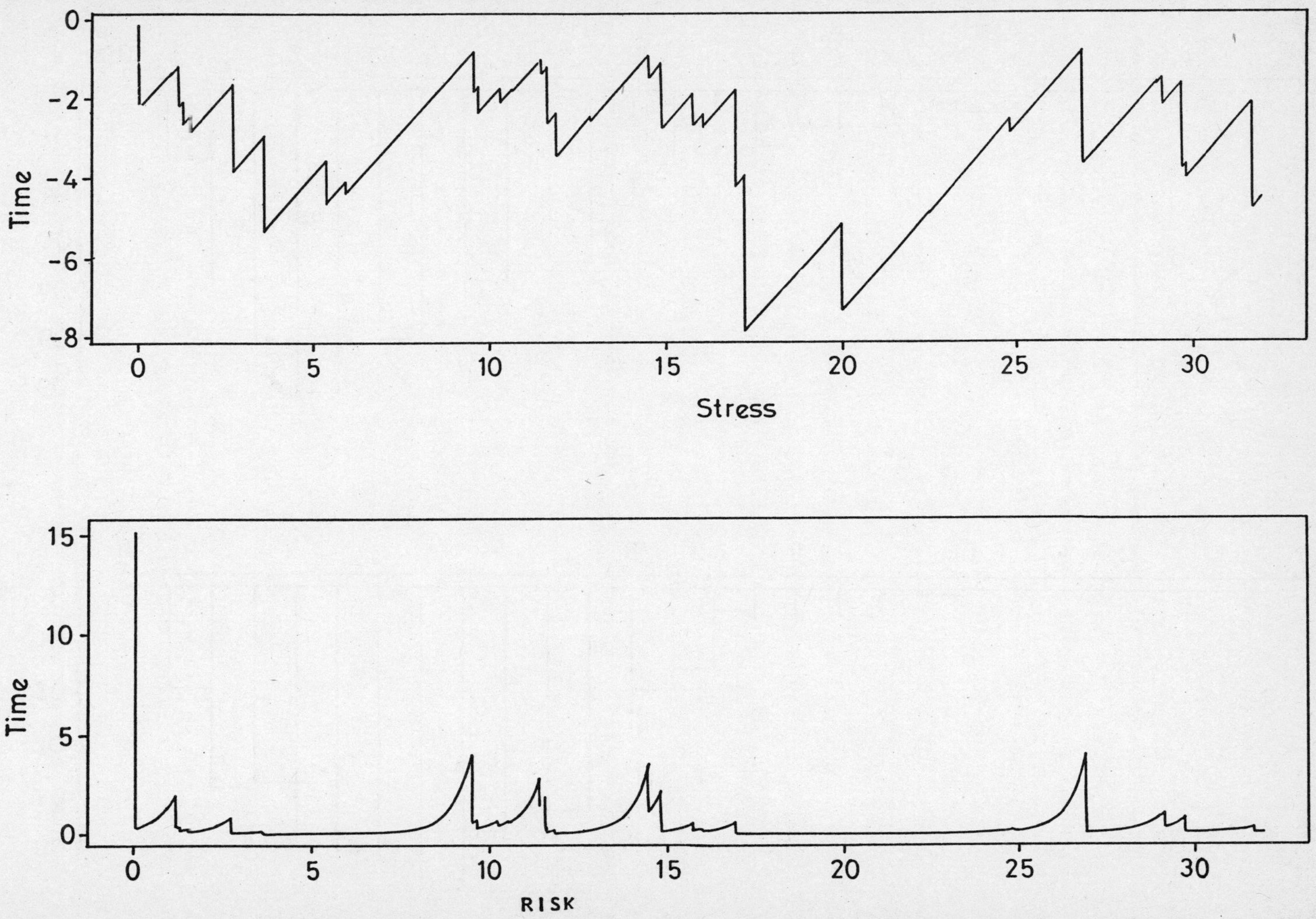

Time
0
-2
-4
-6
-8
0
5
10
15
20
25
30
Stress
Time
15
10
5
0
0
5
10
15
20
25
30
RISK

Figure 1b

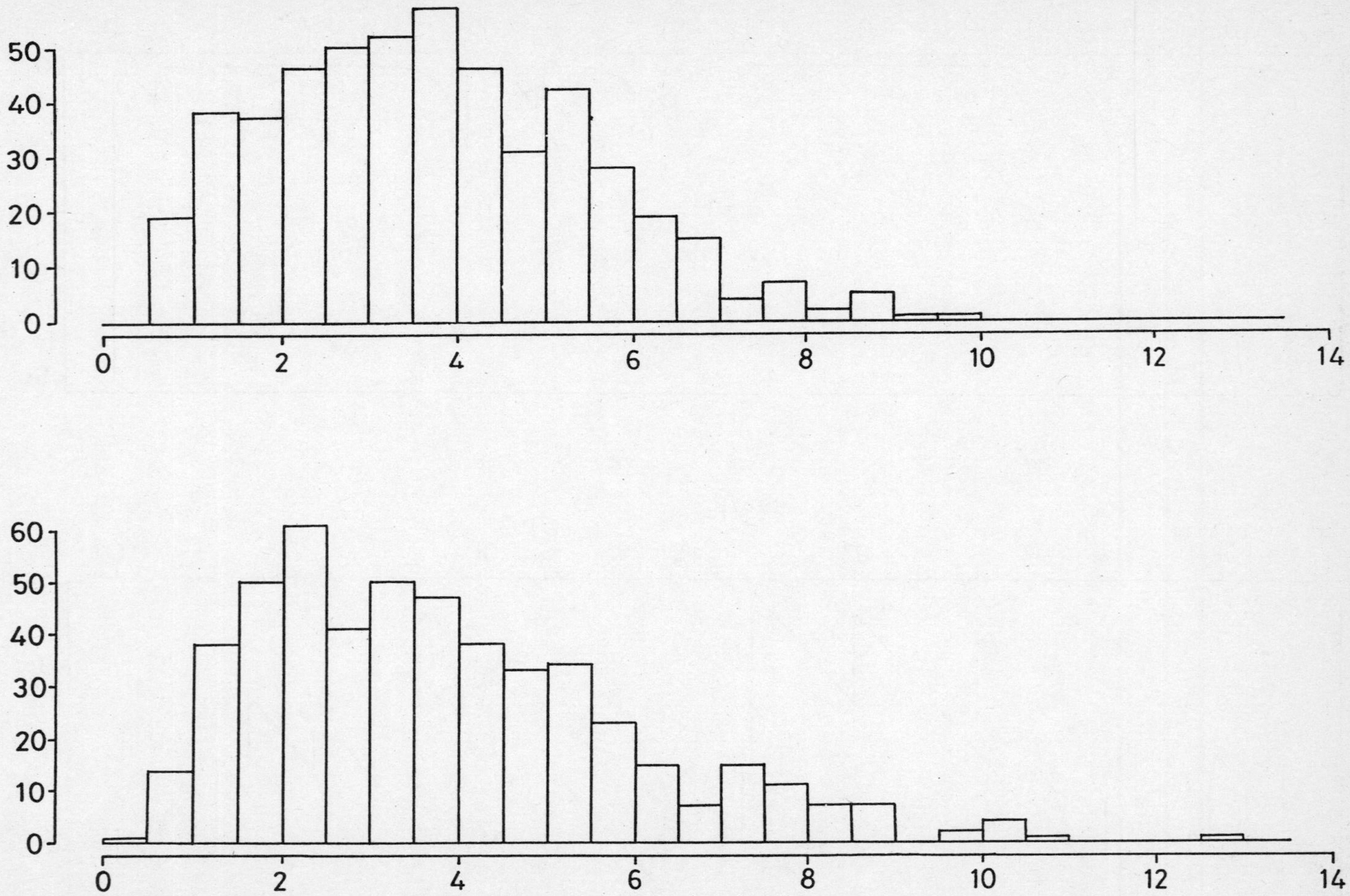

Figure 2a

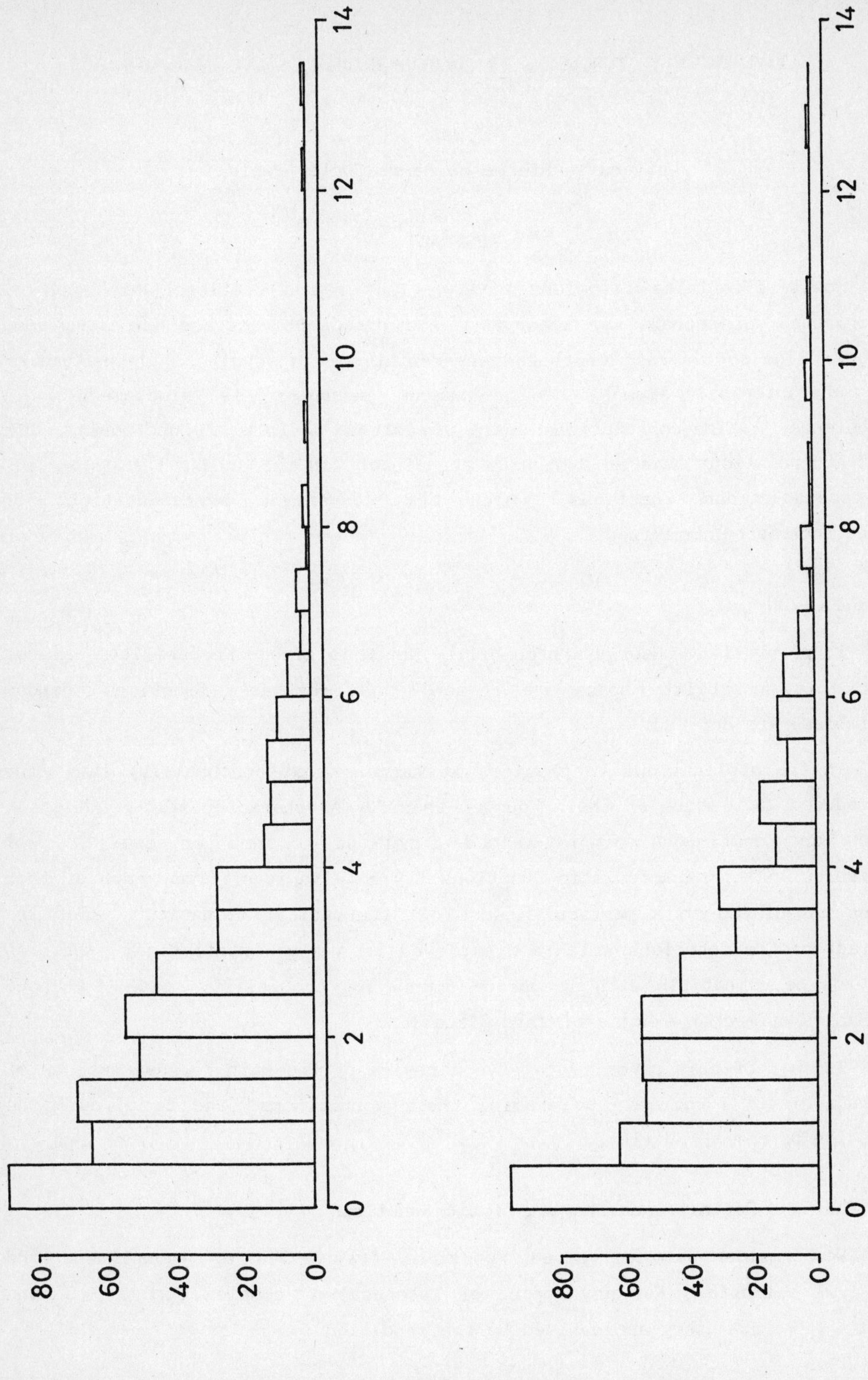

Figure 2b

POSITIVE DEFINITE FUNCTIONS IN QUANTUM MECHANICS AND IN TURBULENCE

J. BASS
Universite Pierre et Marie Curie, Paris

ABSTRACT

Positive-definite functions have various representations which can be interpreted in probability theory, in quantam mechanics, and in turbulence theory. The equivalence of these representations is studied. The nature of the characteristic functions in quantum mechanics is discussed. In turbulence, time-correlations are positive-definite functions, but space-correlations have an ambiguous structure. In the annex, the properties of pseudo-random functions, which are convenient representations of turbulence, are summarized.

INTRODUCTION

In probability theory, a probability space is given. Probability measures over this probability space are chosen and measurable functions (random variables) are defined.

In the applications to physics, it happens that "probability laws" are introduced, by means of their Fourier transforms (characteristic functions), which are continuous complex-valued functions of one or several real variables. The characteristic functions may have various forms, each of them being associated to a particular sort of "statistical mechanics". But the characteristic functions which are employed in a given statistical mechanics may not be compatible with an unique underlying probability space. Quantum mechanics offers an example of this situation.

The aim of this paper is to give a review of the main representations of characteristic functions, to examine their connections and to discuss the question of compatiability.

1 - Positive-Definite and Characteristic Functions

We consider complex-valued functions $f(\lambda)$ which are continuous and *positive definite*. For any choice of integeral n, complex $C_1, C_2, \ldots, C_n$ and real $\lambda_1, \lambda_2, \ldots, \lambda_n$ they are defined by the condition :

$$\Sigma\, \bar{C}_k C_l f(\lambda_k - \lambda_l) \geq 0 \tag{1}$$

These functions have several representations and interpretations as *characteristic functions* and as *correlation functions.*

I - According to the fundamental theorem of Bochner f is the Fourier transform of a positive bounded measure m :

$$f(\lambda) = \int_{-\infty}^{\infty} \exp(i\lambda s)\, dm(s) \tag{2}$$

$f(\lambda)/f(0)$ is the characteristic function of some probability law.

II - Let H be a Hilbert space and ψ an element of H such that $\|\psi\| = 1$. Let A be a hermitian operator over H, bounded or sometimes not bounded. If < > denotes the scalar product in H, the expression

$$f(\lambda) = < \exp(i\lambda A)\psi\ ,\psi > \tag{3}$$

is a positive-definite function. If it is continuous, it is the characteristic function of a probability law.

In *quantum mechanics*, this probability law is interpreted as the law of the quantity associated with the operator A , in the state ψ, by a "rule of correspondance".

III - Let α be a real-valued function defined over $\mathbb{R}$, or over $\mathbb{R}^+$. The mean values

$$f(\lambda) = \lim_{T\to\infty} \frac{1}{2T}\int_{-T}^{T} \exp\,(i\lambda\alpha(t))\, dt, \text{ or } \lim_{T\to\infty} \frac{1}{T}\int_{0}^{T} \exp\,(i\lambda\alpha(t))\, dt \tag{4}$$

are positive-definite functions, if they exist.

They are used in the deterministic theory of *turbulence* and can be interpreted as characteristic functions of probability laws.

IV - Let Ω be a probability space and s an element of Ω. We chose a probability measure m(s). Let p(s,t) be a stationary random function of second order. Then the mathematical expectation

$$E\, \bar{p}(s,t)\; p(s,t+\tau) = \int_{-\infty}^{\infty} \bar{p}(s,t)\; p(s,t+\tau)\, dm(s) \tag{5}$$

is a positive-definite function of τ (independent of t). It is interpreted as the *covariance* of the random function p. This function is the Fourier transform of a "spectral measure".

V - Let g be a complex valued function such that, for every τ ,

$$f(\tau) = \lim_{T\to\infty} \frac{1}{2T} \int_{-T}^{T} \bar{g}(t)\, g(t+\tau)\, dt \qquad (6)$$

exists. Then $f(\tau)$ is a positive definite function. It is interpreted as the correlation function of the function g.

The proofs of all these results are almost obvious. We see that there exist two main interpretations of a positive definite function :

i) characteristic function of a probability law ;

ii) covariance or correlation of a function, random or not random.

2. TWO-DIMENSIONAL CHARACTERISTIC FUNCTIONS

The definitions in section 1 are for the case of one-dimensional characteristic functions. They are easily extended to several dimensions; but, in the cases II and III (quantum mechanics and turbulence), this extension needs precautions.

a) Let A, B be two hermitian operators and consider

$$f(\lambda,\mu) = \langle \exp(i(\lambda A+\mu B)\, \psi,\ \psi \rangle \qquad (7)$$

If A and B commute, it is easy to verify that $f(\lambda,\mu)$ is positive-definite. It is the Fourier transform of a probability measure $m(s,s')$. If A and B do not commute, this result need not be true; $f(\lambda,\mu)$ is not generally a characteristic function. Only some particular points ψ in the Hilbert space have this property [11, 12].

A classical example can be given when H is a L^2-space, and

A = Multiplication by x (operator of position)

$B = \frac{h}{i}\frac{d}{dx}$ (operator of momentum)

It is found that

$$f(\lambda,\mu) = \int_{-\infty}^{\infty} \exp(i\lambda s)\, \psi\left(s+\frac{\mu h}{2}\right) \bar{\psi}\left(s-\frac{\mu h}{2}\right) ds \qquad (8)$$

f is the Fourier transform of

$$\frac{1}{2\pi}\int_{-\infty}^{\infty} \exp(-i\mu s)\, \psi\left(s + \frac{\mu h}{2}\right) \bar{\psi}\left(s - \frac{\mu h}{2}\right) d\mu \qquad (9)$$

This is Wigner's pseudo-density function (called Wigner-Bass pseudo-density by L. de Broglie in [8]).

It is easy to find functions such that (9) is not everywhere positive. For instance, this situation is verified when

$$\psi(s) = \frac{1}{\sqrt{2}} \qquad \text{for } |s| < 1$$
$$= 0 \qquad \text{for } |s| > 1$$

This result shows that the probabilistic structure of quantum mechanics is not compatible with a unique underlying probability space. This remark appears more clearly in terms of operators. The set of all hermitian operators is a vector space. Any family of commutative operators belongs to a subspace. But two non-commutative operators belong to independent subspaces and the intersection of these subspaces is itself a smaller subspace of the whole space.

Let us take the example of the hamiltonian operator H in a L^2-space. It does not commute with x or with $p = \frac{h}{i}\frac{d}{dx}$. The family of unitary operators exp(iHt) (an abelian group) is used for generating the solution of Schrödinger equation. We have

$$\psi(x,t) = \exp\,(iHt)\;\psi(x,0) \tag{10}$$

$\langle \exp(iHt)\psi, \psi \rangle$ can be identified to be the characteristic function of the hamiltonian operator, the time t playing the role of the variable λ .

In the space of operators, the trajectory of exp(iHt) belongs to a subspace ε_1 of commutative operators. But this subspace is different from the spaces ε_2, ε_3 of x or of p. The situation is roughly represented by fig. 1 $\varepsilon_1, \varepsilon_2, \varepsilon_3$ have only the common point 1 (operator identity).

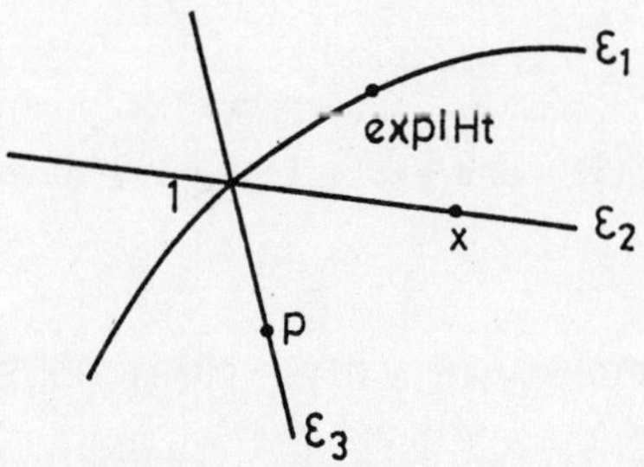

Figure 1

b) The characteristic functions of pairs of functions give rise to very similar structure. The characteristic function of the pair of functions $\alpha(t),\beta(t)$ is the mean value

$$f(\lambda,\mu) = \lim_{T\to\infty} \frac{1}{2T}\int_{-T}^{T} \exp i[\lambda\alpha(t) + \mu\beta(t)]dt \tag{11}$$

if this limit exists. For instance, the characteristic function of (cos t, sin t) is equal to

$$\lim_{T\to\infty} \frac{1}{2T}\int_{-T}^{T} \exp i[\lambda\cos(t) + \beta\sin(t)]dt = J_o\left(\sqrt{\lambda^2 + \mu^2}\right) \tag{12}$$

Of course, the associated probability is concentrated over the unit circle. But it happens that the mean value (11) does not exist, even when $\alpha(t)$ and $\beta(t)$ possesses good characteristic functions. For instance, the formula

$$\frac{1}{T}\int_{0}^{T} \exp(i\lambda \log t)\, dt = \frac{\exp (i\lambda \log T)}{i\lambda+1} \tag{13}$$

shows that log t has no characteristic function. But it is easy to see that t+logt has a characteristic function. That is due to the fact that t/logt is an increasing and invertible function, not essentially different from t. As t possesses a (discontinuous) characteristic function, $\lambda t+\mu(t+\log t)$ has characteristic function if $\lambda+\mu \neq 0$, but not if $\mu = -\lambda$.

We say that two function α, β (real or complex) are *comparable* if the mean value of $\bar{\alpha}\beta$ exists. Comparability of functions is analogous to commutativity of operators. A family of comparable functions constitutes a vector space, a sub-set of the set E of all functions possessing a quadratic mean value. The set E is not a vector space. But it is itself a subset of a vector space, namely the *Marcinkiewiz space*, which is the space of complex valued functions α such that

$$\lim_{T\to\infty} \sup \frac{1}{2T}\int_{-T}^{T} |\alpha(t)|^2 dt < \infty \tag{14}$$

E can be dissociated into vector sub-spaces of comparable functions. As an example, the functions exp(it) and exp[i(t+log t)] have quadratic mean values, but are not comparable.

3. CONSTRUCTION OF FUNCTIONS WITH A GIVEN CHARACTERISTIC FUNCTION

We restrict ourselves to the case of a characteristic function which is the Fourier transform of an absolutely continuous measure. We are given a characteristic function f . According to Bochner's theorem, we have

$$f(\lambda) = \int_{-\infty}^{\infty} \exp[i\lambda s\ \varphi(s)]ds\ , \quad \varphi \geq 0 \tag{15}$$

We can write

$$f(\lambda) = \langle \exp\ i\lambda s\sqrt{\varphi(s)},\ \sqrt{\varphi(s)}\rangle \tag{16}$$

where $\sqrt{\phi} \in L^2$. Therefore there is certainly a representation of f under the form (II) of section 1.

Is it possible to represent f as

$$\lim_{T\to\infty} \frac{1}{T}\int_0^T e^{i\lambda\alpha(t)}dt \tag{17}$$

(we use $\frac{1}{T}\int_0^T$ rather than $\frac{1}{2T}\int_{-T}^T$ for practical reasons).

For this purpose, we shall introduce step functions, defined by *uniformly distributed* sequences (see annex).

Let z_n be a uniformly distributed sequence over]0,1[. If A is a function defined over]0,1[and Riemann-integrable, we know that (Weyl's theorem).

$$\lim_{N\to\infty} \frac{1}{N}\sum_{n=0}^{N} A(z_n) = \int_0^1 A(z)\ dz \tag{18}$$

Therefore

$$\lim_{N\ \infty} \frac{1}{N}\sum_{n=0}^{N} \exp(i\lambda A(z_n)) = \int_0^1 \exp(i\lambda A(z))\ dz \tag{19}$$

Let $\alpha(t)$ be a function equal to $A(z_n)$ for $n<t<n+1$). Then left-side member of (19) is exactly

$$\lim_{T\ \infty} \frac{1}{T}\int_0^T e^{i\lambda\alpha(t)}dt$$

The question reduces to finding A such that

$$f(\lambda) = \int_0^1 e^{i\lambda A(z)}dz \tag{20}$$

It is well known that it is possible. For instance we choose for A a function increasing from $-\infty$ to $+\infty$ when $0<z<1$, and introduce the inverse function A_1 such that

$$A(z) = u, \quad z = A_1(u)$$

If A_1 is differentiable,

$$f(\lambda) = \int_{\infty}^{\infty} e^{i\lambda u} A_1'(u)\, du \tag{21}$$

where A_1' is the Fourier transform of f.

This is nothing but a constructive interpretation of Bochner's theorem.

Let us now study the case of a two-dimensional characteristic function $f(\lambda, \mu)$.

<u>First Method</u>: We introduce a two dimensional sequence (x_n, y_n) uniformly distributed in the square $(0,1) \times (0,1)$. We suppose that the two sequences (x_n) and (y_n) are *independent* (see annex).

In order to solve the equation

$$f(\lambda, \mu) = \lim_{T\ \infty} \frac{1}{T} \int_0^T \exp i(\lambda\alpha(t) + \mu\beta(t))dt \tag{22}$$

we represent α and β by step functions, namely

$$\alpha(t) = A(x_n, y_n), \quad \beta(t) = B(x_n, y_n), \quad \text{for } n<t<n+1 \tag{23}$$

We have

$$f(\lambda, \mu) = \lim_{T\ \infty} \frac{1}{N} \sum_{n=0}^{N} \exp i[\lambda A(x_n, y_n) + \mu B(x_n, y_n)]$$

$$= \iint_C e^{i(\lambda A(x,y) + \mu B(x,y))} dxdy \tag{24}$$

where C is the square $(0,1)\times(0,1)$. (24) has the ordinary form of a characteristic function. If X and Y are two independent random variables, uniformly distributed in the square C , f is the characteristic function of the pair of random variables A(X,Y), B(X,Y) .

If the mapping

$$\xi = A(x,y) \quad \eta = B(x,y)$$

is invertible, and if J is the jacobian, we have

$$f(\lambda, \mu) = \int\int e^{i(\lambda\xi + \mu\eta)} \frac{1}{|J|} d\xi d\eta \tag{25}$$

and $\frac{1}{|J|}$ is the probability density corresponding to f.

We have only to choose A and B such that

$$\frac{D(A,B)}{D(x,y)} = J \qquad (26)$$

If A is a function of x_n only and not of y_n, then $J_n = \frac{\partial A}{\partial x}\frac{\partial B}{\partial y}$ and $\exp[i\lambda A(x_n)]$ is the characteristic function of the function α. This procedure gives an inductive method for constructing an n-dimensional characteristic function.

Second method - We shall show that the introduction of two sequences x_n, y_n is not necessary. Let z_n be a unique uniformly distributed sequence, and

$$\alpha(t) = A(z_n), \quad \beta(t) = B(z_n) \quad \text{for } n < t < n+1 \qquad (27)$$

Can we find A, B such that

$$f(\lambda, \mu) = \lim_{N\ \infty} \frac{1}{N} \sum_{n=0}^{N} e^{i(\lambda A(x_n) + \mu B(x_n))} \qquad (28)$$

$$= \int_0^1 e^{i(\lambda A(x,y) + \mu B(x,y))} dxdy \ ? \qquad (29)$$

Here A(X) and B(X) are two random variables, functions of X, uniformly distributed over (0,1). They are not independent.

Let X be a random variable, uniformly distributed over (0,1). Let us assume that it is possible to find two functions U(X), V(X) such that the two random variables U(X), V(X) are independent and uniformly distributed over (0,1). We define A(x) and B(x) by

$$A(x) = F_1[U(x), V(x)], \quad B(x) = F_2[U(x), V(x)] \qquad (30)$$

Then

$$f(\lambda,\mu) = \int_0^1 \exp i[\lambda F_1(U(x), V(x)] + \mu F_2[U(x), V(x)]\, dx \qquad (31)$$

$f(\lambda,\mu)$ is the stochastic mean value (expectation) of the random variable $\lambda F_1(U,V) + \mu F_2(U,V)$, defined over the probability space constituted by the interval (0,1). Therefore

$$f(\lambda,\mu) = \iint_C e^{i(\lambda F_1(u,v)+\mu F_2(u,v))} du\, dv \qquad (32)$$

where C is the square (0,1) x (0,1) .

As in the first method, we see that it is possible to choose F_1 and F_2 in such a way that f is a given function.

We must now prove that is possible to find two functions U, V defined over (0,1), such that the random variables U(x), V(x) are independent. First let us recall the definition of *Rademacher functions.* We represent a real number x, $0 \le x \le 1$ by its infinite binary expansion :

$$x = \sum_{k=1}^{\infty} \frac{a_k}{2^k} \qquad a_k = 0 \text{ or } 1 \tag{33}$$

Instead of (33), we prefer writing

$$1 - 2x = \sum_{k=1}^{\infty} \frac{r_k}{2^k}, \text{ with } r_k = 1 - 2a_k , \quad r_k = 1 \text{ or } -1 \tag{34}$$

r_k is a function of x, called Rademacher function.

It is easy to verify that $r_k(x)$ is a step function, such that

$$r_k(x) = (-1)^p , \qquad \frac{p-1}{2^k} < x < \frac{p}{2^k} , \qquad p = 1,2, \ldots . \; 2k \tag{35}$$

One has the formula [9]:

$$\int_0^1 \exp \left[i \sum_{k=1}^{n} [\lambda_k r_k(x)]\right] dx = \prod_{k=1}^{n} \int_0^1 \exp [i\lambda_k r_k(x)] \, dx \tag{36}$$

We give the proof in the case where n=2, which corresponds to the two-dimensional characteristic functions. We have to compare

$$f = \int_0^1 \exp [i(\lambda r_1(x) + \mu r_2(x))] \, dx$$

$$f_1 = \int_0^1 \exp [i\lambda r_1(x)] dx ; \qquad f_2 = \int_0^1 \exp [i\mu r_2(x)] dx.$$

It is easy to verify that

$$f = \frac{1}{4} \left[\exp[i(\lambda+\mu)] + \exp[i(\lambda-\mu)] + \exp[-i(\lambda-\mu)] + \exp[-i(\lambda+\mu)]\right]$$

$$f_1 = \frac{1}{2} [\exp(i\lambda) + \exp(-i\lambda)] = \cos \lambda$$

$$f_2 = \frac{1}{4} [\exp(i\mu) + \exp(-i\mu) + \exp(i\mu) + \exp(-i\mu)] = \cos \mu.$$

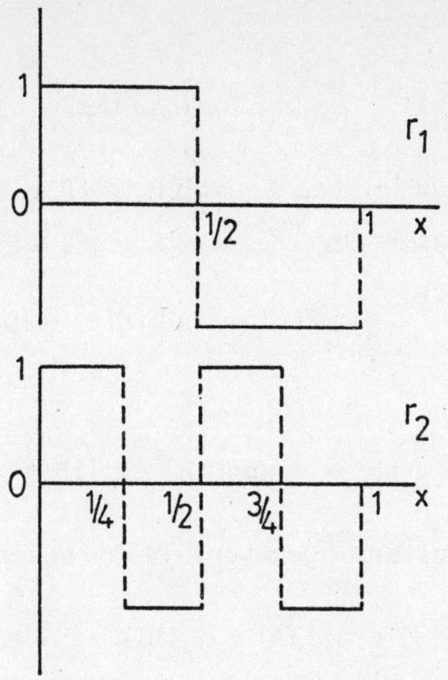

Therefore

$$\int_0^1 \exp\,[i(\lambda r_1(x) + \mu r_2(x))]\,dx = \int_0^1 \exp\,[i\lambda r_1(x)]\,dx \,.\, \int_0^1 \exp\,[i\mu r_2(x)]\,dx \qquad (37)$$

This formula signifies that the random variables $r_1(x)$, $r_2(x)$ are independent. The functions r_1, r_2 have the values -1, 1 . But it is clear that the property (37) is still valid for the functions a_1, a_2, which take the values 0, 1 .

The functions $a_1(X)$, a_2 (X), resulting from the binary expansion of X, constitute two independant random variables. They can be used as functions U(X), V(X) .

4 - SOME QUESTIONS CONCERNING CORRELATION FUNCTIONS.

We have seen that positive-definite functions can be interpreted not only as characteristic functions of a probability law but also as correlation functions.

We shall show that, in quantum mechanics as well as in turbulence, this notion of correlation leads to some contradictions.

a) *In quantum mechanics*, a hermitian operator A, associated with a state φ, generates a probability law and plays the role of a random variable. As ψ is a function of time, this random variable seems to be a random function of time. At two times t, t' , we have the pairs $(A,\psi(t))$ and $(A,\psi(t'))$, equivalent to a pair of random variables X(t), X(t'). Is it possible to attribute a significance to the mean value of the product X(t).X(t')?.

As the mean values are defined by a unique ψ, we must reduce $(A,\psi(t'))$ to the form $(B, \psi(t))$, in such a way that $(A, \psi(t'))$ and $(B, \psi(t))$ correspond to the same statistics ; they must have the same probability law, i.e. the same characteristic function :

$$\langle \exp(i\lambda A)\psi(t') \, , \, \psi(t')\rangle = \langle \exp(i\lambda B)\psi(t) \, , \, \psi(t)\rangle \tag{38}$$

But ψ satisfies Schrödinger equation. If H is the hamiltonian operator, the evolution of ψ is given by

$$\psi(t') = \exp[iH(t'-t)\psi(t)] \tag{39}$$

Therefore

$$\langle \exp(i\lambda A)\psi(t'),\psi(t'))\rangle = \langle \exp(i\lambda A) \exp[iH(t'-t)\psi(t)], \exp[iH(t'-t)\psi(t)]\rangle \tag{40}$$

As exp(iλH) is a unitary operator, (40) is verified if

$$\exp[-iH(t'-t)] \exp(i\lambda A) \exp[iH(t'-t)] = \exp(i\lambda B) \tag{41}$$

This equality for all values of λ is equivalent to

$$B = \exp[-iH(t'-t)].A.\exp[iH(t'-t)] \tag{42}$$

B is well defined. But in general this operator B does not commute with A. Thus even if we interpret (A, $\psi(t)$) as a random function, we cannot attribute any significance to the covariance of this function.

This remark makes it clear that quantum mechanics is different from classical statistical mechanics. The probability laws in quantum mechanics cannot be related to a unique probability space.

b) Let us now consider the problem of *turbulence*. The experiments in fully developed turbulence strongly suggest that turbulence must be represented by deterministic functions, namely pseudo-random functions of time (J.BASS [3,4]). The idea of a deterministic representation of turbulence is also contained in the new theories of dynamic systems, which concern the birth and evolution of turbulence, rather than the study of a stationary state of flow [7] .

The notion of time correlation of turbulence suffers no special difficulty. Let us introduce a symbol for the representation of time mean values. We represent by Mf the mean value

$$Mf = \lim_{T\to\infty} \frac{1}{T}\int_0^T f(t)\, dt \tag{43}$$

If u(x,t) is the velocity of the fluid, as a function of time at a given point x, the time correlation is given by

$$\Gamma(x,\tau) = M\ u(x,t)u(x,t+\tau)\ dt \tag{44}$$

But the "turbulencists" introduced, around the years 1930-40, the notion of *space correlation,* between $u(x,t)$ and $u(x+h,t)$, measured at two different points at the same time. It is still a time mean value, namely

$$R(x,h,t) = M\ u(x,t)\ u(x+h,t)\ dt \tag{45}$$

In general, it is a function of x and h. It has no property of homogeneity, i.e. stationarity in space. As the fluid is confined in a bounded box, clearly its properties at the center of the box are different from those near the walls.

Consequently, there is no reason for $R(x,h,t)$ to be the Fourier transform of a spectral measure. In particular R is not an even function of h.

Of course, we improve the structure by transforming the correlation function (bilinear covariance) into a correlation coefficient :

$$r(x,h,t) = \frac{R(x,h,t)}{\sqrt{Mu^2(x,t)}\sqrt{Mu^2(x+h,t)}} \tag{46}$$

The expansion of r in the vicinity of $h = 0$ is

$$r(x,h,t) = 1 - \frac{h^2}{2}\left[\frac{Mu^2Mu'^2 - (Muu')^2}{(Mu^2)^2}\right] + \ldots . \tag{47}$$

By Schwarz's inequality, we verify that the coefficient of h^2 is positive. The experimental evidence shows that:

1. The *time corrleation* is rapidly decreasing (with small oscillations) and tends to zero when $\tau \to \infty$. In other words, the velocity is a *pseudo-random function* (terminology introduced by J. BASS ([2, 3, 4, 5]).

2. The *space correlation* has, when h increases, a very similar shape. It approaches zero rapidly and this fact is not related to the presence of the boundaries.

A function $u(x,t)$ can then represent the velocity of a turbulent fluid if it satisfies the following requirements :

i) At every point it is a pseudo-random function of time ;
ii) Its space correlation - a time mean value - is rapidly decreasing and approaches zero when h increases.

We say that such a function is a *turbulent function.*

It is easy to give examples of turbulent functions. The most elementary one is a function defined by

$$u(x,t) = f(x+at) \tag{48}$$

where f is a pseudo-random function and a a constant. If γ is the correlation function of f, the time correlation of u is $\gamma(a\tau)$. The space correlation is $\gamma(h)$. This function is pseudo-random in spacee and in time .

More generally, let f_n be a sequence of independent pseudo-random functions :

$$M\, f_n(t)\, f_p(t+\tau) = 0 \quad \text{for} \quad n \neq p\ ,$$

$$= \gamma_n(\tau) \quad \text{for} \quad n = p \tag{49}$$

(correlation function of f_n).

The existencee of independent functions is explained in [5] .
If $c_n(x)$ is a sequence of functions of space, the function

$$u(x,t) = \Sigma\, c_n(x)\, f_n(x+at) \tag{50}$$

is a turbulent function (for an infinite series, trivial conditions of convergence are necessary). In fact, the mean value of $u(x,t)u(x,t+\tau)$ and $u(x,t)u(x+h,t)$ are

$$M\, u(x,t)u(x,t+\tau) = \Sigma\, c_n(x)c_p(x) M\, f_n(x+at) f_p(x+at+a\tau)$$

$$= \Sigma\, |c_n(x)|^2\, \gamma_n(a\tau) \tag{51}$$

$$M\, u(x,t)u(x+h,t) = \Sigma\, c_n(x)c_p(x+h) M f_n(x+at) f_p(x+h+at)$$

$$= \Sigma\, c_n(x)c_n(x+h)\, \gamma_n(h) \tag{52}$$

The decay of the space correlation when h increases is related to the structure of f_n, and does not depend on the coefficients $c_n(x)$. But in this case there is no homogeneity in space.

As a conclusion any theoretical approach of turbulence must take into account the fact that, for a stabilized turbulence, the functions which represent the velocity of the flow must be not only pseudo-random in time, but "turbulent" in space and time.

Annex - Uniformly distributed sequences.

A sequence z_n of real numbers is uniformly distributed over (0,1) if the following properties are satisfied :

i) $0 < z_n < 1$.

Let N′ be the number of points $z_1, z_2, \ldots . z_n$ such that $a<z_n<b$, where (a,b) is an arbitrary sub-inteerval of (0,1). Then

$$\lim_{N\to\infty} \frac{N'}{N} = b-a \, .$$

An analogous definition is available for a sequence of points belonging to the cube $(0, 1)^p$ of $\mathbb{R}^p$. (a,b) is replaced by an arbitrary parallelopiped.

A sequence z_n of numbers is p-uniformly distributed if the sequence $\{z_{n+p-1}, z_{n+p}\}$ is uniformly distributed in the cube $(0,1)^p$. A theorem of H.weyl (1916-see [5]) gives a condition for a sequence to be uniformly distributed :a sequence z_n is uniformly distributed if, for any Riemann-integrable function ϕ, one has

$$\lim_{N\to\infty} \frac{1}{N} \sum_{n=1}^{N} \phi\,(z_n) = \int_0^1 \phi(z)\, dz \tag{54}$$

An equivalent condition is that, for any integer $\imath$ different from zero,

$$\lim_{N\to\infty} \frac{1}{N} \sum_{n=1}^{N} \exp(2i\pi\imath z_n) = 0 \tag{55}$$

Let $P(z) = a_0\, z^p + a_1\, z^{p-1} + \ldots + a_p$ be a polynomial of degree $p \geq \imath$ with real coefficients. If a_0 is an irrational number, the sequence {p(n), modulo 1}, is p-uniformly distributed .

The sequence $\{\Theta^n$, modulo 1$\}$ is uniformly distributed for almost every Θ . But we don't know explicitly the values of Θ having this property. Only counter examples are known: for $\Theta = \frac{1+\sqrt{5}}{2}$, the sequence Θ^n is not uniformly distributed modulo 1.

If the sequence z_n is 2-uniformly distributed, the function equal to $\varnothing\,(z_n)$ for $n<t<n+1$, and 0 for $t<0$, is pseudo-random. Its correlation function is equal to $A(1-|\tau|)\, U(\tau)$ where

$$U(\tau) = 1 \text{ if } |\tau| \leq 1 \text{ , and } 0 \text{ if } |\tau| \geq 1$$

$$A = \int_0^1 |\phi(z)|^2 \, dz.$$

Two uniformly distributed sequences z_n , s_n are independent if the sequence $lz_n + l's_n$ is uniformly distributed, whatever be the integers l, l' not simultaneously equal to zero. It is equivalent to saying that

$$\lim_{N\to\infty} \frac{1}{N} \sum_{n=1}^{N} \exp\,[i(lz_n + l's_n)] = 0.$$

If z_n and s_n are independent,

$$\lim_{N\to\infty} \frac{1}{N} \sum_{n=1}^{N} \phi\,(z_n, s_n) = \iint_C \phi\,(z,s)\,dz\,ds$$

where C is the square $(0,1)^2$.

Examples of independent sequences : two polynomials P,Q of different degrees. Two polynomials $p = a_0 z^p + \ldots, Q = b_0 z^p + \ldots$ of the same degree, when the irrational numbers a_0, b_0 are independent (i.e. $\lambda a_0 + \mu b_0 = \nu$, λ, μ, ν integers, implies $\lambda = \mu = \nu = 0$; $\sqrt{2}$ and $\sqrt{3}$ are independent but $\sqrt{2}$ and $1+\sqrt{2}$ are not).

REFERENCES

1) ARNOUS, E., Quelques applications de la théorie des groupes de transformations unitaires en calcul des probabilitiés et en mécanique quantique, La revue scientifique , n° 3273, 1947.

2) BASS, J., Lois de probabilité, équations hydrodynamiques mécanique quantique, La revue scientifique, n° 3299, 1948.

3) BASS, J., Les fonctions pseudo-alèatories, Mémorial des sciences mathématiques, Gauthier-Villars, 1962 .

4) BASS, J., Stationary functions and their applications to the theory of turbulence, J. of mathematical analysis and applications, 47, No.2 & 3, 1974.

5) BASS, J., Fonctions de corrélation, fonctions pseudo-aléatoires et applications, Masson, 1984.

6) BASS, J., Fonctions de corrélation des fonctions pseudo-aléatoires, Ann. Inst. H. Poincaré, 25, 1989.

7) BERGE,P., POMEAU,Y., VIDAL,Ch.,L'ordre dans le chaos. Vers une approache déterministe de la turbulence, Hermann, 1984.

8) BROGLIE, L. de., Les incertitudes d'Heisenberg et l'interprétation statistique de la mécanique ondulatoire, Gauthier-Villars, 1982.

9) KAC, M., Statistical independance in probability, analysis and number theory, J.Wiley, 1959.

10) KARMAN, Th. Von., and HOWARTH, L., On the statistical theory of isotropic turbulence, Proc. Roy. Soc., A 164, 1938.

11) PIQUET, Cl., Fonctions de type positif associées à deux opérateurs hermitiens, C.R. Acad. Sc., Paris, 279, série A ,1974.

12) URBANIK, K.,Joint probability distributions of observables in quantum mechanics, Studia Math., 21, 1967.

POPULATION MONITORING AND THE QUANTUM INPUT-OUTPUT FORMALISM

J. Jeffers

Physics Department, University of Essex, Wivenhoe Park, Colchester, CO4 3SQ, United Kingdom

and

T. J. Shepherd

Royal Signals and Radar Establishment, St. Andrews Road, Malvern, Worcestershire, WR14 3PS,

United Kingdom

ABSTRACT

The quantum Input-Output formalism of Gardiner and Collett is used to derive the rate equation describing the photoelectron counting statistics of an electromagnetic field output from a high-Q optical cavity. The evolution of the cavity field is assumed to be governed by an arbitrary Hamiltonian for a single mode of the field. The resulting equation is shown to conform with one previously derived on intuitive grounds, using purely population statistical arguments. As an explicit example, the solution for a freely-evolving cavity field is computed.

Glossary of Principal Terms Used in Text

$\hat{a}(t)$, $\{\hat{a}_{in}(t)$, $\hat{a}_{out}(t)\}$	cavity, {input, output} field operator.
$\hat{d}$, $\hat{D}$	detection operators.
FKE	forward Kolmogorov equation.
$\hat{g}(s; t)$	$(1-s)^{\hat{n}(t)}$.
$\hat{h}(s'; t)$	$(1-\eta s')^{\hat{N}(t)}$.
$\hat{H}$	Hamiltonian operator.
$\hbar$	Planck's constant.
IO	Input-Output.
MGF	moment generating function.
$n(t)$	number of photons (individuals) in cavity population at time t.
$\hat{n}$	cavity photon number operator.
$N(t)$	number of photoelectron counts registered by detector in interval $[0, t)$.
$\hat{N}$	output number operator.
$p(n; t)$	probability distribution for n.
$P(n, N; t)$	joint probability distribution for n and N.
$Q(s; t)$	MGF of $p(n; t)$, with expansion parameter s.
$Q_0(s)$	initial cavity MGF.
$\mathbf{Q}(s, s'; t)$	MGF of $P(n, N; t)$, with expansion parameters s and s'.
γ	cavity loss rate.
λ, μ, ν	population birth, death, and immigration rates, respectively.
η	detector quantum efficiency.
$\hat{v}$	vacuum field for detector.
$\hat{\rho}$	density operator.
ω	cavity mode angular frequency.

1. INTRODUCTION

Many quantum optical systems can be described using classical population statistics: indeed, Srinivasan has extensively employed the formalism of classical point processes to analyse a range of systems representing cavity radiation models[1]. When an electromagnetic field is describable in terms of photon number, the evolution of the photon statistics can often conform to those of a population model, and visualisation of the behaviour of the system at the microscopic level is greatly simplified.

For such systems it is comparatively straightforward to include a mechanism for direct detection of the field within the population model[2]; one such process has been dubbed *population monitoring*[3]. To date, however, the justification for this detection formulation has involved purely intuitive arguments, although for some particular examples of field statistics the population monitoring approach has yielded results equivalent to those obtained by purely quantum field theoretic methods[2]. It would be preferable to demonstrate the equivalence *ab initio*.

In the following we shall use the quantum Input-Output (IO) methods formulated by Gardiner and Collett[4, 5] to derive an equation for the photoelectron statistics for a detected field from a high-Q cavity, and show that, in the situations where the cavity field alone can be described as a simple population process, the photoelectron statistics are equivalent to those obtained by population monitoring.

2. POPULATION MONITORING

Consider a population containing $n(t)$ individuals at time t, where $n(t)$ is a Markov process. Let $p(n;\, t)$ represent the probability of there being n individuals in the population at time t. Now define the moment generating function (MGF) for this distribution as $Q(s;\, t)$, where

$$Q(s;\, t) = \sum_{n=0}^{\infty} p(n;\, t)\,(1-s)^n \,. \tag{2.1}$$

We shall assume that the evolution of $p(n;\, t)$ is governed by a forward Kolmogorov equation (FKE) which can be expressed in terms of the MGF as follows,

$$\frac{\partial Q(s;\, t)}{\partial t} = \psi(s,\, \partial/\partial s)\, Q(s;\, t)\,. \tag{2.2}$$

Here, $\psi(s,\, \partial/\partial s)$ is a differential operator appropriate to a given system; for example, if the population evolution is governed by homogeneous birth, death, and immigration processes, with rates λ, μ, and ν, respectively, the FKE is

$$\dot{p}(n;\, t) = \lambda(n-1)\, p(n-1;\, t) + \mu(n+1)\, p(n+1;\, t) + \nu p(n-1;\, t) - (\nu + (\lambda+\mu)\, n)\, p(n;\, t) \tag{2.3}$$

and from (2.1) and (2.2) the corresponding function ψ is

$$\psi(s,\, \partial/\partial s) = -\,\nu s - \mu s\, \frac{\partial}{\partial s} + \lambda s\,(1-s)\, \frac{\partial}{\partial s}\,. \tag{2.4}$$

For a single mode of a quantised cavity electric field, $n(t)$ represents the corresponding photon number. Detection of the field leaking out of the cavity may be incorporated into this picture by extending $p(n;\, t)$ to the *joint* distribution $P(n, N;\, t)$, defined as the joint probability that there are n photons in the cavity at time t, *and* that $N(t)$ photoelectrons have been counted at the detector in the semi-open interval $[0,\, t)$, (i.e. excluding any count at the point t). The manner in which $N(t)$ is determined from the process for $n(t)$ is defined by *population monitoring*[3]. In this, the photons are assumed to leak from the cavity in accordance with a death process of rate γ, say, and are registered at the detector with probability (quantum efficiency) η. The associated extended FKE for $P(n, N;\, t)$ contains terms, associated with the cavity processes, similar to those in the equation for $p(n;\, t)$, but now merely appended by the parameter N, as these processes involve no change in the photoelectron number. The detection processes above introduce the additional contributions to the FKE for $P(n, N;\, t)$: [6]

$$\dot{P}_{additional}(n, N;\, t) = \eta\gamma\,(n+1)\, P(n+1, N-1;\, t) + (1-\eta)\gamma\,(n+1)\, P(n+1, N;\, t) - \gamma n\, P(n, N;\, t). \tag{2.5}$$

Upon introduction of the MGF Ω for the joint distribution $P(n, N;\, t)$,

$$Q(s,s';t) = \sum_{n=0}^{\infty}\sum_{N=0}^{\infty} P(n,N;t)\,(1-s)^n\,(1-s')^N \quad , \tag{2.6}$$

the general FKE for $P(n, N; t)$ is transformed to

$$\frac{\partial Q}{\partial t} = \psi(s, \partial/\partial s)\, Q - \gamma s \frac{\partial Q}{\partial s} + \eta\gamma s' \frac{\partial Q}{\partial s} \; . \tag{2.7}$$

Setting $s'= 0$ in $Q(s, s'; t)$ recovers the MGF $Q(s; t)$ for the cavity population alone. Comparison of eqs. (2.2) and (2.7) clearly reveals that the effect of population monitoring on the undetected cavity population is merely to introduce an additional death process of rate γ. The term $\eta\gamma s'\partial Q/\partial s$ in (2.7) is responsible for the specific detection characteristics[2].

In practice, it has often been found that if (2.2) has an exact analytic solution, then so also does (2.7). For a Markov process, the solution to (2.7) provides all the information needed to determine any multilinear moment of the counting distribution. The population monitoring procedure has also been applied by Srinivasan to several *non-Markovian* population systems[1].

3. THE IO FORMALISM

Gardiner and Collett have developed a formalism describing a single mode of a high-Q cavity quantum electric field coupled to the continuum of field modes *outside* the cavity[4,5]. In the following we consider only a *one-sided* cavity, in which one mirror is transmitting and the other totally reflecting. For an isolated system the field mode is described by the Heisenberg picture annihilation and creation operators $\hat{a}(t)$ and $\hat{a}^\dagger(t)$, which evolve according to the Heisenberg equation

$$\dot{\hat{a}}(t) = -i/\hbar\, [\, \hat{a}(t), \hat{H}\,] \; , \tag{3.1}$$

where $\hat{H}$ is the given system Hamiltonian, and $\hbar$ is Planck's constant divided by 2π. When coupled to the outside world, however, the mode evolves according to the modified equation

$$\dot{\hat{a}}(t) = -i/\hbar\, [\, \hat{a}(t), \hat{H}\,] - \gamma/2\, \hat{a}(t) + \gamma^{1/2} \hat{a}_{in}(t) \quad . \tag{3.2}$$

Here $\hat{a}_{in}(t)$ is the Fourier transform of an input external field mode operator, and γ is the cavity damping constant, proportional to the transmission coefficient of the input-output mirror. The operators $\hat{a}_{in}(t)$, $\hat{a}_{in}^\dagger(t)$ possess the commutator

$$[\hat{a}_{in}(t), \hat{a}_{in}^\dagger(t')] = \delta(t-t') \quad . \tag{3.3}$$

In addition, a further external operator $\hat{a}_{out}(t)$ represents the *output* from the cavity. This operator has a commutator similar to that for $\hat{a}_{in}(t)$,

$$[\hat{a}_{out}(t), \hat{a}_{out}^\dagger(t')] = \delta(t-t') \; , \tag{3.4}$$

and is related to $\hat{a}_{in}(t)$ and $\hat{a}(t)$ through the boundary condition

$$\hat{a}(t) = \gamma^{-1/2}(\, \hat{a}_{in}(t) + \hat{a}_{out}(t)\,) \; . \tag{3.5}$$

Causality is assured by demanding that the following commutators hold:

$$[\, F_1(\hat{a}^\dagger(t), \hat{a}(t)), F_2(\hat{a}_{in}(t'), \hat{a}_{in}^\dagger(t')\,] = 0 \quad , \quad t' > t \; , \tag{3.6}$$

$$[\, F_3(\hat{a}^\dagger(t), \hat{a}(t)), F_4(\hat{a}_{out}(t'), \hat{a}_{out}^\dagger(t')\,] = 0 \; , \quad t > t' \; , \tag{3.7}$$

(where F_1, F_2, F_3, and F_4 are given functions), implying that an input field at a particular time can have no effect upon the cavity field at an *earlier* time, and that a cavity field can have no effect upon an output field at an earlier time. Equations (3.2), (3.5), (3.6) and (3.7) ensure that *unitarity* is maintained in this scheme, i.e.

$$[\, \hat{a}(t), \hat{a}^\dagger(t)\,] = 1 \qquad \forall\, t \, . \tag{3.8}$$

More general system operators $\hat{c}(t)$ (e.g. functions of $\hat{a}(t)$, $\hat{a}^\dagger(t)$) evolve according to the modified Heisenberg equation

$$\dot{\hat{c}}(t) = -i/\hbar\,[\,\hat{c}(t), \hat{H}\,] - \{\,[\hat{c}(t),\ \hat{a}^{\dagger}(t)]\,(\gamma/2\,\hat{a}(t) - \gamma^{1/2}\hat{a}_{in}(t)\,)$$

$$-\,(\gamma/2\,\hat{a}^{\dagger}(t) - \gamma^{1/2}\hat{a}_{in}{}^{\dagger}(t)\,)\,[\hat{c}(t)\,,\hat{a}(t)]\,\}\ , \tag{3.9}$$

and the operators $(\gamma/2\ \hat{a}(t) - \gamma^{1/2}\hat{a}_{in}(t))$ and $(\gamma/2\ \hat{a}(t) - \gamma^{1/2}\hat{a}_{out}(t))$ can be shown to commute with all *system* operators at time t.[5]

4. PHOTODETECTION IN THE IO FORMALISM

In order to relate the IO formalism sketched above to the photodetection problem described earlier, we shall assume that the cavity in question is one-sided and that no light is input to the cavity. Thus, *the input state is assumed to consist solely of the vacuum.* The internal cavity field evolves in accordance with eq. (3.2), and the output field $\hat{a}_{out}$ interacts with the detector, which registers the photoelectron counts. A general detector quantum efficiency η is included by noting that the relevant operator giving the statistics of the counting process is $\hat{d}(t)$, given by the unitary transformation

$$\hat{d}(t) \;=\; \eta^{1/2}\,\hat{a}_{out}(t) + (1-\eta)^{1/2}\,\hat{v}(t)\ , \tag{4.1}$$

where $\hat{v}(t)$ is a vacuum field, contributing nothing to normally-ordered moments of $\hat{d}(t)$ and $\hat{d}^{\dagger}(t)$, and commuting with all other operators[7]. Moments of the photocounting distribution for the interval $[0, t)$ are given by moments of the detection operator $\hat{D}(t)$, where

$$\hat{D}(t) \;=\; \int_0^t dt'\ \hat{d}^{\dagger}(t')\,\hat{d}(t')\ , \tag{4.2}$$

so that, for example, the mean-square of the number of photoelectrons $N(t)$ counted in the interval $[0, t)$ is

$$\langle (N(t))^2 \rangle \;=\; \mathrm{Tr}\,[\,\hat{\rho}\,(\hat{D}(t))^2\,]\ , \tag{4.3}$$

where Tr denotes the quantum mechanical trace over the subsequent operators, and $\hat{\rho}$ is the relevant density operator.

We need to construct from these operators an MGF that generates such moments. The obvious choice corresponding to eq. (2.6) is

$$Q(s, s'; t) \;=\; \langle\,(1-s)^{\hat{n}(t)}\,(1-s')^{\hat{D}(t)}\,\rangle\ , \tag{4.4}$$

where

$$\hat{n}(t) \;=\; \hat{a}^{\dagger}(t)\,\hat{a}(t) \tag{4.5}$$

is the photon number operator for the cavity field.

In principle, $Q(s, s'; t)$ in (4.4) generates the joint distribution $P(n, N; t)$. This quantity is not uniquely defined quantum mechanically, however, until it can be demonstrated that the operators $\hat{n}(t)$ and $\hat{D}(t)$ commute. Now, since $\hat{v}(t')$ commutes with all other operators, and $\hat{D}(t)$ consists of operators $\hat{a}_{out}(t')$ and $\hat{v}(t')$ with $t' < t$, it follows from the causality relations in (3.7) that $\hat{D}(t)$ commutes with all system (cavity) operators at time t. Hence

$$[\,\hat{n}(t), \hat{D}(t)\,] \;=\; 0\ , \tag{4.6}$$

and the MGF in (4.4) is uniquely defined. Note that (4.6) implies that the number of photons in the cavity at time t and the number of photoelectrons counted up to time t are simultaneously measurable.

5. EFFECT OF QUANTUM EFFICIENCY

In order to derive an evolution equation for $Q(s, s'; t)$ we must eliminate the operator $\hat{d}(t)$ in favour of the operator $\hat{a}_{out}(t)$. To do this, we shall employ the normal-order theorem for continuous-mode operators[8]. For arbitrary function $q(t)$, this reads

$$\exp\Big(\int_{-\infty}^{\infty} dt'\ q(t')\,\hat{d}^{\dagger}(t')\,\hat{d}(t')\Big) \;=\; :\exp\Big(\int_{-\infty}^{\infty} dt'\ (\,\exp q(t') - 1\,)\,\hat{d}^{\dagger}(t')\,\hat{d}(t')\Big):\ , \tag{5.1}$$

where the colons denote normal ordering of the enclosed expression, *viz.*, all annihilation operators stand to the right of all creation operators. Setting

$$q(t') = \chi(t, t') \log(1-s') , \tag{5.2}$$

where

$$\chi(t, t') = \begin{cases} 0, & -\infty < t' < 0 \\ 1, & 0 \leq t' < t \\ 0, & t \leq t' < \infty , \end{cases} \tag{5.3}$$

and noting that

$$\{\chi(t, t')\}^2 = \chi(t, t') , \tag{5.4}$$

(5.1) becomes

$$(1-s')^{\hat{D}(t)} = :e^{-s' \hat{D}(t)}: . \tag{5.5}$$

Using the fact that $\hat{v}(t)$ commutes with all other operators, and is in a vacuum state, (5.5) gives

$$Q(s, s'; t) = \langle (1-s)^{\hat{n}(t)} (1-s')^{\hat{D}(t)} \rangle = \langle (1-s)^{\hat{n}(t)} :e^{-\eta s' \hat{N}(t)}: \rangle , \tag{5.6}$$

where

$$\hat{N}(t) = \int_0^t dt' \, \hat{a}_{out}^{\dagger}(t') \, \hat{a}_{out}(t') . \tag{5.7}$$

Using the normal ordering theorem on (5.6), we may write

$$Q(s, s'; t) = \langle (1-s)^{\hat{n}(t)} (1-\eta s')^{\hat{N}(t)} \rangle . \tag{5.8}$$

The MGF for a system involving a perfect detector is clearly given by (5.8) with η=1. The fact that η appears as a multiplier of the expansion parameter s' in (5.8) is an expression of the *Bernoulli* or *binomial sampling* of the output photon distribution by an inefficient detector[9].

6. EVOLUTION EQUATION FOR $Q(s, s'; t)$

The temporal evolution of $Q(s, s'; t)$ is dictated by $\partial Q/\partial t$, and thus we require a partial differential equation for Q to compare with (2.7). First, we shall rationalise notation by defining

$$\hat{g}(s; t) = (1-s)^{\hat{n}(t)} , \tag{6.1}$$

and

$$\hat{h}(s'; t) = (1-\eta s')^{\hat{N}(t)} = :e^{-\eta s' \hat{N}(t)}: , \tag{6.2}$$

so that, from (5.8),

$$Q(s, s'; t) = \langle \hat{g}(s; t) \, \hat{h}(s'; t) \rangle . \tag{6.3}$$

Note that $\partial Q/\partial s$ is easily obtained, for

$$\frac{\partial \hat{g}(s; t)}{\partial s} = -\hat{n}(t) (1-s)^{\hat{n}(t)-1} , \tag{6.4}$$

and hence

$$\frac{\partial Q}{\partial s} = -(1-s)^{-1}\langle \hat{n}(t)\,\hat{g}(s;t)\,\hat{h}(s';t)\rangle . \tag{6.5}$$

A consequence of the canonical commutator (3.8) is the identity, for any function $f(\hat{a}^\dagger(t), \hat{a}(t))$, that[10]

$$e^{x\hat{n}(t)} f(\hat{a}^\dagger(t), \hat{a}(t))\, e^{-x\hat{n}(t)} = f(\hat{a}^\dagger(t)\, e^{x}, \hat{a}(t)\, e^{-x}) . \tag{6.6}$$

In particular, setting

$$e^{x} = 1-s , \tag{6.7}$$

it is trivial to show that

$$[\hat{g}(t), \hat{a}^\dagger(t)] = -s/(1-s)\,\hat{g}(t)\,\hat{a}^\dagger(t) , \tag{6.8a}$$

$$[\hat{g}(t), \hat{a}(t)] = s/(1-s)\,\hat{a}(t)\,\hat{g}(t) . \tag{6.8b}$$

Replacing $\hat{c}(t)$ by $\hat{g}(s;t)$ in (3.9), and using eqs. (6.8), we obtain

$$\begin{aligned}\frac{\partial \hat{g}(s;t)}{\partial t} &= -i/\hbar\,[\hat{g}(s;t), \hat{H}] + \gamma s/(1-s)\,\hat{g}(s;t)\,\hat{n}(t) \\ &\quad - s/(1-s)\,\gamma^{1/2}\left(\hat{g}(s;t)\,\hat{a}^\dagger(t)\,\hat{a}_{in}(t) + \hat{a}_{in}^\dagger(t)\,\hat{a}(t)\,\hat{g}(s;t)\right) .\end{aligned} \tag{6.9}$$

In (6.9) we have used the fact that $\hat{g}$ is a function of $\hat{n}$, and thus commutes with it.

$\partial\hat{h}(s';t)/\partial t$ will be calculated using the normal-order form in (6.2), in which

$$\hat{h}(s';t) = \sum_{r=0}^{\infty} (-\eta s')^r/r! : \hat{N}^r(t) : . \tag{6.10}$$

Differentiating term-by-term, it is straightforward to show that

$$\frac{\partial}{\partial t} :\hat{N}^r(t): = r\,\hat{a}_{out}^\dagger(t) : \hat{N}^{r-1}(t) : \hat{a}_{out}(t) , \tag{6.11}$$

and hence

$$\frac{\partial \hat{h}(s';t)}{\partial t} = -\eta s'\,\hat{a}_{out}^\dagger(t)\,\hat{h}(s';t)\,\hat{a}_{out}(t) . \tag{6.12}$$

Now, from (6.3),

$$\frac{\partial Q(s,s';t)}{\partial t} = \left\langle \frac{\partial \hat{g}(s;t)}{\partial t}\,\hat{h}(s';t)\right\rangle + \left\langle \hat{g}(s;t)\,\frac{\partial \hat{h}(s';t)}{\partial t}\right\rangle , \tag{6.13}$$

and we may substitute from eqs. (6.9) and (6.12) for $\partial\hat{g}/\partial t$ and $\partial\hat{h}/\partial t$. Details of further manipulations are omitted for brevity, but make use of the following properties: (*i*) $\hat{a}_{in}(t)$ is in a vacuum state, and so contributes zero when situated to the *right* of all other operators; (*ii*) similarly, $\hat{a}_{in}^\dagger(t)$ contributes zero when situated to the *left* of all other operators; (*iii*) the boundary condition (3.5); (*iv*) causality, as explained in Section 3; (*v*) commutation of the operators $(\gamma/2\,\hat{a}(t) - \gamma^{1/2}\,\hat{a}_{in}(t))$ and $(\gamma/2\,\hat{a}(t) - \gamma^{1/2}\,\hat{a}_{out}(t))$ with all cavity operators at time t. The result is

$$\frac{\partial Q}{\partial t} = -i/\hbar \left\langle [(1-s)^{\hat{n}(t)} (1-\eta s')^{\hat{N}(t)}, \hat{H}] \right\rangle - \gamma s \frac{\partial Q}{\partial s} + \gamma\eta s' \frac{\partial Q}{\partial s} . \tag{6.14}$$

Eq. (6.14) is our central result. Comparison with eq. (2.7) reveals that a mechanism identical to that for population monitoring governs both the loss from the cavity (in the term $-\gamma s \partial Q/\partial s$) and the detection characteristics (in the term $\eta\gamma s' \partial Q/\partial s$). Clearly, if, for some Hamiltonian $\hat{H}$, the expression $\langle [\hat{g}\hat{h}, \hat{H}] \rangle$ can be written explicitly in terms of Q, then the cavity process might become identifiable as a population process, and the quantum result pictured in terms of population monitoring. The following is a simple example of where this situation holds.

7. AN EXPLICIT EXAMPLE - THE FREE FIELD

Here we consider an example of monitoring a population for which there are no active internal processes in the *isolated* system. In the population picture, this simply implies that the population distribution remains constant in time, or

$$\dot{p}(n; t) = 0, \qquad \frac{\partial Q(s; t)}{\partial t} = 0 , \tag{7.1}$$

and hence, from (2.2),

$$\psi(s, \partial/\partial s) = 0 . \tag{7.2}$$

According to (2.7), the corresponding equation describing the field detection becomes

$$\frac{\partial Q}{\partial t} = -\gamma s \frac{\partial Q}{\partial s} + \gamma \eta s' \frac{\partial Q}{\partial s} , \tag{7.3}$$

for which the solution is[3]

$$Q(s, s'; t) = Q_0(s e^{-\gamma t} + \eta s'(1 - e^{-\gamma t})) , \tag{7.4}$$

where $Q_0(s) = Q(s, 0; 0)$ is the initial *cavity* MGF. Eq. (7.4) describes an exponential loss of photons from the cavity, with loss rate γ, and a Bernoulli sampling of the initial cavity distribution, with sampling factor $\eta(1-e^{-\gamma t})$. This solution was first discussed by Mollow[11].

In the quantum formalism, the Hamiltonian for a single cavity mode free field is given by

$$\hat{H} = \hbar\omega\hat{n} , \tag{7.5}$$

where ω is the mode angular frequency. From the Heisenberg equation for $\hat{g}(s; t)$, we know

$$\frac{\partial Q(s; t)}{\partial t} = \frac{\partial}{\partial t} \langle (1-s)^{\hat{n}(t)} \rangle = -i/\hbar [\hat{g}(s; t), \hat{H}] = 0 , \tag{7.6}$$

which follows since $\hat{g}(s; t)$ and $\hat{H}$ are both explicit functions of $\hat{n}$. Thus, the free quantum field also displays no evolution of the photon probability distribution.

When the interaction with external modes is introduced, however, the internal field mode evolves according to eq. (3.2), which, using (7.5), becomes

$$\dot{\hat{a}}(t) = -i\omega\hat{a}(t) - \gamma/2\, \hat{a}(t) + \gamma^{1/2} \hat{a}_{in}(t) . \tag{7.7}$$

for which the solution is

$$\hat{a}(t) = \hat{a}(0) e^{-(i\omega + \gamma/2)t} + \gamma^{1/2} \int_0^t du\, e^{-(i\omega+\gamma/2)(t-u)} \hat{a}_{in}(u) . \tag{7.8}$$

In order to compute $Q(s, s'; t)$ explicitly, we shall use eq. (5.6) expressed in normal-order form for both external and internal fields[10],

$$Q(s, s'; t) = \langle :e^{-s\hat{n}(t)} :: e^{-\eta s' \hat{N}(t)} : \rangle . \tag{7.9}$$

Expanding in powers of s', we obtain

$$Q(s,s';t) = \sum_{r=0}^{\infty}(-\eta s')^r/r! \int_0^t \cdots \int_0^t \prod_{k=0}^{r} dt_k \langle \hat{a}_{out}^\dagger(t_k) :e^{-s\hat{n}(t)}: \hat{a}_{out}(t_k) \rangle . \tag{7.10}$$

Substituting for $\hat{a}_{out}(t_k)$, $\hat{a}_{out}^\dagger(t_k)$ in favour of $\hat{a}(t_k)$, $\hat{a}_{in}(t_k)$, etc., from (3.5), and noting that $\hat{a}_{in}$ is a vacuum field, results in

$$Q(s,s';t) = \sum_{r=0}^{\infty}(-\gamma\eta s')^r/r! \int_0^t \cdots \int_0^t \prod_{k=0}^{r} dt_k \langle \hat{a}^\dagger(t_k) :e^{-s\hat{n}(t)}: \hat{a}(t_k) \rangle . \tag{7.11}$$

Substituting next for $\hat{a}(t_k)$, $\hat{a}^\dagger(t_k)$ from (7.8), and again using the vacuum property for $\hat{a}_{in}$ gives

$$Q(s,s';t) = \sum_{r=0}^{\infty}(-\gamma\eta s')^r/r! \int_0^t \cdots \int_0^t \prod_{k=0}^{r} dt_k \langle \hat{a}^\dagger(0)^r e^{-(-i\omega+\gamma/2)t_k} :e^{-s\hat{n}(t)}: \hat{a}(0)^r e^{-(i\omega+\gamma/2)t_k} \rangle , \tag{7.12}$$

for which the integrals may be performed to give

$$Q(s,s';t) = \sum_{r=0}^{\infty}(-\eta s'(1-e^{-\gamma t}))^r/r! \langle \hat{a}^\dagger(0)^r \sum_{q=0}^{\infty}(-s)^q/q! \hat{a}^\dagger(t)^q \hat{a}(t)^q \hat{a}(0)^r \rangle . \tag{7.13}$$

After insertion of $\hat{a}(t)$, $\hat{a}^\dagger(t)$ once more from (7.8), together with the causality property that $\hat{a}_{in}(t)$, $\hat{a}_{in}^\dagger(t)$ $(t>0)$ commute with $\hat{a}(0)$, $\hat{a}^\dagger(0)$, a little algebra yields

$$Q(s,s';t) = :\exp\{-(se^{-\gamma t} + \eta s'(1-e^{-\gamma t}))\hat{a}^\dagger(0)\hat{a}(0)\}: . \tag{7.14}$$

However, with the form of the initial cavity distribution MGF,

$$Q_0(s) = Q(s,0;0) = :\exp\{-s\hat{a}^\dagger(0)\hat{a}(0)\}:, \tag{7.15}$$

(7.14) may be expressed exactly in the form (7.4). Thus, the population monitoring and quantum detection formulations coincide precisely for a free cavity field.

8. CONCLUSIONS

In the foregoing sections we have shown that the intuitive picture of population monitoring is supported by the quantum IO formalism of Gardiner and Collett, when the external field input to the cavity is in a vacuum state. We have also shown the equivalence explicitly with reference to the example of a freely-evolving single-mode cavity field. This tends to justify the use of the population monitoring techniques in population statistical treatments of cavity radiation.

Examples of systems, other than that of a free field, for which the commutator in eq. (6.16) has an ensemble average expressible in terms of $Q(s,s';t)$ are relatively few. In general, evolution of the photon density operator (in the number representation) involves exchange between different diagonals, i.e. non-trivial phase evolution, and the system can not easily be described in population-statistical terms without the introduction of reservoirs into the model to render the cavity field statistics Markovian. It is, however, possible to extend the analysis above to include the relevant phase evolution, and this aspect will be explored in a future paper.

ACKNOWLEDGEMENTS

The authors wish to thank Professor Rodney Loudon and Dr. Matthew Collett for useful conversations and correspondence. One of us (J. J.) thanks the Science and Engineering Research Council and the Royal Signals and Radar Establishment for a C.A.S.E. studentship award.

REFERENCES

1. S. K. Srinivasan, *Point Process Models of Cavity Radiation and Detection*, (Charles Griffin and Co., Ltd., London, 1988).
2. T. J. Shepherd, *A model for photodetection of single-mode cavity radiation*, Optica Acta, **28**, 567, (1981).
3. T. J. Shepherd, *Photoelectron counting - semiclassical and population monitoring approaches*, Optica Acta, **31**, 1399, (1984).
4. M. J. Collett and C. W. Gardiner, *Squeezing of intracavity and travelling-wave light fields produced in parametric amplification*, Phys. Rev., **30**, 1386, (1984).
5. C. W. Gardiner and M. J. Collett, *Input and output in damped quantum systems: quantum stochastic differential equations and the master equation*, Phys. Rev. **31**, 3761, (1985).
6. E. Jakeman and T. J. Shepherd, *Population statistics and the counting process*, J. Phys. A **17**, L 745, (1984).
7. H. P. Yuen and J. H. Shapiro, *Optical communication with two-photon coherent states - Part III: quantum measurements realizable with photoemissive detectors*, IEEE Trans. Inf. Theory, **26**, 78, (1980).
8. K. J. Blow, R. Loudon, S. J. D. Phoenix, and T. J. Shepherd, *Continuum fields in quantum optics*, Phys. Rev. A **42**, 4102, (1990).
9. F. Ghielmetti, *Some comments on the relation between photoelectron and photon statistics and on the scaling properties of laser light*, Nuovo Cim. B **35**, 243, (1976).
10. W. H. Louisell, *Quantum Statistical Properties of Radiation*, (John Wiley and Sons, New York, 1973).
11. B. R. Mollow, *Quantum theory of field attenuation*, Phys. Rev. **168**, 1896, (1968).

AN APPLICATION OF THE KALMAN FILTER IN GEOASTRONOMY

P.M. Mathews
Department of Theoretical Physics, University of Madras
Guindy Campus, Madras 600 025, India

ABSTRACT

This paper is presented as a contribution to the Symposium in honour of my distinguished colleague and long-time friend, S.K. Srinivasan. My intention in this paper is to provide a brief overview of an application of stochastic theory to geoastronomy, an area which Professor Srinivasan has not touched, but one in which there have been some remarkable advances in the past decade. Stochastic processes are inextricably mixed up in the recordings of signals from celestial bodies which constitute the basic raw material of geoastronomy. The Kalman filter (*Kalman* [1960], *Kalman and Bucy* [1960]), developed as an efficient method of including in an estimation process parameters whose values change during a period over which data are collected, uses stochastic-process models to predict their changes between epochs of observation. The use of the Kalman filter in extracting results of astonishingly high accuracy from the data contaminated by the stochastic "noise" is detailed in a recent paper by *Herring et al.* [1990], on which the present outline is largely based. The kinds of results that can be obtained relating to the surface and the interior of the Earth are illustrated herein by a few examples.

INTRODUCTION

Geoastronomy is the study of the Earth with the aid of astronomical observations. That the Earth is for ever engaged in *pradikshana* around the Sun, pirouetting all the while, is a fact inferred from the apparent motions of celestial objects in the heavens, known from the earliest times. We have come a long way from this elementary stage of geoastronomy. Theoretical developments based on Newton's dynamics and his theory of gravitation made it possible, for instance, to infer from the astronomically observed precession of the figure axis of the Earth that the Earth's dynamical ellipticity (i.e., the fractional difference between the moments of inertia about polar and equatorial axes) is about 1/300. During the last decade it has become possible not only to monitor such phenomena as the motions of the tectonic plates of which the Earth's crust is composed, but even to "see" certain features of the Earth's interior "reflected" in the skies. One of the clearest to emerge is the shape of the boundary between the fluid core of the Earth and the solid mantle within which it is contained.

Information of this kind has been obtained through a combination of two types of data: the elastic characteristics of the Earth's interior and the relative moments of inertia of the different regions (mantle, fluid outer core and solid inner core), as deduced from seismological observations, on the one hand, and values of very high accuracy for the amplitudes of certain Fourier components of the nutations of the Earth, on the other hand. The latter are determined from data obtained by Very Long Baseline Interferometry (VLBI), which involves simultaneous recordings of radio signals from quasars - the furthest radio sources available - by a number of radio telescopes whose separations are of the order of thousands of kilometres. The observed nutations, which are motions of the direction of the (polar) symmetry axis of the Earth in space about a mean direction, are a superposition of a number of circular motions at different frequencies contained in the perturbing gravitational forces of the Sun and the Moon on the Earth. The VLBI technique has made it possible to measure the angular amplitudes of the individual Fourier components with an accuracy which would have been unimaginable even a decade ago. The uncertainty in a typical nutation amplitude is now only about 0.04 milliarcseconds. A tilt of the Earth through this angle would move a point on the surface by just about 1.3 millimetres! Along with advances in instrumentation and technology, sophisticated methods of data analysis have played a crucial role in bringing down the uncertainty to this level. The advantages of the Kalman filter in this context have been adequately recognised only in the last couple of years.

The times of arrival of a signal from a source at two widely separated radio telescopes are different, since the telescopes are not, in general, equidistant from the source. The difference in arrival times, the so-called group delay τ_g, is D/c, where c is the velocity of light in vacuum, and D, the difference in the distances of the two telescopes from the quasar, could be up to several thousand kilometres. For a given pair of telescopes, τ_g will depend on the length of the baseline vector connecting the telescopes and on the orientation of this vector relative to the direction towards the quasar. Simultaneous recording of signals from a particular quasar by several telescopes (for typical durations of 100 to 400 seconds) enables the delays corresponding to all the baselines connecting the various telescopes to be determined. These delays constitute the basic data from which the orientation of the Earth during the observation session is determined, besides information about the baselines. After each session, the telescopes are switched simultaneously to another quasar for the next session; recordings are continued in this fashion, for 24-28 hours. (The next day-long experiment is a few days later, according to a regular schedule.) The nutation of the Earth manifests itself in the changes in Earth orientation from session to session during a day's experiment; amplitudes of individual Fourier components of the nutation, which have periods very close to 24 hours as perceived from the rotating Earth, are computed from the Earth orientation changes. Since the delays form the "raw" data from which all geophysical information is extracted, the fundamental problem in geoastronomy with VLBI is the accurate determination of τ_g.

Ideally, the recordings at two stations should coincide exactly if shifted relative to each other by a time τ_g (after compensating for the variation of τ_g on account of the spin of the Earth); they should exhibit perfect correlation for a relative shift $\tau = \tau_g$. In reality, the radio signals arrive at a telescope after suffering delays which are subject to random fluctuations in their passage through the Earth's atmosphere; noise in the receiver gets superposed on the signal; and the rate of the clock which provides the time reference for the recorded signals is subject to small changes during the observing session. The problem then is to achieve high accuracy in the determination of the group delay τ_g relating to a given pair of radio telescopes (as also related quantities such as the delay rate, i.e, the time rate of change of the delay) through the correlation of the contaminated signals recorded at the two stations. The Kalman filter has been used as an effective means of dealing with the stochastic part of the contamination.

THE KALMAN FILTER

The delays and delay rates for the various baselines, radio sources, observing sessions, and days of observation, are the "observables" from which various parameters of interest such as nutation amplitudes, baseline lengths, etc., are to be determined. After making corrections for known deterministic effects (e.g., changes in baselines caused by tidal deformations of the Earth's surface, and the reduced speed of the signals during the traversal of a standard atmosphere), one sets up the equations to be solved in the form

$$y_n = \mathbf{A}_n x_n + v_n \tag{1}$$

wherein all quantities refer to the nth epoch of time. For each n, y_n is a vector made up of the differences between observations and their theoretical values computed from *á priori* values of the parameters involved; x_n is the vector of corrections (to be estimated) to the *á priori* values; $\mathbf{A}_n$ is the matrix of partial derivatives of the observed quantities with respect to the parameter values; and v_n, the vector of residuals attributable to noise in the observations.

The evolution of the parameters from the nth epoch to the next is expressed through the equation

$$x_{n+1} = \mathbf{S}_n x_n + w_n \tag{2}$$

where $\mathbf{S}_n$ is the transition matrix representing the dynamics of the evolution, and w_n is the vector of stochastic contributions to the parameters during the interval between the two epochs. Those elements of w_n which correspond to nonstochastic parameters (e.g., the positions of the radio telescopes and of the sources) are zero. Stochastic parameters include the clock rates and atmospheric delays.

The problem is to estimate the vectors x_n, given the observations y_n, the transitions matrices $\mathbf{S}_n$ and the matrices $\mathbf{A}_n$, as well as the characteristics of the stochastic processes represented by v_n and w_n, for all n. The following assumptions are made about the expectation values of the stochastic variables:

$$\langle v_n \rangle = \langle w_n \rangle = 0 \quad \text{for all } n \tag{3}$$

$$\langle v_n v_{n+j} \rangle = \langle w_n w_{n+j} \rangle = 0 \quad \text{for all } j \neq 0 \tag{4}$$

$$\langle v_n x_{n+j} \rangle = \langle v_n w_{n+j} \rangle = 0 \quad \text{for all } n \text{ and } j \tag{5}$$

$$\langle x_n w_{n+j} \rangle = 0 \quad \text{for all } j \geq 0 \tag{6}$$

Eq.(6) states that the state at epoch n, represented by x_n, does not affect the random part of the evolution to later epochs; and according to eq.(5), the noise v_n at a given epoch is not correlated with the state at, or with the stochastic part of the evolution from, the same or any other epoch. Further characterisation of the stochastic parameters that is needed for our purposes is provided by the variance-covariance matrices $\mathbf{V}_n$ and $\mathbf{W}_n$ of the elements of v_n and w_n respectively:

$$\mathbf{V}_n = \langle v_n v_n^T \rangle, \quad \mathbf{W}_n = \langle w_n w_n^T \rangle \tag{7}$$

where T denotes the transpose. Determination of these matrices to match the stochastic properties of the various quantities which play a role in the VLBI method is one of the problems in the implememtation of the Kalman filter. Once $\mathbf{V}_n$ and $\mathbf{W}_n$ are specified, a sequential procedure is employed to estimate the x_n at each epoch of observation, starting from the earliest epoch, and adding data pertaining to successive epochs one at a time.

Let x_n^n be the estimate of x_n made using all the data up to and including the nth epoch; and let $\mathbf{C}_n^n$ be the variance-covariance matrix for the elements of x_n^n. Two steps are involved in determining the corresponding quantities at epoch $(n+1)$. The first is a prediction for epoch $(n+1)$ on the basis of the already computed x_n^n and $\mathbf{C}_n^n$; and the second updates the estimates by taking account of the data pertaining to the epoch $(n+1)$ also.

Prediction:

$$x_{n+1}^n = \mathbf{S}_n x_n^n \tag{8a}$$

$$\mathbf{C}_{n+1}^n = \mathbf{S}_n \mathbf{C}_n^n S_n^T + \mathbf{W}_n \tag{8b}$$

Updating:

$$x_{n+1}^{n+1} = x_{n+1}^n + \mathbf{K}(y_{n+1} - \mathbf{A}_{n+1} x_{n+1}^n) \tag{9a}$$

$$\mathbf{C}_{n+1}^{n+1} = \mathbf{C}_{n+1}^n - \mathbf{K}\mathbf{A}_{n+1}\mathbf{C}_{n+1}^n \tag{9b}$$

where K is the Kalman gain, given by

$$\mathbf{K} = \mathbf{C}_{n+1}^n \mathbf{A}_{n+1}^T (\mathbf{V}_{n+1} + \mathbf{A}_{n+1}\mathbf{C}_{n+1}^n \mathbf{A}_{n+1}^T) \tag{10}$$

The two-step process is now repeated, starting with x_{n+1}^{n+1} and $\mathbf{C}_{n+1}^{n+1}$, to obtain the estimates pertaining to the epoch $(n+2)$, and so on, till all the observations are exhausted. The estimate x_f^f (with f referring to the final epoch) is based on all the available observations, and provides the best estimates of the corrections to be made to the *á priori* values of the parameters. This statement applies to all the "global" parameters, which are epoch-independent, e.g., the positions of the radio sources in a space-fixed frame, the coefficients in expressions linear in time for the base-line lengths, etc.

As for the stochastic parameters (e.g., the clock parameters) and other parameters such as the pole positions (orientations of the Earth in space) which are time-dependent, the estimated values at a particular epoch n are contained in x_n^n; however, this estimate has not taken account of any of the data relating to epochs after the nth. In order to obtain final estimates based on all the available data, a backward or smoothing run of the Kalman filter needs to be performed. The BRF (backward-running filter) starts with the last epoch of observations and adds data from successive earlier epochs. In this run, one "predicts" $\hat{x}_{n-1}^n$ and $\hat{\mathbf{C}}_{n-1}^n$ from $\hat{x}_n^n$ and $\hat{\mathbf{C}}_n^n$ and "updates" to $\hat{x}_{n-1}^{n-1}$ and $\hat{\mathbf{C}}_{n-1}^{n-1}$. If x_+, $\mathbf{C}_+$ denote the updated FRF (forward-running filter) results at a particular stage (say the epoch n) and x_-, $\mathbf{C}_-$ the predicted but not updated results of the BRF for the same epoch, a weighted average of these two sets provides the final estimates. The weighted averages, according to the *Gelb* [1974] algorithm, are given by

$$x_n^s = x_+ + \mathbf{B}(x_- - x_+) \tag{11a}$$

$$\mathbf{C}_n^s = \mathbf{C}_+ - \mathbf{B}\mathbf{C}_+ \tag{11b}$$

where

$$\mathbf{B} = \mathbf{C}_{+}(\mathbf{C}_{-} + \mathbf{C}_{+})^{-1} \tag{11c}$$

In order to implement the above procedure, one needs to know the nature of the stochastic processes involved in w_n so that $\mathbf{W}_n$ can be specified. The processes which contribute to w_n are those which represent the stochastic behaviour of the clocks at the various stations and the stochastic aspects of the atmospheric contribution to the time taken by signals from the radio source to reach the various telescopes. Models have to be employed to represent these processes, partly because the underlying physics is not well enough known to identify on theoretical grounds the stochastic processes involved; in any case, exact implementation would be too complicated to be practicable even if the process itself were adequately known.

STOCHASTIC MODELS

The characterisation of noise in frequency standards such as hydrogen masers which serve as clocks in VLBI experiments is almost always in terms of the Allan standard deviation σ_A. For a stochastic process $p(t)$, the Allan variance σ_A^2 at a sampling interval τ is defined by

$$\sigma_A^2(\tau) = \frac{1}{2}\langle[p(t) - 2p(t+\tau) + p(t+2\tau)]^2\rangle \tag{12}$$

For a random walk process $p_r(t)$ and an integrated random walk $p_i(t)$, generated from a white noise process $w(t)$ with power spectral density Φ through the equations

$$\frac{dp_r}{dt} = w(t) \quad \text{and} \quad \frac{d^2p_i}{dt^2} = w(t) \tag{13}$$

the Allan variances are Φ/τ and $\Phi\tau/3$ respectively, in consequence of the forms of the covariance function $R(t_1, t_2) \equiv \langle p(t_1)p(t_2)\rangle$ in the two cases: $R(t_1, t_2) = \Phi \min(t_1, t_2)$ for the random walk and $\Phi t_1^2(3t_2 - t_1)/6$ for the integrated random walk. Studies on the performances of hydrogen masers show that, typically, the Allan variance decreases like τ^{-1} for τ up to a few thousand seconds, while for larger sampling intervals, it is proportional to τ. This behaviour implies that the process can be modelled by a combination of a random walk and an integrated random walk; and it is this mo[illegible] is employed for the clock statistics in the processing of VLBI data using the Kalman filter.

The stochastic part of the atmospheric delays originates primarily from fluctuations in the water vapour content of the atmosphere. The structure function $D(\tau) \equiv \langle[p(t+\tau) - p(t)]^2\rangle$ for this process has been theoretically determined on the basis of the Kolmogorov turbulence theory (*Treuhaft and Lanyi*) [1987]; its dependence on τ, expressed in terms of a power law τ^α, has α varying from 5/3 for small τ to 2/3 for large τ. However it has been found adequate to model the fluctuations by a random walk process which has a structure function with a constant index 1 (i.e., $D(\tau)$ proportional to τ).

With the stochastic models chosen thus, the next step is to assign appropriate values to the parameters of the models. Results of performance studies on a number of hydrogen masers indicate that the Allan standard deviation (ASD) of the clock process is lowest at a sampling time τ of around 50 minutes. It is found that a conservative upper bound on the observed ASD for all τ would be provided by a superposition of a random walk and an integrated random walk if the PSDs of the white noise processes which drive them are chosen to be 0.15 $(\text{picosec})^2/\text{sec}$ and 0.05 $(\text{femtosec})^2/\text{sec}$ respectively. The ratio of the two PSDs is so chosen as to make the ASD a minimum at $\tau = 50$ min. As regards the atmospheric delays, the PSD of the white noise which drives the random walk model may be inferred from the VLBI data themselves. This is done by taking advantage of the observation that it is this white noise process (the time derivative of the random walk) which appears as the noise in the delay rate which is one of the observables in the VLBI experiment. The r.m.s. scatter of the delay rates then provides a measure of the PSD. Estimates of the PSD obtained in this fashion shows strong dependence on the station and the season, but dominates over the clock noise in almost all cases.

The Kalman filter is run with the parameters of the clock and atmospheric delay processes thus specified.

GEOASTRONOMY

We now present a few of the results of studies in geodesy and geophysics based on the output from VLBI data analysis. Incredibly accurate determinations of the lengths of baselines connecting pairs of radio telescopes and of the rates of change of components of the baseline vectors are among the most striking of these results. A few examples of intercontinental baseline lengths at a particular epoch, determined from measurements made during the eighties, are given in Table I; it will be noted that accuracies to within a few millimeteres in baselines as long as a few thousand kilometres have been achieved. Another set of results, given in Table II, exhibits the relative motions of three radio telescopes located near the San Andreas Fault in California. Mojave and Owens Valley Radio Observatory (OVRO) are on one side of the Fault, and the Jet Propulsion Laboratory station (JPL) on the other, Mojave being almost directly east of JPL and OVRO to the north-east, almost directly north of Mojave. The large magnitude of the relative motions between stations on opposite sides of the Fault, compared to the much smaller magnitude pertaining to the two stations on the same side (Mojave and OVRO) is evident. The relevance of an accurate knowledge of such motions to studies on the genesis of, and to programmes for prediction of, earthquakes, hardly needs emphasis. (The results given in both the tables are by courtesy of T.A. Herring.)

Amplitudes of various Fourier components of the Earth's nutations are among the parameters which are estimated using the Kalman filter. These amplitudes are the rotational responses of the Earth to the corresponding Fourier components of the gravitational torques of the Sun and the Moon on the Earth, and are dependent on certain mechanical and elastic properties of the Earth. Such properties include the dynamical ellipticities and the ratios of the moments of inertia of the various regions of the Earth (mantle, fluid outer core, and solid inner core), as well as the compliances which represent the deformability of the different regions in response to the rotation of one or the other of the regions relative to the others. All such parameters are computed from standard Earth models constructed using seismological data supplemented by the assumption of *hydrostatic equilibrium.* (The deviation from sphericity of the shape of the Earth and of the boundaries between different regions and other constant density surfaces within, under a condition of steady rotation, is taken to be as in a fluid body in equilibrium under the combination of self gravitation and the centrifugal forces due to rotation.) Solution of the coupled equations of angular momentum balance for the different regions reveals the existence of a nutational eigenmode (caused by the presence of the fluid core) whose eigenfrequency turns out to be of magnitude $\approx (1+1/460)$ cycles per sidereal day (cpsd) when values of the Earth parameters computed from a widely accepted model such as PREM are used. This natural frequency is quite close to the frequency $\approx (1+1/366)$ of the so-called *retrograde annual* component of the perturbing gravitational torque. The near resonance which results from the proximity of the eigenfrequency causes the nutational response at this frequency to be significantly different from what it would have been if the Earth had been wholly solid and rigid. The theoretical value for this amplitude, computed from PREM, is 33.00 milliarcseconds (mas) as against 24.89 mas for the rigid case. But the VLBI-determined value is 31.38 mas, with an uncertainty of only 0.04 mas. The large discrepancy between the observed and theoretical values of this amplitude, whcih far exceeds the uncertainties, is in contrast with the relatively good agreement in the amplitudes of forced nutations at other frequencies, and indicates strongly that the resonance frequency as computed from the standard Earth model is in error. It was found that the theoretical value of the retrograde annual amplitude could be matched to the observed amplitude (corrected for effects not included in the theory) by choosing the resonance frequency to be $(1+1/430)$ cpsd. From a consideration of the dependence of the frequency, as deduced from theory, on the Earth parameters (*Sasao et al.* [1980], *Mathews et al* [1991*a*]), it has been concluded that the difference between this value inferred from observations and that obtained from the Earth model is to be understood as reflecting a higher value of the dynamical ellipticity of the fluid core than in the hydrostatic model. This, in turn, has been explained as a consequence of an extra flattening of the core-mantle boundary compared to that in the hydrostatic model, the difference between the equatorial and polar radii of the core-mantle boundary (mean radius 3480 km) being determined to be about 9.3 km instead of a little under 8.9 km (*Gwinn et al.* [1986], *Mathews et al* [1991*b*]). What is most remarkable about this finding is not that a deviation from hydrostatic equilibrium is shown to exist, but that values 5 to 20 times higher for such deviations which had been claimed from seismological studies are firmly ruled out.

CONCLUDING REMARKS

The key to the new ability to make quantitative statements regarding such phenomena as the extremely slow relative motions between tectonic plates (which have literally earth-shaking consequences), and about the structure and properties of the Earth's interior, lies in the extremely high accuracy of of the observational parameters extacted from the raw data. The role of the methods of stochastic theory in making still more accurate determinations possible will become more crucial in future, inasmuch as noise from certain sources (e.g., atmospheric fluctuations) remain outside of human control even as the capabilities of the equipment and instruments used keep advancing.

REFERENCES

Gelb, A., Applied optimal estimation, MIT Press, Cambridge, 1974.

Gwinn, C.R., T.A. Herring, and I.I. Shapiro, Geodesy by radio interferometry: Studies of the forced nutations of the Earth 2. Interpretation, *J. Geophys. Res.,* 91, 4755–4765, 1986.

Herring, T.A., J.L.Davis, and I.I. Shapiro, Geodesy by radio interferometry: The application of Kalman filtering to the analysis of VLBI data, 1990, *J. Geophys. Res.,* 95, 12561–12581, 1990.

Kalman, R.E., A new approach to linear filtering and prediction problems, *J. Basic Engg., (ASME Trans.), 82D*, 1960.

Kalman, R.E., and R.S. Bucy, New results in linear filtering and prediction theory, *J. Basic Engg.*, *83D*, 1961.

Mathews, P.M., B.A. Buffett, T.A. Herring, and I.I. Shapiro, Forced nutations of the Earth: Influence of inner core dynamics I. Theory, *J. Geophys. Res.*, 1991*a*, (In Press).

Mathews, P.M., B.A. Buffett, T.A. Herring, and I.I. Shapiro, Forced nutations of the Earth: Influence of inner core dynamics II. Numerical results and comparisons, *J. Geophys. Res.*, 1991*b*, (In Press).

Sasao, T., S. Okubo, and M. Saito, A simple theory on the dynamical effects of a stratified fluid core upon nutational motion of the Earth, in *Proceedings of IAU Symposium No.78*, Ed. E.P. Federov, M.L. Smith, and P.L. Bender, pp.165–183, D. Reidel, Hingham, Massachusetts, 1980.

Treuhaft, R.N., and G.E. Lanyi, The effect of the dynamic wet troposphere on radio interferometric measurements, *Radio Science*, *22*, 251–265, 1987.

Table I

Typical Baselines Measured by VLBI[1]

Baseline	Number	Length (m)
Onsala – Wettzell	27	$919,661.0012 \pm .0010$
Wettzell – Westford	166	$5,998,325.3514 \pm .0015$
Westford – Mojave	19	$3,903,767.7429 \pm .0021$
Mojave – Kwajalein	6	$7,576,938.5297 \pm .0086$

[1] The stations shown are located as follows: Onsala in Sweden; Wettzell in Germany; Westford in Eastern U.S.; Mojave in Western U.S.; Kwajalein in Japan. The second column shows the number of one-day experiments from which the baseline is determined.

Table II

Velocities of Relative Motions of Telescope Stations

Baseline		Velocity Components (cm/yr)	
From	To	East	North
Mojave	JPL	0.3 ± 0.2	-0.3 ± 0.2
Mojave	OVRO	1.8 ± 0.3	-2.1 ± 0.3
JPL	OVRO	-1.5 ± 0.3	1.8 ± 0.3

CONFORMAL MARTINGALES IN STOCHASTIC MECHANICS

K.V. PARTHASARATHY
Department of Mathematics
Ramakrishna Mission Vivekananda College
MADRAS 600 004, INDIA.

ABSTRACT

We study the probabilistic description of quantum systems in terms of stochastic differential equations, subject to multiplicative noise. The martingale properties of the complex Brownian motion are exploited. The conformal stochastic calculus is the key concept in arriving at the new result in terms of diffusions.

1. INTRODUCTION

Nelson's stochastic mechanics (Nelson, 1976, 1984) deals with quantum phenomena described in terms of diffusions instead of wave functions. The kinematics of diffusion processes involves extensive use of stochastic calculus based on Ito's martingale integrals. The core of Ito calculus is the integration by parts formula, with an additional quadratic variation term indicating time asymmetry in the Ito integral. Hence, Nelson uses the two-sided symmetric stochastic integral of Ito (1978), which gives the chain rule as in classical calculus. The construction of diffusions and their martingale equations is the main theme in the works of Carlen (1985) and Durrett (1984). Lavenda and Santamato (1984) study the stochastic formalism of quantum theory by extending the probability measure density for fluctuating paths to the complex domain.

Getoor and Sharpe (1972) established the existence of a class of complex martingales called conformal martingales, playing the role similar to that of analytic functions in the classical two dimensional theory. A conformal martingale is a complex Brownian motion obeying the rules of ordinary calculus, in contrast to real Brownian motion. It can also be represented by a time change of complex Brownian motion. Bessel processes and skew product representations of Brownian motions, enrich the study of conformal martingales, from the point of view of stochastic differential equations. (Ito and McKean, 1965; Pitman and Yor, 1981; Rogers and Williams, 1987).

We feel that the concept of conformal martingale plays a vital part in analysing some problems of stochastic dynamics.

The plan of the paper is as follows: Section 2 is a short review and reference on conformal martingales and skew products. We devote section 3 to highlight the advantage of windings of plane Brownian motion with topological constraints in the context of Aharonov-Bohm effect in quantum dynamics. This note ends with some concluding remarks in section 4.

2. CONFORMAL MARTINGALES AND SKEW PRODUCTS

A complex valued process $W_t = U_t + iV_t$ is a conformal martingale if U_t and V_t are each continuous local martingales such that $\langle U \rangle = \langle V \rangle$ and $\langle U,V \rangle = 0$. $\langle W \rangle$ is the quadratic variation of the process W_t and $\langle W,Q \rangle$ is the process defined by complex polarization. $\langle W,Q \rangle$ is linear in W and conjugate linear in Q.

It is an important property of complex Brownian motion that integrals of complex differential forms along Brownian paths may be rewritten as Ito type stochastic integrals.

The following theorem is the key result for our application.

Let f be an analytic function on a domain D. Then

$f(Z_{t\wedge\tau}) = f(Z_0) + \int_0^{t\wedge\tau} f'(Z_s)\, dZ_s$ and so $f(Z_t)$ is a conformal martingale until Z_t exists in D. (Here $t \wedge \tau$ is $\min(t,\tau)$).

<u>Proof:</u> We write $f(Z) = u(x,y) + iv(x,y)$ and apply Ito formula to get

$$df(Z_t) = [u_x\, dx_t + u_y\, dy_t] + i[v_x\, dx_t + v_y\, dy_t], \tag{2.1}$$

the second order terms vanishing since u and v are harmonic. By Cauchy-Riemann equations this is just $f'(Z_t)dZ_t$. Thus we find that complex Brownian motion obeys the rules of ordinary calculus.

To see that we have a conformal martingale, we find the real and imaginary parts have increasing processes $\int_0^t (u_x^2 + u_y^2)ds$ and $\int_0^t (v_x^2 + v_y^2)ds$ and these are equal by Cauchy-Riemann conditions. Their joint quadratic variation $\int_0^t (u_x v_x + u_y v_y)ds$ also vanishes by the same reason.

Next, we introduce the concepts of windings and skew product representations.

We write a complex Brownian motion $Z = (Z_t, t \geq 0) = (X_t + iY_t, t \geq 0)$ in polar coordinates as

$$Z_t = R_t \exp(i\phi_t) \tag{2.2}$$

where $R = |Z|$ is the radial part of Z and ϕ_t is the total winding of Z about 0 upto time t. The total winding upto t may be considered as the integral along the path of Z of the closed differential form on $C - \{0\}$

$$\mathrm{Im}\left(\frac{dZ}{Z}\right) = \frac{xdy - ydx}{x^2 + y^2} \tag{2.3}$$

If we define

$$\log Z_t = \int_{Z[0,t]} \frac{dz}{z} \tag{2.4}$$

then $(\log Z_t, t \geq 0)$ is the unique continuous determination of $\log Z_t$ starting at $\log Z_0 = 0$ and

$$\log Z_t = \log R_t + i\phi_t . \tag{2.5}$$

Using the above theorem for conformal martingales for the example $f(z) = \frac{1}{z}$, we have

$$\log Z_t = \int_0^t \frac{dZ_s}{Z_s} , \tag{2.6}$$

with the stochastic integral representations

$$\log R_t = \int_0^t \frac{X_s dX_s + Y_s dY_s}{R_s^2}, \quad \phi_t = \int_0^t \frac{X_s dY_s - Y_s dX_s}{R_s^2} .$$

The log radial process $\log R_t$ and the winding process ϕ_t are orthogonal local martingales with common increasing process U_t, called the log clock, defined by

$$U_t = \int_0^t \frac{ds}{R_s^2} . \tag{2.7}$$

A conformal martingale may be represented by a time change of a complex Brownian motion. Hence the conformal martingale $\log Z_t = \log R_t + i\phi_t$ may be represented as $\log Z_t = \beta + i\theta = \xi(U_t)$, where β and θ are the independent, log radial and angular Brownian motions respectively.

We also have

$$\log R_t = \beta(U_t) \text{ and } \phi_t = \theta(U_t), \ t \geq 0 \tag{2.8}$$

The angular Brownian motion is independent of the whole radial process and in particular independent of log clock U_t. The representation of the winding process ϕ_t in (2.8) as a Brownian motion run with an independent clock U_t, determined by the radial motion is called the skew product representation of ϕ_t. With these remarks, we are ready to present our main result in the next section.

3. AHARONOV-BOHM EFFECT

The notion of total windings as stochastic integrals plays a key role in the investigation of constraint problems of topological nature. An interesting example is the Aharonov-Bohm problem for which the singularity represents an impenetrable solenoid confining a magnetic flux. Even in the absence of flux, the topological effect due to windings will remain. Schulman (1971, 1981) recognised this effect and found the path integral solution for the motion of the particle in a multiply connected space, exploiting homotopy theory.

This is the problem of a free particle moving in a plane where the origin is a singular point. We consider the paths beginning from R_0 at $t = 0$ and ending at R at time t. Since the origin has a singularity, the space becomes multiply-connected. The path from R_0 to R cannot cross the singularity. It can only encircle a singular point or go straight without encircling. The above paths are grouped according to the number of turns they make around the singular point. The different homotopy classes of paths may be classified according to the number of turns n around the singularity. We refer to Khandekar and Lawande (1986), Khandekar, Bhagwat and Wiegel (1988), for a lucid account on the path integral representation of the nth propagator, imposing periodic constraints.

The skew product representation of the conformal martingale gives the solution immediately, when the winding process ϕ_t is considered. The homotopic and homological aspects of this process in the universal covering space have been studied at great length in Lyons and McKean (1984). Ito and McKean (1965) use the skew product representation to find the characteristic function of this process, viz $E[\exp(i\alpha\phi_t)]$.

Now ϕ_t is a centred Gaussian random variable with variance equal to

$$U_t = \int_0^t \frac{ds}{R_s^2} \, . \tag{3.1}$$

This is also equal to $E[\exp(-\frac{\alpha^2}{2} U_t)]$ by the law of total probability on conditional expectation.

We bring out our analysis through martingale approach, (McKean, 1969; McGill, 1981).

We concentrate on the skew product representation of the complex Brownian motion (R_t, ϕ_t). Let $\phi_t(0) = \theta_0$ and $p_t(R,R_0;\ \theta,\theta_0)$ be the transition density of (R_t,θ_t) where $\theta_t=\phi_t - \theta_0$. We also define the smooth function

$$g_t(R,\theta) = E[e^{i\mu(\phi_t - \theta_0)} | R_t = R, \theta_t = \theta] \tag{3.2}$$

We note that (with $J_\gamma(x)$, the Bessel function)

$$J_{|\gamma|}(\alpha R_t) \exp\left\{ i\gamma\phi_t + \frac{\alpha^2 t}{2} \right\} \tag{3.3}$$

is a local martingale for $\alpha \geqslant 0$, γ real.

We replace γ by $\mu + n$ where μ is real and n an integer, and use the local martingale property. We have

$$E[e^{i(\mu+n)\phi_t} J_{|\mu+n|}(\alpha R_t)] = J_{|\mu+n|}(\alpha R_0) \exp\left\{ i(\mu+n)\theta_0 - \frac{\alpha^2 t}{2}\right\} \tag{3.4}$$

Now the left hand side of (3.4) is

$$\int_0^\infty R\ J_{|\mu+n|}(\alpha R)\ [e^{i(\mu+n)\theta_0} \int_{-\infty}^{\infty} e^{in\theta}\ g_t(R,\theta)\ p_t(R,R_0;\theta,\theta_0)d\theta]dR \tag{3.5}$$

and this can be considered as the Hankel transform of the quantity in the square bracket. Inverting this and through right hand side of (3.4), we get this quantity as

$$e^{i(\mu+n)\theta_0} \int_0^\infty \alpha \exp(-\frac{\alpha^2}{2}t)\ J_{|\mu+n|}(\alpha R)\ J_{|\mu+n|}(\alpha R_0)d\alpha \tag{3.6}$$

$$= e^{i(\mu+n)\theta_0}\ I_{(\mu+n)}(\frac{RR_0}{t})\ \frac{1}{t} \exp(-\frac{R^2 + R_0^2}{2t}) \tag{3.7}$$

using Weber's formula, $I_\gamma(x)$ being the modified Bessel function. The right hand side of (3.7) is the Fourier transform of $e^{i(\mu+n)\theta_0}\ g_t(R,\theta)\ p_t(R,R_0;\theta,\theta_0)$. Fourier inversion gives $g_t(R,\theta)\ p_t(R,R_0;\theta,\theta_0)$ and with this function we get the probability of the event

$$[(\phi_t - \theta_0 = 2k\pi) \cap (R_t = R;\ \theta_t = \theta)],$$

k being an integer. This agrees with the results in the existing literature, with due changes in the arguments of the functions concerned;

4 CONCLUSIONS AND OUTLOOK

The core of the paper is the evaluation of the probability distribution function for the winding process of a particle moving in a multiply-connected space, as encountered in AB effect analysis. The discussion of the paper is based on the use of conformal stochastic calculus and skew product representations. To our knowledge this approach is entirely different from the various other methods used in the literature.

The theory of conformal martingales, with conceptual extensions in higher dimensions, plays a significant role in the description of Bessel bridges and other processes closely connected with Brownian motion in R^d. Its relevance in stochastic dynamics forms the content of a future contribution.

Thus we observe that the analysis of the structure of the inherent characteristics of some stochastic phenomena, involves the use of stochastic calculus. An important advantage for the model builders was that the martingale property for the noise processes involved was not insisted. But the modelling of stochastic differential equations driven by martingales with both jump and continuous components has a natural choice in our study. A fruitful and enlightening discussion on these concepts can be had in Srinivasan (1982).

ACKNOWLEDGEMENTS

I wish to express my sincere gratitude to Professor S.K. Srinivasan for stimulating discussions which kindled my interest in Nelson's stochastic mechanics. It is a pleasure to thank Professor R. Vasudevan for introducing me to the joys of Ito Calculus. I thank the organisers of this symposium for giving me an opportunity to present this paper.

REFERENCES

Nelson, E. (1976). Dynamical Theories of Brownian Motion. Princeton, U.P., Princeton, New Jersey.

Nelson, E. (1984). Quantum Fluctuations. Princeton Univ. Press. Princeton, New Jersey.

Ito, K. (1978). "Extension of Stochastic Integrals" in K.Ito, ed., Proc. of Intern. Symp. SDE, Kyoto 1976, Wiley, New York.

Carlen, E. (1985). "Existence and Sample Path Properties of the Diffusions in Nelson's Stochastic Mechanics", in Stochastic Processes in Mathematics and Physics, S. Albeverio, et. al., ed., Lecture Notes in Mathematics 1158, Springer, Berlin, p.25.

Durrett, R. (1984). Brownian Motion and Martingales in Analysis. Wadsworth, Monterey, California.

Lavenda, B.H. and Santamato, E. (1984). "Stochastic Interpretations of Nonrelativistic Quantum Theory", Int. Jour. of Theore. Phy 23 (7), 585.

Getoor, R.K. and Sharpe, M.J. (1972). "Conformal Martingales", Inventions. Math 16, 271.

Ito, K. and McKean H.P. (1965). Diffusion Processes and Their Sample Paths. Springer, Berlin.

Pitman, J. and Yor, M. (1981). "Bessel Processes and infinitely divisible laws", in Stochastic Integrals, D. Williams, ed., Lecture Notes in Mathematics 851, Springer, Berlin, p.285.

Rogers, L.C.G. and Williams, D. (1987). Diffusions, Markov-Processes, and Martingales(Vol.2), Ito Calculus. John Wiley and Sons, New York.

Schulman, L.S. (1971). "Approximate Topologies", J. Math. Phys., 12, 304.

Schulman, L.S. (1981). Techniques and Applications of Path Integration, Wiley Interscience, John Wiley & Sons, N.Y.

Khandekar, D.C. and Lawande, S.V. (1986), "Feynman Path Integrals: Some exact Results and Applications", Phys. Reports 137, 115.

Khandekar, D.C. Bhagwat, K.V. and Wiegel, F.W. (1988). "On a path Integral with a topological constraint", Phys. Lett. 127A, 379.

Lyons, T. and McKean, H.P. (1984), "Windings of the Plane Brownian Motion", Adv. in Math. 51, 212.

McKean, H.P. (1969) Stochastic Integrals, A.P, NY.

McGill., P. (1981). "Brownian Motion in Hyperspherical Coordinates", Stochastics, 6, 57.

Srinivasan, S.K. (1982), Stochastic Calculus and Models and their Applications to some problems of Physics, Mathematics Department Lecture Notes No 15, National University of Singapore.

-oOo-

PROBABILITY DISTRIBUTIONS OVER NONCOMMUTING VARIABLES

E.C.G. Sudarshan
Center for Particle Theory and Department of Physics
University of Texas
Austin, Texas 78712
U.S.A.

ABSTRACT

The notion of probability distributions and associated characteristic functions over a set of noncommuting variables are studied. With the provision that the marginal distributions restricted to a commuting set should give the nonnegative probability distributions over those variables we find that such generalized master distributions exist which, while incorporating implicit positivity requirements, are not themselves pointwise necessarily nonnegative. The cumulant generating functions are studied and the applicability of the Marcinkiewicz theorem explored.

INTRODUCTION : STATES AS LINEAR MAPS

Given a system with primary dynamical variables $x = \{x_1, x_2, \ldots\}$ in terms of which a generic dynamical variable may be displayed as

$$f(x) \equiv f(x_1, x_2, \ldots)$$

the generic notion of a state or configuration is the linear homogeneous map

$$f(x) \longrightarrow \langle f(x) \rangle$$

of the dynamical variables into numbers with the constraints:

$$1 \longrightarrow \langle 1 \rangle = 1$$

$$a_1 f_1(x) + a_2 f_2(x) \longrightarrow a_1 \langle f_1 \rangle + a_2 \langle f_2 \rangle$$

$$f(x) \geq 0 \longrightarrow \langle f(x) \rangle \geq 0.$$

These are, respectively, normalization, linearity and positivity. This definition of generalized states [1] is applicable to both classical and quantum systems.

In general the map is such that functional relationships are not preserved :

$$\langle f(x)\rangle \neq f(\langle x\rangle).$$

If the inequality is replaced by an equality for all functions f(x) then the states are dispersion-free: such states may or may not exist. It is sufficient to compute the dispersions of dynamical variables defined by

$$\sigma_r^2 = \langle x_r^2\rangle - \langle x_r\rangle^2 \equiv \langle(x_r - \langle x_r\rangle)^2\rangle .$$

These quantities are nonnegative. If they all vanish the state is dispersion-free; and the state is a pure state with sharp values for the dynamical variables. The map is then merely substitution of the values for the primary dynamical variables:

$$f(x) \longrightarrow \langle f(x)\rangle \equiv f(\langle x\rangle)$$

$$x \longrightarrow \langle x\rangle$$

For quantum systems with noncommuting dynamical variables there are no dispersion-free states. Yet the convexity requirement that

$$\rho = \rho_1 \cos^2\theta + \rho_2 \sin^2\theta$$

is a state whenever ρ_1, ρ_2 are states holds [2]. Under very mild restrictions one could find the generating elements of the convex set of states and identify them as "pure states". *The generic pure quantum state is not dispersion-free in contrast to the classical states (where the pure states are dispersion-free) that correspond to definite values for all dynamical variables.*

The value obtained from the map is the expectation value and may be denoted by

$$f(x) \longrightarrow \langle f(x)\rangle = E\,\{f(x)\}.$$

A particularly important quantity which always exists and characterizes the state completely is the characteristic function:

$$\chi\{\lambda\} = \chi(\lambda_1, \lambda_2, \ldots) = E\{\exp(i\lambda x)\}$$

$$\equiv \langle \exp[i(\lambda_1 x_1 + \lambda_2 x_2 + \ldots)]\rangle.$$

Clearly

$$|\chi(\lambda)| \leq 1$$

with the equality being realized only for dispersion free states.

For classical systems it is usual to introduce the probability measure dP(x) so that

$$E\{f(x)\} = \int f(x)dP(x)$$

and the differential probability density p(x) by

$$p(x)dx = dP(x) \geq 0 .$$

In terms of these the characteristic function becomes

$$\chi(\lambda) = \int \exp(i\lambda x)\, dP(x) \equiv \int \exp(i\lambda x)\, p(x)dx$$

so that $\chi(\lambda)$ is the (multidimensional) Fourier transform of the probability density.

For a quantum system, on the other hand one introduces the density operator ρ in terms of which expectation values are calculated according to the rule

$$E\{f(x)\} = \mathrm{Tr}\,\{f(x)\ \rho\}$$

the trace being taken in the vector space of the quantum system. The requirements of normalization and positivity are transcribed into the requirements

$$\mathrm{Tr}\,\{\rho\} = 1\ ;\ \rho \geq 0 .$$

Linearity is assured by properties of the trace operation. Extremal states are obtained by choosing density matrices which are projections of rank one

$$\rho^2 = \rho\ ;\ \mathrm{Tr}\,\rho = 1 .$$

Then there exists a state vector ψ , defined modulo a phase, such that

$$\rho = \psi\,\psi^\dagger\ ;\ \psi^\dagger\psi = 1$$

Since there are any number of operators which fail to commute with any such ρ these states will not be dispersion-free for those operators.

Classically all dynamical variables are capable of simultaneous measurement, the extremal states are dispersion-free. The complete specification of a statistical state is then the specification of the probability measure for all the variables, the "master probability distribution". If the observation is limited to the functions on a subset on primary variables the results can be expressed in terms of a "marginal distribution". All marginal distributions are obtained by partial integration of the master probability distribution. Needless to say all these are nonnegative normalized distributions.

In quantum theory not all dynamical variables can be simultaneously measured since they do not commute with each other. We must restrict simultaneous measurements to all or any of a complete commuting set of dynamical variables. With each such choice of complete commuting set there is a marginal probability distribution [3] which is nonnegative and normalized. The state of quantum system is thus associated with a probability distribution for each complete commuting set. Not all of them are independent; but they completely determine the state [4]. If $\{\hat{\psi}_1, \hat{\psi}_2, \ldots\}$ is a basis for the quantum system then the probability distributions associated with the complete commuting sets

$$\left\{\varphi_1\varphi_1^\dagger \ , \ \varphi_2\varphi_2^\dagger \ , \ \ldots\right\}$$

$$\left\{\frac{1}{2}(\varphi_1+\varphi_2)(\varphi_1+\varphi_2)^\dagger \ , \ \frac{1}{2}(\varphi_3+\varphi_4)(\varphi_3+\varphi_4)^\dagger \ , \ \ldots\right\}$$

$$\left\{\frac{1}{2}(\varphi_2+\varphi_3)(\varphi_2+\varphi_3)^\dagger \ , \ \frac{1}{2}(\varphi_4+\varphi_5)(\varphi_4+\varphi_5)^\dagger \ , \ \ldots\right\}$$

$$\left\{\frac{1}{2}(\varphi_1+i\varphi_2)(\varphi_1+i\varphi_2)^\dagger \ , \ \frac{1}{2}(\varphi_3+i\varphi_4)(\varphi_3+i\varphi_4)^\dagger \ , \ \ldots\right\} \text{ and}$$

$$\left\{\frac{1}{2}(\varphi_2+i\varphi_3)(\varphi_2+i\varphi_3)^\dagger \ , \ \frac{1}{2}(\varphi_4+i\varphi_5)(\varphi_4+i\varphi_5)^\dagger, \ \ldots\right\}$$

completely determine the state. All other marginal probability distributions are expressible in terms of these.

GENERALIZED QUANTUM PROBABILITY DISTRIBUTIONS; BELL'S INEQUALITIES

Can there be a master probability distribution from which these marginal probabilities can be obtained? Is quantum theory a statistical theory with possibly undetected (and undetectable?) "hidden variables"? We would require that all the above marginal distributions may be obtained from this master distribution by integrating over the unwanted variables

An answer in the negative about the existence of a probability distribution for a set of variables larger than a complete commuting set was given by Bell [5]. Bell dealt with a quantum system with two spin $\frac{1}{2}$ particles in a total spin 0 state. By making independent measurements on the two particles after they have spatially separated we can evaluate virtually correlations of spin components in any two definite directions. The pairs of such measurement would be expected to obey a triangle inequality

$$P(\mathbf{a},\mathbf{b}) \leq P(\mathbf{a},\mathbf{c}) + P(\mathbf{c},\mathbf{b})$$

where **a, b, c** are any three directions. However, quantum mechanical calculations show that these triangle inequalities can be violated; and experiments confirm the quantum predictions.

A generalization of Bell's conclusions applicable to any system with a noncommutative set of dynamical variables has been developed by Jordan and Sudarshan [6].

Is there any way out? Can we find generalized master probability distributions for which the requirements of positivity is relaxed with the proviso that the marginal distribution to any commuting set is nonnegative? The answer is in the affirmative provided the operators have point spectra. We proceed as follows :

Construct the projection operator $\Pi(x_r = \zeta_{r,v})$, to the υth eigenvalue $\zeta_{r,v}$ of a definite variable x_r. If $\phi(x_r) = 0$ is the characteristic equation of x_r the projection operator may be written

$$\Pi\,(x_r = \zeta_{r,v}) = \frac{\phi(x_r)}{(x_r - \zeta_{r,v})\left.\frac{\partial\phi}{\partial x}\right|_{x_r = \zeta_{r,v}}} .$$

In general the Π's for different x_r do not commute and therefore the product of the projections have to be defined with a rule for ordering the factors. One convenient way is to define a completely symmetrized product of the various projection operators; this would always be Hermitian, but there are other Hermitian choices like taking an arbitrary product and taking its Hermitian part. The expectation value of the product of projection operators gives the generalized probability [7] for that configuration. By explicit construction the marginal distributions are given by summing over all possibilities for the irrelevant ones; but the sum of the projection operators for all possible values is the identity operator! Finally when we come to a commuting set of operators their projections all commute and yield nonnegative operators; consequently their expectations would be nonnegative.

For the special cases considered by Bell we have spin $\frac{1}{2}$ particles. The dynamical variables here are $\underline{\sigma}.\mathbf{a}$, $\underline{\sigma}.\mathbf{b}$ etc., where **a, b** ... unit vectors. Two correlated measurements $\underline{\sigma}$. **a** and $\underline{\sigma}$. **b** are made by measuring $\underline{\sigma}.\mathbf{a}$ on the first particle and $-\underline{\sigma}.\mathbf{b}$ on the second particle when the two particles are in a total spin 0 state. Consider the possibility of measuring the spin of the first particle on these directions **a**, **b**, **c** . In each experiment only two of them can be measured but we make three distinct set of measurements for the three pairs of direction. If the measurement of $\underline{\sigma}.\mathbf{a}$ gives +1 and $\underline{\sigma}.\mathbf{b}$ gives -1 we denote the probability for this occurrence by P(**a**+, **b**−) etc. If there is a master probability distribution P(**a**+, **b**±, **c**±, **d**±, ...) then P(**a**+, **b**−) for example could be obtained by summing over ± for **c**, **d**, ... from P(**a**+, **b**−, **c**±, **d**±, ...). If a (nonnegative) master probability distribution of this kind existed then P(**a**±, **b**±, **c**±) would be nonnegative. Then

$$P(\mathbf{a}+,\mathbf{b}+) = \frac{P(\mathbf{a}+,\mathbf{b}+,\mathbf{c}+) + P(\mathbf{a}+,\mathbf{b}+,\mathbf{c}-)}{\sum P(\mathbf{a}+,\mathbf{b}+,\mathbf{c}+)}$$

and similar expressions hold for P(**a**+, **c**+) and P(**a**+, **c**+). The triangle inequality holds:

$$P(\mathbf{a}{+}\mathbf{b}{+}) \le P(\mathbf{a}+,\mathbf{c}+) + P(\mathbf{c}+,\mathbf{b}+).$$

There are more such relations involving four directions **a**, **b**, **c**, **d**. These are the Bell inequalities.

For a quantum mechanical system these inequalities are violated. If we define the projection operators

$$\Pi(\mathbf{a}+) = \frac{1}{2}\,(1 \pm \underline{\sigma}.\mathbf{a}),$$

the joint probability P(**a**+, **b**±) is given by

$$P(\mathbf{a}+,\mathbf{b}\pm) = \frac{1}{2}\,\mathrm{Tr}\left\{\Pi(\mathbf{a}+)\ \Pi(\mathbf{b}\pm)\right\}$$

$$= \frac{1}{2}\,\frac{1}{2!}\,\mathrm{Tr}\left\{\Pi(\mathbf{a}+)\ \Pi(\mathbf{b}\pm) + \Pi(\mathbf{b}\pm)\ \Pi(\mathbf{a}\pm)\right\} = \frac{1}{4}\,(1 \pm \mathbf{a}.\mathbf{b}).$$

If we choose **a**, **b**, **c** in a plane with **a**.**c** = **b**.**c** = cos 2α, **a**.**b** = cos 4α then the triangle inequality becomes

$$\sin^2 2\alpha \le 2\sin^2\alpha$$

which is, in general, not true!

But, surely, we can define a generalized master probability distribution:

$$P(\mathbf{a}\lambda,\mathbf{b}\mu,\mathbf{c}\nu) = \frac{1}{3!}\ \frac{1}{2}\,\mathrm{Tr}\left\{\Pi(\mathbf{a}\lambda)\ \Pi(\mathbf{b}\mu)\ \Pi(\mathbf{c}\nu) + \text{permutations}\right\}$$

which can be computed to give

$$P(\mathbf{a}\lambda,\mathbf{b}\mu,\mathbf{c}\nu) = \frac{1}{8}\left\{1 + \lambda\mu\ \mathbf{a}.\mathbf{b} + \mu\nu\ \mathbf{b}.\mathbf{c} + \nu\lambda\ \mathbf{c}.\mathbf{a}\right\}.$$

This is not always nonnegative. For the configuration

$$\mathbf{a}.\mathbf{b} = \cos 4\alpha,\quad \mathbf{b}.\mathbf{c} = \mathbf{c}.\mathbf{a} = \cos 2\alpha$$

we get

$$P(+\ +\ +) = P(-\ -\ -) = \frac{1}{4}\cos\alpha\ (1 + \cos\alpha)$$

$$P(+\ +\ -) = P(-\ -\ +) = \frac{1}{4}\cos\alpha\ (-1 + \cos\alpha)$$

$$P(+\ -\ +) = P(-\ +\ -) = \frac{1}{4}\sin^2\alpha$$

$$P(+\ -\ -) = P(-\ +\ +) = \frac{1}{4}\sin^2\alpha$$

Clearly the first two lines can become negative under suitable conditions.

For two spin 1 particles coupled to zero angular momentum we can again demonstrate quantum mechanical violations of triangle equalities deduced from the anticipated nonnegativity of a master probability distribution. In this case the results of the measurement of a spin component $\mathbf{s}\,.\mathbf{a}$ can give three distinct values +1,0,-1. We have, accordingly the projection operators

$$\Pi(\mathbf{a}+) = \frac{1}{2}\,\mathbf{s}.\mathbf{a}\ (1+\mathbf{s}.\mathbf{a})$$

$$\Pi(\mathbf{a}0) = 1-(\mathbf{s}.\mathbf{a})^2$$

$$\Pi(\mathbf{a}-) = \frac{1}{2}\,\mathbf{s}.\mathbf{a}\ (-1+\mathbf{s}.\mathbf{a}).$$

We denote by $P(\mathbf{a},\mathbf{b})$ the quantity

$$P(\mathbf{a},\mathbf{b}) = P(\mathbf{s}.\mathbf{a} = +,\ \mathbf{s}.\mathbf{b} = -) + P(\mathbf{s}.\mathbf{a} = +,\ \mathbf{s}.\mathbf{b} = 0) + P(\mathbf{s}.\mathbf{a} = 0, \mathbf{s}.\mathbf{b} = +).$$

Classically we expect the triangle inequalities. The Bell's inequality for the configuration $\mathbf{a}.\mathbf{b} = \cos 4\alpha$, $\mathbf{b}.\mathbf{c} = \mathbf{c}.\mathbf{a} = \cos 2\alpha$ is equivalent to

$$6\cos^4\alpha - 7\cos^2\alpha - 2\cos\alpha + 3 \geq 0.$$

When $\cos\alpha \geq 0.611$ this inequality is violated.

Note that the quantities $P(\mathbf{a}\lambda, \mathbf{b}\mu, \mathbf{c}\nu)$ are inferred probabilities and not susceptible of direct measurement. In every possible measurement at most two of the spin components can be measured; for them the expected correlations are always nonnegative.

A THEOREM ON PROBABILITY DISTRIBUTIONS OVER NONCOMMUTING VARIABLES

Jordan and Sudarshan [6] have shown very simply that for any two observables represented by Hermitian operators that do not commute, there is a state for which there is no nonnegative joint probability distribution for these two observables.

If A and B are two noncommuting observables there are projections E and F associated with A and B respectively which do not commute. The eigenvalues of the projections are 0,1. If there were (nonnegative) joint probability distributions for E, F there are nonnegative probabilities p(1,1), p(1,0), p(0,1), p(0,0) with marginal probabilities

$$p_E(1) = p(1,1) + p(1,0)$$

$$p_E(0) = p(0,1) + p(0,0)$$

$$p_E(1) = p(1,1) + p(0,1)$$

$$p_E(0) = p(1,0) + p(0,0)$$

If E,F do not commute there must be a state ψ such that

$$E\psi = 0 \ , \ F\psi \neq 0 \ , \ EF\psi \neq 0 \ .$$

To see this we observe that since

$$EF \neq FE$$

$$EF(1 - E) \neq 0$$

Since EF = EFE implies EF = (EF) = FE. If for all $E\psi = 0$, $F\psi = 0$ then it is true for every ϕ that

$$EF(1-E)\phi = 0.$$

The probability that the measurement of F for the state ψ gives 1 is

$$p_F(1) = \langle F \rangle = || F\psi ||^2;$$

and the probability that a subsequent measurement of E gives 1 is

$$p_E(1) = \left|\left| \frac{1}{|| F\psi ||} EF\psi \right|\right|^2 = \frac{|| EF\psi ||^2}{|| F\psi ||^2} \neq 0 \ .$$

But then p(1,1) ≠ 0 contrary to our stated assumption.

This demonstration does not require A,B to have discrete spectra; but the associated set of spectral projectors have the discrete eigenvalues 1,0 only. In particular for canonical commutation relations in orthodox quantum mechanics or for the set of generators of a (noncompact) Lie group characterizing a quantum system these considerations apply.

For the canonical position and momentum operators Wigner gave a distribution function almost six decades ago [8]. This "Wigner density" is a phase space function

$$\rho(q,p) = \int \psi^*(q + \tfrac{x}{2})\, e^{ipx}\, \psi(q - \tfrac{x}{2})\, dx .$$

This function is real, bounded by unity, integrable and square integrable. The marginal distributions give true nonnegative probability distributions:

$$\int \rho(q,p)dp = \psi^*(q)\, \psi(q)$$

$$\int \rho(q,p)dp = \tilde{\psi}^*(p)\, \tilde{\psi}(p) \; ; \quad \tilde{\psi}(p) = \frac{1}{\sqrt{2\pi}} \int \psi(x)\, e^{-ipx}\, dx$$

But ρ itself is not always nonnegative. For any of an orthonormal set of states at most one density can be nonnegative. For example for the harmonic oscillator the ground state Wigner density is nonnegative everywhere but for the excited states the densities take on negative values also.

Nevertheless the Wigner density is a complete dynamical description of the state; and the dynamics formulated in terms of the Wigner density involves the Moyal brackets [9].

The Wigner-Moyal method can be extended to other systems with noncommuting dynamical variables. One such is the application to Wolf functions [10] which gives the density of light rays in a generic pencil, recognizing the fact that direction and location of a ray of light cannot be simultaneously specified. When correlated ray distributions are taken the Bose nature of the quantum field is reflected in a positive correlation of light rays [11].

CHARACTERISTIC FUNCTIONS AND THE MARCINKIEWICZ THEOREM

For classical multivariate probability distributions it is useful to define the characteristic function $\chi(\lambda)$ which is the expectation value of $\exp(i\lambda x)$:

$$\chi(\lambda) = \langle e^{i\lambda x} \rangle = \int p(x)\, \exp(i\lambda x)\, dx.$$

Clearly

$$|\chi(\lambda)| \le 1$$

with equality only for probability distribution concentrated at a point. If $\chi(\lambda)$ is analytic in λ at the origin then it can be expanded in power series with coefficients which are the moments:

$$\chi(\lambda) = \sum_{0}^{\infty} \frac{(i\lambda)^n}{n!} \langle x^n \rangle .$$

The moments may not exist but $\chi(\lambda)$ will exist; except that it maynot be analytic at $\lambda=0$.

More useful is the cumulant generating function

$$\zeta(x) = \log \chi(\lambda)$$

whose power series expansion around $\lambda = 0$ gives the cumulants :

$$\zeta(\lambda) = \log \left[1 + \sum_1^\infty \frac{(i\lambda)^n}{n!} \langle x^n \rangle \right]$$

$$= i\lambda\langle x\rangle + \frac{(i\lambda)^2}{2!} \left[\langle x^2\rangle - \langle x\rangle^2 \right] + \ldots$$

the first two terms giving the mean and the variance. For a Gaussian distribution, the characteristic function is also a Gaussian and the cumulant generating function is a quadratic polynomial. In a remarkable theorem Marcinkiewicz [12] has shown that the positivity of $p(x)$ demands that if $\zeta(x)$ is not a quadratic polynomial it cannot be a polynomial at all.

The question naturally arises whether for distributions over noncommuting variables where the distribution function is not everywhere nonnegative the Marcinkiewicz theorem [12] can be overcome. In particular we would like to know whether the Wigner-Moyal phase space densities could lead to cumulant functions which are nonquadratic polynomials. To answer this question we must define the characteristic function for a distribution of noncommuting variables. For the case of canonical variables in quantum mechanics we define

$$\chi(\lambda,\mu) = \langle \exp(i\lambda q + i\mu p)\rangle \equiv \int \psi^*(x)\, e^{(i\lambda x + \mu \frac{\partial}{\partial x})}\, \psi(x)\, dx .$$

Using the identity

$$\exp(i\lambda q + i\mu p) = \exp\left[\frac{i\mu p}{2}\right] \exp(i\lambda q) \exp\left[\frac{i\mu p}{2}\right]$$

we compute

$$\chi(\lambda,\mu) = \int \psi^*\left[x - \frac{\mu}{2}\right] e^{i\lambda x}\, \psi\left[x + \frac{\mu}{2}\right] dx .$$

This is seen to be the double Fourier transform of the Wigner density

$$\rho(q,p) = \int \psi^*\left[x - \frac{1}{2}y\right] e^{-ipy}\, \psi\left[x + \frac{1}{2}y\right] dy$$

Can $\log \chi(\lambda,\mu)$ be a nonquadratic polynomial? After all $\rho(q,p)$ is not positive definite. The answer is in the negative. We proceed to show this based on the fact that while $\rho(q,p)$ is not nonnegative, its marginal distributions are nonnegative. Write

$$\lambda = \nu \cos\theta \;,\; \mu = \nu \sin\theta$$

so that,

$$\lambda q + \mu p = \nu(q \cos\theta + p \sin\theta) \ .$$

The change of variables from (q,p) to (q',p') according to

$$q' = q \cos\theta + p \sin\theta$$

$$p' = -q \sin\theta + p \cos\theta$$

is a canonical transformation. Then the characteristic function $\chi(\nu \cos\theta, \nu \sin\theta)$ is the characteristic function for the marginal distribution in q'. But if $\chi(\lambda,\mu)$ were a nonquadratic polynomial in λ,μ so would $\chi(\nu \cos\theta, \nu \sin\theta)$ be a polynomial of the same degree in ν. But all such polynomials are restricted to be quadratic. Hence the same applies to $\chi(\lambda,\mu)$. The Marcinkiewicz theorem applies to the Wigner densities. Consequently it applies also to the quantum (or semiclassical) correlation functions in optics; for Gaussian fields all correlations are expressible as functions of the two-point correlations and the linked clusters beyond the two point vanish: all order linked clusters are nontrivial. Hence the complete description of a non-Gaussian optical field involves an infinity of primitive correlation functions.

When the dynamical variable takes on only a discrete number of values the characteristic function is a sum of exponentials; and cannot be the exponentials of polynomials!

CONCLUDING REMARKS

In this brief survey we have shown how to define and study probability distributions over noncommuting variables. In many cases a characteristic function can be defined. It is the expectation value of the exponential of a (pure imaginary) linear sum of the noncommuting variables. When the variables are generators of a noncompact group this is a convenient definition: we compute the characteristic function as the expectation value for the generic group element. For systems in which the variables take on only a finite number of values this is cumbersome: a more direct method is to compile the probabilities for each of the (finite number of) configurations. This is what we have done for the case of spin systems.

Much of the machinery of classical probability theory can be extended to this case also except for the possible violation of point wise nonnegativity. There are still implicit positivity requirements, but whether it be for the Wigner density or the Bell densities the positivity requirements are more subtle.

We have dealt with the "operator ordering" problem for noncommuting variables by symmetrizing with respect to the basic variables. There are other possible ways of doing this which give rise to other kinds of distributions. For quantum optics the diagonal coherent state density (the Sudarshan density [13]) and the diagonal matrix elements of the density operator in the coherent states (the Kano density [14]) correspond, respectively, to the normal and the antinormal ordering of creation and annihilation operators. While these have excellent properties, they do not have the property that their marginal distributions represent the probability distributions over commuting variables. Because of the simple form of the Baker-Campbell-Hausdorff identities for exponentials of canonical operators the characteristic functions of the Sudarshan and the Kano distributions are Gaussian multiples of the characteristic function of the Wigner distribution; therefore they too satisfy the Marcinkiewicz theorem.

This paper is written on the occasion of the sixtieth birthday of Professor Sumangali Kidambi Srinivasan who has contributed so decisively to probability theory and its applications. The author is honored to acknowledge the continuing stimulation of discussions with Professor Srinivasan over the decades of their mutual acquaintance.

This work was supported in part by research contract DOE-FG05-85 ER-40-200.

REFERENCES

1. I.E. Segal, Annals of 48, 930 (1947)
   E.C.G.Sudarshan, Lectures on Theoretical Physics. Brandeis, 1959.
2. P.M. Mathews, J.Rau and E.C.G. Sudarshan, Phys. Rev. 2, 920, (1961).
3. I.E. Segal, Illinois J. Math, 6, 500 (1962).
4. E. Hopf, I. Rat. Mech. and Anal 1, 87 (1952)
   I.E. Segal, Can. J.Math 13, 1 (1961)
   E.C.G.Sudarshan, J.Math and Phys. Sci. (Madras) 3, 121 (1969).
5. J.S. Bell, Physics, 1, 195 (1964).
6. T.F. Jordan and E.C.G.Sudarshan, "Simply No Hidden Variables", University of Minnesota preprint, submitted to Am. J.Physics.
7. E.C.G.Sudarshan and Tony Rothman, "A New Interpretation of Bell's Inequalities", University of Texas report DOE-199 .
8. E.P. Wigner, Phys. Rev. 40, 749 (1932).
9. J.E.Moyal, Proc. Cambr. Phil. Soc. 45, 99 (1948).
10. E.C.G.Sudarshan, Physica 96A, 315 (1979); Phys. Rev. A23, 2802, (1981).

11. E.C.G.Sudarshan, N.Mukunda and R.Simon, "Coherence, propagation and Fluctuations of Light", University of Texas report, presented at the University of Rochester Symposium "The Optical Field", Oct.1987.

12. J.Marcinkiewicz, Math. Z. 44, 612 (1939).

13. E.C.G.Sudarshan, Phys. Rev. Lett. ID, 27 (1963)
C.L. Mehta and E.C.G.Sudarshan, Phys. Rev. 138B, 274 (1965).

14. Y. Kano, J. Math. Phys. 6, 1913 (1965).

STOCHASTIC QUANTUM MECHANICS

R. Vasudevan
The Institute of Mathematical Sciences
Madras 600 113
INDIA

ABSTRACT

The need for answering the question whether quantum mechanics is complete by itself and whether quantum fluctuations and thermal fluctuations are of the same genus, is discussed in this report. The hydrodynamic Hamilton Jacobi equations for the quantum Madelung fluid reveal the existence of a quantum potential which corresponds to the mysterious dependence of the individual on the statistical ensemble of which it is a member. Nelson's analysis, starting from the Brownian motion of a particle moving in the field of a background white noise, leads to coupled equations for the velocity fields under certain conditions; these are described in the first three sections.

Section 4 deals with indeterminacy relation in a novel way based on the two velocities envisaged in Nelson's work and some examples are given. The concluding section deals with the criticism of these two approaches and promises the unification of the three approaches including the derivation of the Feynman's path formalism obtained by complexifying the phase accumulation and assigning a proper measure for it. The extension of this approach to relativistic and spinning particles will form the content of part II of this article to be published elsewhere.

DEDICATION

This contribution is in honour of Professor S.K.Srinivasan, my esteemed colleague and lifelong collaborator, who is crossing his sixtieth birth day and has covered himself with great repute as an eminent stochastic theorist. I am proud that we were both initiated into the wonders of stochastic processes by our great teacher, innovator, and famous educationist of our country, Professor Alladi Ramakrishnan thirty years ago. He gave to the stochastic theorists the magnificent tool of product densities in the early fifties, with which problems in different fields are tackled by our group extensively.

Professor Srinivasan used sophisticated and creative ideas to solve problems thrown up by a wide spectrum of fields such as operation research, population processes, neural spike trains, doubly stochastic processes, quantum optics and cascade theories. By his endeavour in teaching and research he has carved a "niche" for himself in the annals of great probabilists and I am happy to be associated with this symposium.

1. THE NON-LOCALITY PROBLEM

This is a subject which will make even a phlegmatic physicist to sit up and argue about the conventional interpretation of quantum mechanics. The orthodox view that quantum mechanics gives the complete description of a quantum system suffers from the fact that it lacks a definite answer to the question: What is the distinction between quantum fluctuations and thermal fluctuations? The first one refers to the uncertainty in the outcome of measurement of a single system, while the latter refers to fluctuations within the members of the ensemble. This distinction must remain absolute though consistent under restrictions. The effort now is to modify the interpretation of quantum state so that it refers to an ensemble of physical systems. This is an attempt to understand the probabilistic nature of quantum mechanics by invoking a more precise description of the world at a deeper level; such a description would complete quantum mechanics as statistical mechanics completes thermodynamics. In a famous paper Einstein et.al. [1] using the reasonings based on gadenken experiments called EPR - paradox, spelled out that quantum mechanics must be completed. Bohr disagreed with his conclusion and others did not commit themselves by saying it was a matter of taste [2]. It was Bell's [3] paper that opened the route to real experiments designed by the 1970's and thanks to the technological advances, these findings are getting refined all the time. We refer the reader to the experiments of Clauser et. al. [4] and Aspect and others [5] to test Bell's inequality using a pair of low energy photons.

These experiments have emphasised the principle of non-locality in quantum mechanics. Let us summarise briefly the experiment (which mimics the spirit of the so called EPRB gadenken experiment) with photons. Clauser et al. [6] in atomic cascades $4P^2S_o$ - $4S4P^1P_1 \longrightarrow 4S^{21}S_o$ in Calcium, used two photons with energies γ_1 and γ_2 going in opposite directions. The two photons from source S counter propagate along OZ and -OZ and impinge on linear polarisation analysers I and II with the orientations of polarisers along unit vectors a and b. We can write the state vector

$$|\psi\,(r_1,r_2)\rangle = 2^{-1/2}(|x_1,x_2\rangle + |y_1,y_2\rangle) \tag{1.1}$$

where $|x_1\rangle$ refers to ν_1 photon and $|y_1\rangle$ to the ν_2 photon. A photon can be found in one of the two exit channels labelled +1, or -1 of the analysers I and II at the two ends according as the linear polarisations of the photons are parallel or perpendicular to the orientations a or b of the two analysers.

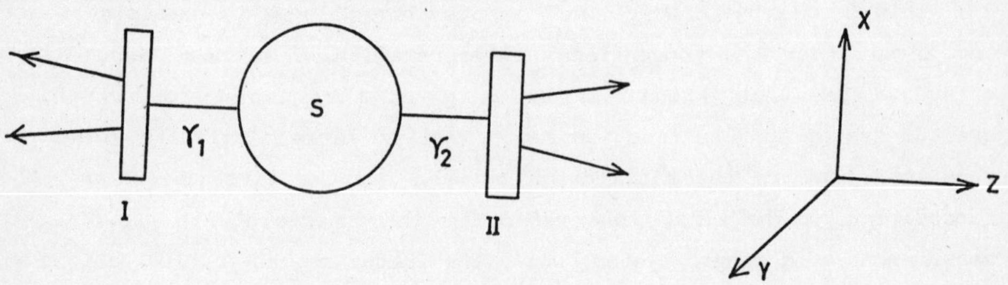

Fig. 1

atomic cascade $J = 0 \longrightarrow J = 1 \longrightarrow J = 0$ (atomic angular momentum states)

This is like the Stern and Gerlach filter for spin 1/2 particles, and for single measurements at analyser I or at II we obtain

$$P_+(\vec{a}) = P_-(\vec{a}) = 1/2\ ,\ P_+(\vec{b}) = P_-(\vec{b}) = 1/2 \tag{1.2}$$

For joint measurement, a simple quantum mechanical calculation will give [5]

$$P_{++}(\vec{a},\vec{b}) = P_{--}(\vec{a},\vec{b}) = (1/2)\cos^2(\vec{a}.\vec{b})$$

$$P_{+-}(\vec{a},\vec{b}) = P_{-+}(\vec{a},\vec{b}) = (1/2)\sin^2(\vec{a}.\vec{b}) \tag{1.3}$$

For $\vec{a}\ .\ \vec{b} = 0$ the correlation related function [5] is calculated from experiments.

$$E(\vec{a}.\vec{b}) = P_{++}(\vec{a},\vec{b}) + P_{--}(\vec{a},\vec{b}) - P_{+-}(\vec{a},\vec{b}) - P_{-+}(\vec{a},\vec{b}) = \cos^2(\vec{a}.\vec{b})\ . \tag{1.4}$$

According to this photon γ_1 has equal chance of going into channel +1 or channel -1, so also for photon ν_2. However if photon γ_1 is experimentally found to be in +1, ν_2 goes into channel +1 and conversely. It is difficult to understand these correlations according to standard interpretation of quantum mechanics. How does ν_2 know that ν_1 has entered into a particular channel? It is easy to understand strong correlation between the pair distant from each other if they had interacted at the source and share a common property from the *classical stand point of the EPR* experiments. Hence there is an element of objective physical reality determining this result. Einstein therefore concluded that quantum mechanics is not complete.

J.S.Bell using the idea of local supplementary variables for the ensemble of pairs with orientations $\vec{a}$, $\vec{b}$ and $\vec{a}'$, $\vec{b}'$ for the analysers measured with a set of various polarisations constructed for the function S defined by

$$S = E(\vec{a}, \vec{b}) - E(\vec{a}, \vec{b}') + E(\vec{a}', \vec{b}) + E(\vec{a}', \vec{b}') \qquad (1.5)$$

the inequality

$$-2 \le S \le 2 .$$

However quantum mechanical calculation leads to values of S that do not obey Bell's inequality in many situations. Bell's inequality also implies locality conditions. The locality conditions of "hidden variable" [15] theories may lead to violations of Bell's inequality. This means that there is no classical looking picture in the spirit of Einstein's idea that could mimic all the predictions of quantum mechanics. The question now arises whether it is not possible to consider quantum mechanics *as an average of a deeper level theory* if we can improve classical looking features at this level ? [1]

Thus the EPR paradox and the experiments referred above bring out the fact that when one particle is experimented upon, the information travels to the correlated particle instantaneously. In the "many world theories" of the Dewitt [7] the second particle collapses into a corresponding quantum state. This strengthens our idea of essential non-locality in the processes. Wigner's idea [8] of attributing this collapse to the interaction of matter and mind (or universal consciousness) points out the essential non-locality in sensing events. Feynmann [9], writes in answer to the question "Is there a real problem?" that "It has not become obvious to me that there is no real problem. I cannot define the real problem and therefore I suspect there is no real problem. But I am not sure that there is no real problem". We will see in the sequel the different kinds of attempts to resolve this problem of non-locality.

2. THE MADELUNG FLUID AND CAUSAL INTERPRETATION

The causal interpretation of quantum theory [9] brought out the question of non-locality into sharp focus and this introduces the concept of a new kind of quantum potential allowing for immediate interaction of distant particles thus explaining the EPR- paradox. The path of a particle in non-relativistic quantum mechanics is given by the wave function ψ obeying the Schrodinger equation

$$i\hbar \frac{\partial\psi}{\partial t} = -\frac{\hbar^2}{2m}\nabla^2\psi + V\psi . \tag{2.1}$$

Considering the solution $\psi = Re^{iS/\hbar}$ we obtain

$$\frac{\partial S}{\partial t} + \frac{1}{2m}(\vec{\nabla}S)^2 + V - \frac{\hbar^2}{2m}\frac{\nabla^2 R}{R} = 0 \tag{2.2}$$

$$\frac{\partial R^2}{\partial t} + \vec{\nabla} . \left(R^2 \frac{\nabla S}{m}\right) = 0 \tag{2.3}$$

which are called the flow equations of the Madelung [18] fluid with density $\rho(x,t)=R^2$. If the term with $\hbar^2$ is neglected, (2.2) gives the Hamilton-Jacobi equation for a classical motion or the WKB approximation for the Quantum Jacobi equation. This is interpreted as the path of a particle normal to wave front S = const at each time and moving with momentum $\vec{P}= \vec{\nabla}S$. In fact this defines an ensemble of trajectories whose density is R^2 at each x and t and the equation (2.3) is the conservation of probability or continuity equation. The last term in (2.2) is an additional quantum potential $Q = -\frac{\hbar^2\nabla^2 R}{2m\,R}$ which depends not on ψ, but on the form of the solution and corresponds to "active information" for the motion. This is likened to a ship in motion by an automatic pilot propelled by information in radio waves [10]. When there are other particles in the system this quantum potential is altered so that non-local interaction is possible since the guidance condition and the quantum potential belong to the whole system or the "pool of active information". As an immediate application we can *figure out stationary states* with real wave functions, when $\vec{p} = \vec{\nabla}S = 0$ due to cancellation of V by the quantum potential in the opposite direction. There is no net force on the particle. But this idea is not liked by many physicists.

As explained by Bohm and Hiley [9] and derived by Nelson [11],on the Brownian motion model, a steady background field is postulated giving rise to another drift velocity $\vec{u}_o$, called osmotic velocity; $\vec{u}_o = \frac{D\vec{\nabla}\rho}{\rho}$, D being a constant. The diffusion current $\vec{J} = -D\,\vec{\nabla}\pi$. The random component of the motion which is responsible for diffusion may be due to vacuum fluctuations or

Brownian noise with spectrum coefficient D. The total current is

$$\vec{j} = \pi \frac{\vec{\nabla} S}{m} + \pi D \frac{\vec{\nabla}\rho}{\rho} - D \vec{\nabla}\pi \tag{2.4}$$

where $\frac{\vec{\nabla} S}{m} = \vec{v}$ is the velocity of the particle while $\rho = |\psi|^2$, and $\pi(x,t)$ is the probability of the particle being at x at t. The conservation equation for $\pi(x,t)$, the probability density, is

$$\frac{\partial \pi}{\partial t} + \vec{\nabla} \cdot \vec{J} = 0 \tag{2.5}$$

The osmotic part of the velocity arises from the other source and the systematic velocity is

$$\vec{v}_s = \frac{\vec{\nabla} S}{m} + D \frac{\vec{\nabla}\rho}{\rho} \tag{2.6}$$

where ρ is the solution of the continuity equation $\frac{d\rho}{dt} + \vec{\nabla}.(\rho \frac{\vec{\nabla} S}{m}) = 0$. In equilibrium situations π reduces to $\rho(x,t)$. This can be shown to be so for $\vec{\nabla} S = 0$ and also for $\vec{\nabla} S \neq 0$ [10] and $\rho = |\psi|^2$.

We shall consider a many body system or even a two body problem. The wave function $\psi(r_1, r_2, t)$ satisfies the Schrodinger'sequation

$$i\hbar \frac{\partial \psi}{\partial t} = - \frac{\hbar^2}{2m} \left[\nabla_1^2 + \nabla_2^2 \right] \psi + V \psi \tag{2.7}$$

and writing $\psi = Re^{iS/\hbar}$ and defining $R^2 = |\psi|^2$ we obtain

$$\frac{\partial S}{\partial t} + \frac{(\vec{\nabla}_1 S)^2}{2m} + \frac{(\vec{\nabla}_2 S)^2}{2m} + V + Q = 0 \tag{2.8}$$

where

$$Q = - \frac{\hbar^2}{2mR} \left[\nabla_1^2 + \nabla_2^2 \right] R \ , \ \rho = R^2 \tag{2.9}$$

and

$$\frac{\partial \rho}{\partial t} + \vec{\nabla}_1 \cdot \left[\rho \frac{\nabla_1 S}{m} \right] + \vec{\nabla}_2 \cdot \left[\rho \frac{\nabla_2 S}{m} \right] = 0 \tag{2.10}$$

The Hamilton-Jacobi equation for the system of two particles depends not only on V, the classical potential between them, but also responds to the quantum potential Q which depends on the particles in a way that does not fall off with distance. Thus we have a non-local interaction through this action-at-a distance. Of course this is the peculiar feature of quantum mechanics which is responsible for dependence of the whole system with the parts being guided by the shape of the entire wave function. This dynamical interrelationship is the most fundamentally new ontological significance of quantum theory. This suggests that quantum potential in a novel way organises the activity of entire set of particles depending on the state of the whole and may become relevant to explaining the features of living organisms. Also it is claimed that this causal interpretation of quantum theory explains the measurement processes in quantum theory [10].

3. STOCHASTIC QUANTUM MECHANICS OF NELSON

As we discussed in the introduction we need to find a concept which can encompass both the ideas of virtual quantum fluctuations and that of statistical fluctuations. The fundamen'al tenet of the statistical interpretation of the wave function is that the wave function is not a description of an individual system but rather a description of an ensemble of similarly prepared systems. One must therefore assume a universal source of fluctuations and this source must impart some unusual properties to the correlations of these quantum fluctuations (including existence of non-local correlations).

Nelson [11] assumed that the particle is performing a Markovian motion in a background of a noise field, the particle being endowed with a deterministic drift.

Let x(t) be such a Markov process denoting the coordinate of the particle in continuous time t; then the motion is described by

$$dx(t) = b(x,t)dt + dW(t) \tag{3.1}$$

dW is a white noise Gaussian process with independent and identically distributed increments, and ν is the diffusion coefficient which occurs in the mean square value of the noise process:

$$\langle \Delta W \rangle = 0, \quad \langle \Delta W \,.\, \Delta W \rangle = 2\nu dt \tag{3.2}$$

Hence the attempt to explain quantum mechanics as arising from a probabilistic description at some more fundamental level of dynamics must be the stochastic motion of the particle described by equation (3.1). However the trajectories belong to an ensemble endowed with special features to be described below.

Let us now turn our attention to Nelson's [11] derivation of quantum motion from a Brownian motion view point by considering the Markovian process for X(t) as in eqn. (3.1) where W is a Weiner process and is a difference martingale satisfying $E\{[W(a)-W(B)]^2|P_a\} = b-a$. Let us define the forward velocity for the particle:

$$Dx(t) = \lim_{\Delta t \to 0} E\left\{ \frac{x(t+\Delta) - x(t)}{\Delta} \middle| P_t \right\} \tag{3.3}$$

where E stands for expectation for all $t \in I$.

Let I be an open interval on the right and P_t for all $t \in I$ be an increasing family of σ algebras of measureable sets on a probability space. X(t) is such a stochastic process having continuous sample paths with probability one such that X(t) is P_t measurable. DX(t) is continuous. Similarly Nelson defines the velocity in negative sense of t, since the process remains a Markov process when looked at either along the positive sense of time or along the negative sense. We have

$$D_*x(t) = \lim_{\Delta t \to 0} E\left\{ \frac{x(t) - x(t-\Delta)}{\Delta} \middle| A_t \right\}$$

and

$$dx(t) = b_*(x, t)dt + dW_* \tag{3.4}$$

Let A_t be a decreasing family of σ algebras such that X(t) is A_t measurable. While P_t refers to the past, A_t represents the future.

Nelson postulates that nature treats past and future on the same footing and hence defines mean acceleration as

$$a = (1/2)\ [DD_* + D_*D\ ]\ x(t) \tag{3.5}$$

We can satisfy ourselves that this is not incorrect by checking with Markov equations for harmonic oscillators. We can obtain for acceleration

$$a(t) = (1/2)\ [DD_* + D_*D]x(t) = -\omega^2 x(t) \tag{3.6}$$

We know that $d^2x/dt^2 + \omega^2 x(t) = 0$ represents the oscillator average equation. Considering Ito [13] equations (3.1) and (3.4) we use the Ito chain rule to obtain

$$D\, f(x,t) = \left[\frac{\partial}{\partial t} + \vec{b}.\vec{\nabla} + \nu\, \nabla^2\right] f(x,t) \tag{3.7}$$

$$D_* f(x,t) = \left[\frac{\partial}{\partial t} + \vec{b}_*.\vec{\nabla} - \nu_* \nabla^2\right] f(x,t) \tag{3.8}$$

If $\nu = \nu_*$ we have the two types of Fokker-Plank equations :

$$\frac{\partial \rho}{\partial t} = - \vec{\nabla} \,.\, (\vec{b}\, \rho) + \nu \nabla^2 \rho$$

$$\frac{\partial \rho}{\partial t} = - \vec{\nabla} \,.\, (\vec{b}_*\, \rho) - \nu \nabla^2 \rho \tag{3.9}$$

$\rho(x,t)$ in the backward and forward equations refers to the probability density for the position x at time t of the particle.

Let us define the current velocity as $\vec{v} = \frac{1}{2}(\vec{b}+\vec{b}_*)$ and the osmotic velocity as $\vec{u} = \frac{1}{2}(\vec{b}-\vec{b}_*)$. Then we obtain from (3.9) the continuity equation

$$\frac{\partial \rho}{\partial t} + \vec{\nabla} \,.\, (\rho \vec{v}) = 0 \tag{3.10}$$

Also by the definition of acceleration (3.5) adopted by Nelson to take care of time symmetry, we have, using (3.7) and (3.8) the equation for the acceleration a:

$$E\left[\frac{1}{2}(DD_* + D_*D)x(t)\right] = \vec{a} = \frac{\partial \vec{v}}{\partial t} + (\vec{v} \,.\, \vec{\nabla})\vec{v} - (\vec{u} \,.\vec{\nabla})\vec{u} + \nu \nabla^2 \vec{u} \tag{3.11}$$

That means

$$\vec{a} = -\frac{\vec{\nabla} V}{m} = \frac{\partial \vec{v}}{\partial t} + (\vec{v} \,.\, \vec{\nabla})\vec{v} - (\vec{u} \,.\vec{\nabla})\vec{u} + \nu \nabla^2 \vec{u}$$

$$= \frac{\partial \vec{v}}{\partial t} + \frac{1}{2}\vec{\nabla}\,(v^2 - u^2) + \left[\vec{\nabla} \,.(\nu \vec{\nabla})\right]\vec{u} \tag{3.12}$$

where V is the external potential. It has been shown by Nelson in an elegant fashion that

$$\frac{\partial \vec{u}}{\partial t} = \frac{\partial}{\partial t}\left[\nu\, \frac{\vec{\nabla} \rho}{\rho}\right] = \nu \vec{\nabla}\left[\vec{\nabla}.\vec{v}\right] + \vec{\nabla}\left[\vec{v}.\vec{u}\right] \tag{3.13}$$

in view of the continuity equation (3.10) with

$$\vec{u} = \frac{\vec{b}-\vec{b}_*}{2} = \nu\, \frac{\vec{\nabla} \rho}{\rho}\;, \quad \vec{v} = \frac{\vec{b}+\vec{b}_*}{2}$$

Remembering that we started with the assumption that x(t) satisfies the

Langevin equation (3.1), under time reversal

$$t \to -t,\ x \to x\ ,\ \nabla_x \to \nabla_x \text{ and } \nabla_x^2 \to \nabla_x^2$$

along with

$$\partial_t \to -\partial_t,\ \vec{\nabla}_v \to -\vec{\nabla}_v \quad \text{and} \quad \vec{\nabla}\,\partial_v^2 \to \partial_v^2 \tag{3.14}$$

we deduce from (3.12) that

$$\vec{\nabla}\left[\partial_t S + \frac{(\vec{\nabla}S)^2}{2m} - \frac{\hbar^2}{2m}\ \frac{\nabla^2\sqrt{\rho}}{\sqrt{\rho}}\right] = -\vec{\nabla}\,V \tag{3.15}$$

where $\nu = \frac{\hbar}{2m}$ is the assumed diffusion coefficient of noise. Since (3.15) is equivalent to equation (3.11) for u and v we obtain easily

$$\partial_t S + \frac{(\vec{\nabla}S)^2}{2m} + V - \frac{\hbar^2}{2m}\ \frac{\nabla^2\sqrt{\rho}}{\sqrt{\rho}} = 0 \tag{3.16}$$

If $\nu > 0$, the drift does not only depend on actual values of the process but also on density distribution. Thus with the stochastic definition of mean acceleration we have given a particle picture since Hamilton-Jacobi equations of the Madelung fluid are obtained with an added term to the external potential. With the definition ψ of the wave solution of Schrodinger equation as

$$\psi = \exp\left[R + i\ \frac{S}{\hbar}\right] = \sqrt{\rho}\ \ \exp\left[i\ \frac{S}{\hbar}\right] \tag{3.17}$$

and

$$\frac{\hbar}{m}\vec{\nabla}R = \vec{u} = \nu\,\frac{\vec{\nabla}\rho}{\rho}\ ,\ \vec{v} = \frac{\vec{\nabla}S}{m}$$

we can easily obtain from Madelung equations or the analogous particle equations (3.13) and (3.15) the well-known Schrodinger equation

$$ih\,\frac{\partial\psi}{\partial t} = -\frac{\hbar^2}{2m}\,\nabla^2\psi + V\psi \tag{3.18}$$

And this equivalence is well-known. This can be extended to include the description of particles with spin and velocity-dependent potentials such as the one arising from the interaction of a charged particle in a magnetic field. Hence we have the two coupled sets of differential equations (3.16) and (3.10) to be solved successively. These successive solutions of the coupled equations give rise to the continuing WKB approximations. This means that

$$-E + V - \frac{(\vec{\nabla}S)^2}{2m} - \frac{\hbar^2}{2m}\frac{1}{\sqrt{\rho}}\nabla^2\sqrt{\rho} = 0$$

and

$$\vec{\nabla}.(\rho\vec{v}) = 0 \tag{3.19}$$

for stationary states with

$$\frac{\partial\rho}{\partial t} = 0 \quad \text{and} \quad \frac{\partial S}{\partial t} = -E \tag{3.20}$$

Neglecting the term with $\hbar^2$ in (3.19), we obtain the usual WKB solutions from the coupled equations for v and ρ

$$v = \pm\sqrt{\frac{2(E - V)}{m}} = \pm k, \quad \rho = \frac{\text{constant}}{v} \text{ and } \vec{\nabla}S = m\vec{v} \tag{3.21}$$

and the WKB wave function

$$\psi_o \sim \frac{\text{const}}{\sqrt{k}} \exp\left\{\pm \frac{im}{\hbar}\int k dx\right\} \tag{3.22}$$

The second iterated solution yields

$$v_1 = \left\{k^2 + \frac{\hbar^2}{2m^2}\frac{\sqrt{k}}{x}\frac{\partial^2}{\partial x^2}\left[\frac{\alpha}{\sqrt{k}}\right]\right\}^{1/2} = K\left\{1 + \frac{\hbar^2}{2m^2k^2}\left[\sqrt{k}\frac{d^2}{dx^2}\sqrt{k}\right]\right\}^{1/2} \tag{3.23}$$

and $\rho_1 = \alpha/v_1$ and so on similar to the usual iterations of WKB treatment of the Schrodinger solutions.

4. POSITION-MOMENTUM UNCERTAINTY IN STOCHASTIC MECHANICS [13]

Before deriving $\Delta x \, \Delta p \geq \frac{\hbar^2}{4}$ we will consider the following identities. If A and B are hermitian, operator AB is not hermitian but can be decomposed as

$$AB = \frac{1}{2}(AB + BA) + \frac{1}{2}[A, B] \tag{4.1}$$

and

$$\text{Re}\,\langle AB\rangle = \frac{1}{2}\langle AB + BA\rangle$$

$$\text{Im}\,\langle AB\rangle = \frac{1}{2i}\langle [A, B]\rangle \tag{4.2}$$

and

$$\text{Variance of } A = \langle A^2\rangle - \langle A\rangle^2$$

$$\text{cov}\,(AB) = \frac{1}{2}\langle AB + BA\rangle - \langle A\rangle\langle B\rangle$$

If we set a = A - <A> and b = B - <B> then due to Schwarz inequality

$$\langle a^2\rangle \langle b^2\rangle \geq |\langle ab\rangle|^2$$

$$\geq \frac{1}{4}\left[\langle ab + ba\rangle^2 + \langle [a,\ b]\rangle^2\right] \tag{4.3}$$

This was the relation derived by Schrodinger in (1930). We specialise these relations to quantum operators X and P by repeating S.D.E. for the motions of x process as

$$d\xi = b(\xi, t)dt + dW$$

where ξ is the random variable corresponding to the eigenvalues of operator X and also obtain its time reversed version. As we have already seen

$$u = \frac{1}{2}(b - b_*) = \nu \frac{\partial}{\partial x} \log \rho = \frac{\hbar}{m} \operatorname{Re} \frac{\partial}{\partial x} \log \psi$$

$$v = \frac{1}{2}(b + b_*) = \frac{P}{m} = \frac{\hbar}{m} \operatorname{Im} \frac{d}{dx} \log \psi \tag{4.4}$$

The diffusion process ξ and velocity u can be found to satisfy

$$(\mathrm{Var}\ \xi)\ (\mathrm{Var}\ u) = E\left[(\xi - \langle\xi\rangle)^2\right] E\left[(u^2 - \langle u\rangle^2)\right]$$

$$\geq E\left[\left[(\xi - \langle\xi\rangle)(u - \langle u\rangle)\right]^2\right]$$

$$\geq \mathrm{cov}^2(\xi u) \tag{4.5}$$

It is easily seen from the definitions of u and $\rho = |\psi|^2$ that

$$E(u) = 0, \quad E(\xi u) = \nu \int dx\ , \quad x \frac{\partial \rho}{\partial x} = - \nu$$

and

$$\mathrm{cov}^2\ (\xi u) = \nu^2 \tag{4.6}$$

Our attempt is now to calculate the variances of ξ and P :

$$\langle P\rangle = \frac{\hbar}{i} \int dx \left(\psi^* \frac{\partial \psi}{\partial x}\right) = \frac{m}{i} \int dx\ \rho\ (u + iv) = \langle mv\rangle \tag{4.7}$$

$$\langle P^2\rangle = -\hbar^2 \int dx \left(\psi^* \frac{d^2\psi}{\partial x^2}\right) = m^2 \int dx\ \rho(u^2 + v^2) = E(m^2u^2) + E(m^2v^2)$$

$$= \mathrm{Var(mu)} + E(m^2v^2) = m^2\nu^2 + E(m^2v^2) \tag{4.8}$$

$$\mathrm{Var}\ P = \mathrm{Var(mu)} + E(m^2v^2) - [E(mv)]^2$$

$$E(m^2v^2) - [E(mv)]^2 = \mathrm{Var}\ mv \tag{4.9}$$

Hence Var ξ Var P = Var ξ (Var mu + Var mv). In view of equations (4.3) and (4.9)

$$\begin{aligned}(\mathrm{Var}\xi)(\mathrm{Var}\ P) &= \mathrm{Var}\ \xi\ \mathrm{Var\ mu} + \mathrm{Var}\ \xi\ \mathrm{Var(mv)} \\ &\geq \mathrm{cov}^2(\xi,\ mu) + \mathrm{cov}^2(\xi, mv) \\ &\geq m^2\upsilon^2 + \mathrm{cov}^2\ (\xi, mv)\end{aligned} \tag{4.10}$$

Hence the uncertainty relations

$$(\mathrm{Var}\xi)(\mathrm{Var}\ P) \geq \frac{\hbar^2}{4} + \mathrm{cov}^2(\xi, mv) \tag{4.11}$$

We thus have a tighter lower bound in the place of the usual Heisenberg relations $\Delta X\ \Delta P \geq \frac{\hbar^2}{4}$.

As a simple example we can take the solution of the simple harmonic oscillator coherent state problem and write down the solution for ψ as the overcomplete set of displaced oscillator solutions of corresponding to the classical Hamiltonian with classical solutions $P_{cl}(t)$ and q_{cl} (t).The solution for ground state is

$$\psi_{coh}(x,\ t) = \frac{1}{(2\pi\sigma)^{1/4}} \exp\left\{- \frac{1}{4\sigma}\left[x - q_{cl}\right] + \frac{i}{\hbar}\, xP_{cl} - \frac{i}{2\hbar}\, P_{cl}q_{cl} - \frac{i}{2}\,\hbar\,\omega t\right\} \tag{4.12}$$

with $\sigma = \hbar/2m\omega$.

Hence we easily see that

$$\rho\ (x,t) = \frac{1}{\sqrt{2\pi}} \exp\left\{ - \frac{1}{2\sigma}\,(x-q_{cl})^2\right\}$$

$$S(x,t) = x\ P_{cl} - \frac{1}{2}\, P_{cl}q_{cl} - \frac{1}{2}\,\hbar\omega t \tag{4.13}$$

and $(\psi, q_s\psi) = x_s$, "q_s" being the operator for position and

$$(\psi\ , q_s - \langle\psi\ , q_s\psi\rangle^2\ \psi) = \sigma \tag{4.14}$$

We also deduce

$$v(x,t) = \frac{P_{cl}}{m}$$

$$u = -\omega\,(x - q_{cl}) = \frac{\hbar}{m}\,\mathrm{Re}\,\frac{d}{dx}\log\psi$$

$$v_{\pm} = \frac{P_{cl}}{m} \mp \omega\,(x - q_{cl}) \tag{4.15}$$

Therefore the drift and diffusion equation for the coherent state is

$$dx = \left[\frac{P_{cl}}{m} \mp \omega\,(x - q_{cl})\right]dt + dW \tag{4.16}$$

We have already seen in (4.9) and (4.10) that

$$\mathrm{Var}\,\xi\ \mathrm{Var}\,P = \mathrm{Var}\,\xi\ (\mathrm{Var}\ mu + \mathrm{Var}\ mv) \tag{4.17}$$

will yield the required inequality. We also know that

$$\mathrm{Var}\,\xi = \sigma \text{ and } \mathrm{Var}\ mv = \mathrm{Var}\ P_{cl} = 0 \tag{4.18}$$

Since $u = -\omega(x-q_{cl})$, we find

$$\Delta x\ \Delta P = \mathrm{Var}\,\xi\ \mathrm{var}\ mu = \sigma \times \omega^2\ m^2\sigma = \omega^2\ m^2\ \sigma^2 \tag{4.19}$$

Hence

$$\Delta x\ \Delta P = \sigma^2\ m^2\ \omega^2 = \frac{\hbar^2}{4} \tag{4.20}$$

Thus the uncertainty yields the equality as expected. In a similar fashion for the squeezed coherent states $D(\alpha)\ S(z)|0\rangle$, where the $S(z)$ is the squeezing operator and $D(\alpha)$ is the displacement operator. We note that z being a complex number can be written as $z=re^{i\Theta}$. With $\Theta=0$ the wave function can be represented as $|\alpha,r\rangle$. In phase space this can be represented by $\psi_{r\alpha}$

$$\psi_{r\alpha} = \frac{1}{\left[2\pi(\Delta x)'^2\right]^{1/4}}\exp\left\{-\frac{(x-\langle x\rangle)^2}{4(\Delta x)'^2} + \frac{i}{\hbar}\langle P\rangle\left(x - \frac{x_{cl}}{2}\right) - \frac{iEt}{\hbar}\right\} \tag{4.21}$$

with $(\Delta x)' = \sqrt{\hbar/2m\omega}\cdot\imath^{-1}$, $\imath = e^{-r}$.

Hence

$$S(x,t) = \langle P\rangle x - 1/2\ \langle P\rangle\ \langle x\rangle - Et \tag{4.22}$$

where E is energy corresponding to $\psi_{r\alpha}$.

$$v = \frac{P_{cl}}{m} = \frac{\langle P\rangle}{m}$$

$$u = \left(\frac{\hbar}{2m}\right)\frac{\nabla\rho}{\rho} = -\,\omega\,\imath^2\left[x - \langle x\rangle_{cl}\right] \tag{4.23}$$

and hence the corresponding Brownian motion can be generated by the SDE

$$d\xi = \left[\frac{P_{cl}}{m} \mp \omega\imath^2(x - x_{cl})\right]dt + dw \tag{4.24}$$

where $\imath^2 = e^{-2r}$

5. CONCLUSIONS

Thus we have described in a pedestrian fashion the three approaches leading to the resolution of the conceptual difficulties and demystify some aspects of quantum behaviour. The causal interpretation gave rise to the Hamilton-Jacobi and continuity equations with two coupled velocity fields. However the potential appearing in these equations are non-classical and quantum effects are attributed to them. But the density of the particle was taken to be proportional to $\rho = |\psi|^2$, the density of the background fluid in equilibrium situations. If it did not hold initially then it would be brought about in time by fluctuations. Also a prior knowledge of the Schrodinger equation was necessary [16].

The second formulation by Nelson described the equivalence of classical Markov process and non-relativistic schrodinger equation while the kinematics of the process was described by Brownian motion concept, the dynamics of the process is governed by Newton's law of motion. Also the simultaneous existence of the reverse process had to be invoked and the need for the existence of both forward and backward Fokker-Plank equations was found necessary. The fundamental limit to physical measurements is caused by the underlying noise. Quantum fluctuations like classical fluctuations have basically the same origin - the chaotically complicated nature of the motion at the microscopic level. This means that the fluctuations are of the same kind as thermal fluctuations [16].

The velocity field in this stochastic analysis is an average quantity and concepts relating to *complex measured real diffusion* processes were thought of and the result was similar to Feynman path formulation [17] starting from classical accelerated diffusion process. We will describe this method in a later contribution wherein unification of these three approaches are studied systematically.

The analysis of Dirac's relativistic equation and spinning particles have been analysed in this light and we will present later some of the intrinsic quantum mechanical effects in different contexts.

REFERENCES

1. Einstein, A. Podolosky, B. and Rosen, N., (1935), Can quantum mechanical description of reality be considered complete Phys. Rev. 47, p 777.
2. Bohr, N, (1935) can Quantum Mechanical description of reality be considered complete? Phys Rev. 48 p. 695.
3. Bell, J.S. (1964) On the E.P.R Pardox, Physics I, p. 195.
4. Clauser, J.F. (1976) Experimental Investigation of a Polarisation Correlation Anomaly, Phys Rev Letters 36 p. 1223.
5. Aspect, A. (1975) Proposed experiment to test separable hidden variable theories, Phys. Letters 54A p. 117.

   Aspect, A. and Grangier (1986) in Quantum concepts in space and time, (Ed) R.Penrose, clarendon Press.

   David Mermin, N., Microphysical reality and Quantum formalism, Vol. II, Kluwer Acad Pub, p. 339.

   Sudarshan, E.C.G. (1988) Negative probabilities underlying the violation of Bell's inequalities, Preprint.
6. Clauser, J., Horne, M.A., Shimnoy, A., Holt, R.A. (1969) Proposed experiment to test local hidden variable theories, Phys. Rev. Lett. 23 p. 880.
7. Dewitt, B.S., and Graham, N. (1955) Many World Interpretation of Quantum Mechanics, Princeton Univ. Press.
8. Wigner, E.P. (1967) Symmetries and Reflections, Indiana Press, p. 171

   Wigner, E.P. (1970) On hidden variables and Quantum mechanical probabilities, Am. J. of Physics, 38B p. 1005.
9. Bohm, D. and Hiley, B.J (1975) Foundations of Physics, 5, p. 93.

   Feynman, R.P. (1982) Simulating Physics with Computers, International Jour. Th. Phys. 21 p. 487.
10. Bohm, D. and Hiley, B.J. (1987) An Ontological Basis or Quantum Theory, Phys. Reports 144 p. 321.
11. Nelson, E. (1967) Dynamical theories of Brownian motion, Princeton Univ. Press.
12. Albeverio, S., Blanchard, ph. and Hoegh, R., Krohn (1983) in Lecture Notes in Mathematics, 1109 Proceedings Marseille, Springer Verlag.

13. Ito, K. and Mckean, H.P. Jr. (1965) Diffusion processes and their sample paths, Vol. 125. Acad. press Inc., N.Y.

14. Lde Lapena and Auerbauh and Macelto, A. (1972) Stronger form for the position-momentum uncertainty relations, Phys. Letters A 39 p. 65.

15. Guerra, F. (1981) Structural aspects of the stochastic mechanics, Phys. Reports 77 p. 263.

16. Bohm, D. and Vigier, I.P. (1954) Model of the causal interpretation of Quantum theory in terms of a fluid with irregular fluctuations, Phys. Rev. 96 p. 208.

Vigier, J.P. (1989) Lectures in Proc. 3rd Symposium on Foundations of Quantum Mechanics, Tokyo.

17. Schulman, S. (1981) Techniques and applications of path integration, Wiley, N.Y.

18.. Madelung, E. (1926) Z.Phys. 40 p. 322.

A NEW APPROACH TO THE SOLUTION OF NEUROLOGICAL MODELS: APPLICATION TO THE HODGKIN-HUXLEY AND THE FITZHUGH-NAGUMO EQUATIONS

George Adomian
Center For Applied Mathematics
University of Georgia

Matthew Witten
Department of Applications Research and Development
University of Texas System
Center for High Performance Computing
Balcones Research Center
Austin, TX 78758-4497

Gerald E. Adomian
Alfred E.Mann Foundation for Scientific Research
Sylmar, CA 91342
Harbor-UCLA Medical Center and UCLA School of Medicine

ABSTRACT

The solution of the Hodgkin-Huxley and the Fitzhugh-Nagumo equations are demonstrated as applications of the decomposition method [1-3] which can be used as a new and useful approach obtaining analytical and physically realistic solutions to neurological models and other biological problems without perturbation, linearization, discretization, or massive computation.

1.0 INTRODUCTION

Mathematical modeling and analysis of a number of biological problems, involving interactions of physiological systems, can benefit significantly from new advances in mathematical methodology. These advances allow for the solution of dynamical systems involving coupled systems, nonlinear ordinary or partial differential equations, and delay equations. All dynamical problems are basically nonlinear. Neural networks, for example, in their processing and transmission of information display nonlinearity as well as couplings and fluctuations. Commonly used mathematical approaches that involve some form of linearization or approximation are too limited in scope and involve gross and unsatisfactory approximations changing essentially the nonlinear nature of the actual system being analyzed. As a consequence, the mathematical system actually solved is quite different from the desired original nonlinear model.

Randomness is, of course, another factor present in real systems due to a variety of causes. This can be randomness or fluctuations either in parameters of an individual system or involving variations from one individual to another. Randomness, or stochasticity, in physical (or biological) systems, like nonlinearity, is generally dealt with by perturbative methods, averagings which are not generally valid, or other specialized and restrictive assumptions that do not realistically represent fluctuations, especially if fluctuations are not relatively negligible. We emphasize that the methods used here do not require assumption of small nonlinearity or small randomness, linearization or closure approximations. A comprehensive discussions of the classical numerical algorithms/approaches to neuronal modeling can be found in Mascagni [4].

This method - the decomposition method [1-3] - is an approximation method, but all modeling is approximation and this methodology approximates (accurately and in an easily computable manner) the solution of the real nonlinear and possibly stochastic problem rather than a grossly simplified linearized or averaged problem.

Mathematical modeling must represent behavior realistically. An analytic solution of a model that deviates significantly from the actual physical problem being modeled can convey a false sense of understanding, unjustified by experimental or physical results. These simplifications made, of course, for tractability of analysis of equations and the use of well-understood mathematical methods, can often neglect (quite seriously) the essentially nonlinear and stochastic nature of physical and biological phenomena. Linearity is a special case and linearization of nonlinear phenomena can change the problem to a different problem. It may be adequate if the nonlinearity is "weak" so perturbative methods become adequate. If we can deal with "strong" nonlinearities - as we can - then the "weakly non-linear" or the "linear" cases derive from the same theory as well. Random fluctuations are always present in real phenomena and perturbative or hierarchy methods and their various adaptations will be adequate only when randomness is relatively insignificant (Witten [38]; Adomian [2], for example). We wish, therefore, to deal with 'strongly' stochastic cases and to derive the special cases as well without the averaging procedures, closure approximations and truncations, or assumptions of special nature for the processes such as Markov or Gaussian white noise, etc.

It is to be noted that once we realise that we can be less constrained by the mathematics, we are then able to develop more realistic and sophisticated models, since modeling physical phenomena involves retention of essential features while striving for simplicity so resulting equations can be solved. Modeling is always a compromise between realistic representation and mathematical tractability. With fewer limitations imposed to achieve tractability, we can make our models more realistic. We are now able to include delayed effects rather than assuming changes take place instantaneously and can make these delays constant, time-dependent, or random. We are able to deal with coupled nonlinear equations, random initial or boundary conditions, and systems of nonlinear differential equations. The results are easily obtained and accurate. Let us now briefly discuss the decomposition method before we consider the Hodgkin-Huxley equation.

2.0 Mathematical Background

Having discussed the issues of approximation, we now discuss the new methodology. In this new approach we will assume u(t) is the solution of the equation and that u(t) may be written

$$u(t) = \sum_{n=0}^{\infty} u_n \tag{1a}$$

We will begin with the (deterministic) form Fu = g(t) where F is a nonlinear ordinary differential operator with linear and nonlinear terms. Assume that the linear term is written Lu + Ru where L is invertible. In addition, in order to avoid difficult integrations we will choose L as only the highest ordered derivative in the equation. R is then the remainder of the linear operator (the linear portion of the equation). The nonlinear term is represented by Nu. Thus we have that $Fu \equiv Lu + Ru + Nu = g$ and we write

$$Lu = g - Ru - Nu$$

Operating on both sides of the above equation by L^{-1} (which we have assumed exists), we have

$$L^{-1}\,Lu = L^{-1}\,g - L^{-1}Ru - L^{-1}Nu \tag{1b}$$

For initial-value problems, we conveniently define L^{-1} for $L = d^n/dt^n$ as the n-fold definite integration operator from 0 to t. As an example for the operator $L = d^2/dt^2$, we have

$$L^{-1}\,Lu = \int_o^t \int_o^t \frac{d^2u(\alpha)}{d\alpha^2}\,d\alpha\,d\alpha = \int_o^t [u'(t) - u'(0)]d\alpha$$

$$= \int_o^t u'(t)d\alpha - \int_o^t u'(0)d\alpha = u(t) - u(0) - tu'(0)$$

Combining this result with equation (1b), we obtain

$$u = u(0) + tu'(0) + L^{-1}g - L^{-1}Ru - L^{-1}Nu \qquad (1c)$$

Should the problem be a boundary value problem, we let L^{-1} be an indefinite integral. The first three terms of (1c) are identified as u_o in the assumed decomposition equation (1a),

$$u = \sum_{n=o}^{\infty} u_n = u_o - L^{-1} R \sum_{n=o}^{\infty} u_n - L^{-1} \sum_{n=o}^{\infty} A_n \qquad (1d)$$

We will assume that Nu, the nonlinear term, can be written in the form

$$Nu = \sum_{n=o}^{\infty} A_n \qquad (2)$$

where $A_n = f(u_o, \ldots, u_n)$. That is, the A_n's are specially generated polynomials dependent upon $u_o, \ldots, u_n$ (see [1-3, 34, 35], for details). It follows that

$$\begin{aligned} u_1 &= - L^{-1}Ru_o - L^{-1}A_o \\ u_2 &= - L^{-1}Ru_1 - L^{-1}A_1 \\ &\ldots \\ u_n &= - L^{-1}Ru_{n-1} - L^{-1}A_{n-1} \end{aligned} \qquad (3)$$

All components are determinable as A_o depends only upon u_o, A_1 depends only on u_o, u_1, etc. The practical solution will be the n-term approximation $\phi_n(t) = \sum_{i=o}^{n-1} u_i(t)$ and the $\lim_{n\,\infty} \phi_n(t) = \sum_{i=o}^{\infty} u_i(t) = u(t)$ by definition. ϕ_n must, of course, satisfy the initial/boundary conditions of the problem.

In the linear case, where Nu vanishes, we have that

$u = u_o - L^{-1}Ru_o - L^{-1}Ru_1 - \ldots$ or $u_o + (L^{-1}R)(L^{-1}R)u_o - \ldots$ thus,

$$u(t) = \sum_{n=0}^{\infty} (-1)^n \, (L^{-1}R)^n \, u_o \tag{4}$$

If and only if the given conditions are zero, we have the following additional simplification:

$u_o = L^{-1}g$, and $u = \sum_{n=0}^{\infty} (-1)^n (L^{-1}R)^n L^{-1}g$. Thus, our equation $Fu = g$ becomes $u = F^{-1}g$ where the inverse is $F^{-1} = \sum_{n=0}^{\infty} (-1)^n (L^{-1}R)^n L^{-1}$. Let us now consider a slightly more complex mathematical model.

For boundary value problems L^{-1} is an indefinite integral operator. We then have that

$$u = A + Bt + L^{-1}g - L^{-1}Ru - L^{-1}Nu \tag{5}$$

where A and B are to be determined. Applying the decomposition $u=\sum_{m=o}^{\infty} u_m$ where $u_o = A + Bt + L^{-1}g$, we consider successive approximations

$$\varphi_n = \sum_{m=o}^{n-1} u_m \tag{6}$$

Each approximant for $n = 1,2,3, \ldots$ must satisfy the given conditions, beginning with $\varphi_1=u_0$. Suppose that the conditions are $u(b_1)=\beta_1$ and $u(b_2)=\beta_2$. Then $\varphi_1(b_1) = \beta_1$ and $\varphi_1(b_2) = \beta_2$ so that

$$\begin{aligned} A + Bb_1 + L^{-1}g &= \beta_1 \\ A + Bb_2 + L^{-1}g &= \beta_2 \end{aligned} \tag{7}$$

or

$$\begin{pmatrix} 1 & b_1 \\ 1 & b_2 \end{pmatrix} \cdot \begin{pmatrix} A \\ B \end{pmatrix} = \begin{pmatrix} \beta_1 - L^{-1}g \\ \beta_2 - L^{-1}g \end{pmatrix} \tag{8}$$

determining A and B and consequently φ_1. To obtain $\varphi_2 = u_o+u_1 = \varphi_1 + u_1$, we need u_1 defined by

$$u_1 = - L^{-1}Ru_o - L^{-1}A_o \tag{9}$$

Forming φ_2 we satisfy the boundary conditions letting $\varphi_2(b_1)=\beta_1$ and $\varphi_2(b_2)= \beta_2$ thus re-evaluating A and B to yield an evaluated φ_2 approximation. We then calculate $u_2= -L^{-1}Ru_1-L^{-1}A_1$, form φ_3, satisfy the conditions thus evaluating A and B and consequently φ_3. We can continue this process to an arbitrary and satisfactory φ_n.

A further improved, but somewhat more complex procedure has been recently discussed by Adomian and Rach [37]. This procedure uses a double decomposition with the advantage of being the fastest converging and a global procedure applicable to linear or nonlinear, ordinary or partial differential equations.

For a linear ordinary differential equation, we need not use the general procedures and simply carry along the constants of integration and evaluate only once for the general φ_n.

Suppose, now, that Fu = g is a multivariate equation such as $\nabla^2 u + u_t = g$, with g = g(x,y,z,t). Express the operator $\nabla^2 u$ as

$$[L_x + L_y + L_z + L_t]u = g \tag{10}$$

where L_x, L_y, L_z are $\partial^2/\partial x^2$, $\partial^2/\partial y^2$, $\partial^2/\partial z^2$, respectively, and $L_t = \partial/\partial t$. We can express equation (10) for each linear operator term; operating with the appropriate inverse, and solving for u. Thus we have that

$$L_t u = g - L_x u - L_y u - L_z u$$

$$L_x u = g - L_y u - L_z u - L_t u$$

$$L_y u = g - L_z u - L_t u - L_x u$$

$$L_z u = g - L_t u - L_x u - L_y u \tag{11}$$

Applying L_t^{-1} to the first, L_x^{-1} to the second, etc., we obtain the following series of expressions for u(t)

$$u = \Phi_t + L_t^{-1} g - L_t^{-1}(L_x + L_y + L_z)u$$

$$u = \Phi_x + L_x^{-1} g - L_x^{-1}(L_y + L_z + L_t)u$$

$$u = \Phi_y + L_y^{-1} g - L_y^{-1}(L_z + L_t + L_x)u$$

$$u = \Phi_z + L_z^{-1} g - L_z^{-1}(L_t + L_x + L_y)u \tag{12}$$

where Φ_t, Φ_x, Φ_y, Φ_z denote the homogeneous solutions.

Each of these equations is solvable by decomposition [1-3] letting $u = \sum_{n=0}^{\infty} u_n$ and identifying the $\Phi + L^{-1}g$ in each equation as u_o. We now have

$$u = u_o - L_t^{-1}(L_x + L_y + L_z) \sum_{n=0}^{\infty} u_n \tag{13a}$$

$$u = u_o - L_x^{-1}(L_y + L_z + L_t) \sum_{n=0}^{\infty} u_n \tag{13b}$$

$$u = u_o - L_y^{-1}(L_z + L_t + L_x) \sum_{n=0}^{\infty} u_n \tag{13c}$$

$$u = u_o - L_z^{-1}(L_t + L_x + L_y) \sum_{n=0}^{\infty} u_n \tag{13d}$$

In (13a) we have that

$$u_o = \Phi_t + L_t^{-1}g \tag{14a}$$

and

$$u_{n+1} = - L_t^{-1} (L_x + L_y + L_z) u_n \tag{14b}$$

In (13b) we have that

$$u_o = \Phi_x + L_x^{-1}g \tag{15a}$$

and

$$u_{n+1} = - L_x^{-1} (L_y + L_z + L_t)u_n \tag{15b}$$

In (13c) we have that

$$u_o = \Phi_y + L_y^{-1}g \tag{16a}$$

and

$$u_{n+1} = - L_y^{-1} (L_z + L_t + L_x)u_n \tag{16b}$$

Finally, in (13d) we have that

$$u_o = \Phi_z + L_z^{-1}g \tag{17a}$$

and

$$u_{n+1} = - L_z^{-1} (L_t + L_x + L_y)u_n \tag{17b}$$

From equations (14-17), it follows that all components of u can be determined. We have called the solutions of equations (14) to (17) partial solutions. G. Adomian and R. Rach [31] show that the partial solutions are all actual solutions when Φ_t depends on x, y, z, Φ_x depends on y, z, t, etc., i.e., when the conditions are general. This means only one of equations (14) - (17) need be solved. Or, of course, we add them and divide by four as discussed in earlier work on decomposition (Adomian [1-3]). If an equation doesn't contribute because u_o is zero, it is not considered; the solution is determined from the remaining equations. In this case the partial solutions are asymptotically equal.

Now suppose our equations are coupled differential or partial differential equations. Suppose we have, for example, a system of equations in u and v. It is only necessary to define the pair u_o, v_o, then find u_1, v_1 in terms of u_o, v_o etc. For n equations we have an n-vector of terms for the first component. Then an n-vector of second components is found in terms of the first. The procedure is discussed completely in [3] and we will only illustrate the procedure here.

The inclusion of stochastic processes is dealt with in [1-3] and elsewhere. The approximation $\phi_n = \sum_{i=o}^{n-1} u_i$ becomes a stochastic series, and no statistical independence problems [1] are encountered in obtaining statistics from the ϕ_n.

Stochastic ordinary and partial differential equations, both linear and nonlinear, arise in a number of neurobiological applications. We emphasize that deterministic equations are merely a special case of stochastic equations, just as linear equations are a special case of nonlinear equations.

3.0 Applications to the Hodgkin-Huxley Equation

Consider the stochastic Hodgkin Huxley equation (18).

$$C_m \partial v/\partial t = (a/2\rho_1)\partial^2 V/\partial x^2 + \bar{g}_k n^4(V_k - V) + \bar{g}_{Na} m^3 h\,(V_{Na} - V)$$

$$+ g_1(V_1 - V) + \imath(x,t) \tag{18a}$$

and

$$\partial n/\partial t = \alpha_n(V)\,(1-n) - \beta_n(V)_n$$

$$\partial m/\partial t = \alpha_m(V)\ (1-m) - \beta_m(V)_m$$

$$\partial h/\partial t = \alpha_h(V)\ (1-h) - \beta_h(V)_h \tag{18b}$$

Where C_m is membrane capacitance per unit area, a is nerve fibre radius, ρ_i is intracellular resistivity, $\bar{g}_k$ is constant maximal available potassium conductance per unit area, V_k is potassium equilibrium potential relative to resting potential, $\bar{g}_{Na}$ and V_{Na} are the corresponding quantities for sodium. g_l is the leakage conductance per unit area, and V_l is the equilibrium potential for the leakage current. Finally, the coefficients α and β in the last three equations are given by:

$$\alpha_n = \frac{10 - V}{100(e^{10-V\sqrt{10}}-1)} \qquad \beta_n = (1/8)e^{-V/80}$$

$$\alpha_m = \frac{25 - V}{10(e^{(25-V)/10}-1)} \qquad \beta_m = 4e^{-V/18}$$

$$\alpha_h = (7/100)e^{-V/20} \qquad \beta_h = \frac{1}{e^{(30-V)/10}+1} \tag{19}$$

If $L_t=\partial/\partial t$ and $L_x=\partial^2/\partial x^2$, we can write

$$C_m L_t V = (a/2\rho_i)\ L_x V + \bar{g}_k n^4(V_k - V)$$

$$+ \bar{g}_{Na} m^3 h(V_{Na} - V) + g_l(V_l - V) + (x,t) \tag{20}$$

$$L_t n = \alpha_n(1-n) - \beta_n n \tag{21}$$

$$L_t m = \alpha_m(1-m) - \beta_m m \tag{22}$$

$$L_t h = \alpha_h(1-h) - \beta_h h \tag{23}$$

Equations (20)-(23) are extremely difficult because the α's and β's are functions of V; the dependent variables are V, n, m, h. consider (20)

$$L_t V = (a/2\rho_i C_m)\ L_x V - (\bar{g}_k/C_m)n^4 V + (\bar{g}_{Na} V_{Na}/C_m)M^3 h$$

$$- (\bar{g}_{Na}/C_m)n^3 hV - (\bar{g}_l/C_m)V + (\bar{g}_k V_k/C_m)n^4 - (1/C_m)I - (g_l V_l/C_m).$$

By the decomposition method,

$$V = V(0) - L_t^{-1}(1/C_m)\imath - L_t^{-1}(g_l V_l/C_m) - L_t^{-1}(a/2\rho_i C_m)L_x V - L_t^{-1}(\bar{g}_k/C_m)n^4V$$

$$- L_t^{-1}(\bar{g}_{Na} V_{Na}/C_m)m^3h + L_t^{-1}(\bar{g}_{Na}/C_m)n^3hV + L_t^{-1}(g/C_m)V - L_t^{-1}(\bar{g}_k V_k/C_m)n^4 \qquad (24)$$

Let $V_o=V(0) - L_t^{-1}(1/C_m)\ \imath - L_t^{-1}(g_l V_l/C_m)$and $V+\sum_{n=o}^{\infty} V_n$, where V_o is now defined. Thus, $\sum_{n=o}^{\infty} V_m$ is substituted for V in all linear terms. Represent the nonlinear terms by the A_n polynomials. Thus, if f(u) is a nonlinear term, $f(u) = \sum_{n=o}^{\infty} A_n$, and since there are several nonlinear terms (for which the A_n are appropriately different), we write $\sum_{n=o}^{\infty} A_n\{f(u)\}$. The nonlinear terms are n^4V, m^3h, n^3hV, and n^4. For these we write $\sum_{n=o}^{\infty} A_n\{n^4V\}$, $\sum_{n=o}^{\infty} A_n\{m^3h\}$, $\sum_{n=o}^{\infty} A_n\{n^3hV\}$, and $\sum_{n=o}^{\infty} A_n\{n^4\}$. These can, of course, be evaluated [1-3, 5]. Now (24) can be written:

$$V = V_o - L_t^{-1}(a/2\rho_i C_m)L_x \sum_{n=o}^{\infty} V_n - L_t^{-1}(\bar{g}_k/C_m) \sum_{n=o}^{\infty} A_n\{n^4V\}$$

$$- L_t^{-1}(\bar{g}_{Na} V_{Na}/C_m) \sum_{n=o}^{\infty} A_n(m^3h) + L_t^{-1}(\bar{g}_{Na}/C_m) \sum_{n=o}^{\infty} A_n\{n^3hV\}$$

$$+ L_t^{-1}(g/C_m) \sum_{n=o}^{\infty} V_n - L_1^{-1}(\bar{g}V_k/C_m) \sum_{n=o}^{\infty} A_n\{n^4\} \qquad (25)$$

Thus

$$V_1 = - L_t^{-1}(a/2\rho_i C_m)L_x V_o - L_t^{-1}(\bar{g}_k/C_m)A_o\{n^4V\}$$

$$- L_t^{-1}(\bar{g}_{Na} V_{Na}/C_m)A_o\{m^3h\} + L_t^{-1}(\bar{g}_{Na}/C_m)A_o\{n^3hV\}$$

$$+ L_t^{-1}(g/C_m)V_o - L_t^{-1}(\bar{g}V_k/C_m)A_o\{n^4\} \qquad (26)$$

etc., for other components as discussed in [1-3]. Normally a set of four equations such as this will be solved by similarly decomposing n, m, h, determining the set n_o, m_o, h_o, as well as V_o. Then n_1, m_1, h_1, V_1 will be found in terms of the first set ([2]). Because of the complexity of the α's and β's we will proceed somewhat differently. We evaluate the α's and β's by using only V_o. The first terms of V, n, m, h are evaluated. Then V_1 is evaluated from (26) and we evaluate the α's and β's again with the improved approximation $V = V_o + V_1$ and our calculation of components of n, m, h, then of V_1 and V_2. Evidently we can continue this process using $V \simeq V_o + V_1 + V_2$, etc. This procedure yields improving solutions of V, n, m, h which can be substituted into (14-17) for verification. If ι is stochastic, then V_o is stochastic. Normally, stochastic coefficients or inputs merely mean our solutions are stochastic series, i.e., they involve stochastic processes, and from our n-term approximations we can calculate statistics. Here, the complicated dependences of the α, β coefficients on V is a severe complication. To evaluate the α, β coefficients, we must use $\langle V_o \rangle$.

4.0 The Fitzhugh-Nagumo Equations.

The general Fitzhugh-Nagumo equations modeling conduction of nerve impulses are given by

$$u_t = u_{xx} + u(u-a)(1-u) - z$$

$$z = \sigma u - \gamma z \tag{27}$$

where $a \in (0,1)$ and $\sigma, \gamma \geq 0$. If we write $L_t = \partial/\partial t$ and $L_{xx} = \partial^2/\partial x^2$ we rewrite these as

$$L_t u - L_{xx} u = - au + (1+a)u^2 - u^3 + z \tag{28}$$

$$L_t z = u$$

which we can combine into

$$L_t u - L_{xx} u = -au + (1+a)u^2 - u^3 + L_t^{-1} u \tag{29}$$

Solve for $L_t u$ and also for $L_{xx} u$

$$L_t u = - au + (1+a)u^2 - u^3 + L_t^{-1} u + L_{xx} u$$

$$L_{xx} u = au - (1+a)u^2 - u^3 - L_t^{-1} u + L_t u \tag{30}$$

Applying the appropriate inverses

$$u = u(x,0) - L_t^{-1}au + L_t^{-1}(1+a)u^2 - L_t^{-1}u^3 + L_t^{-1}L_t^{-1}u + L_t^{-1}L_{xx}u \tag{31}$$

$$u = A + Bx + L_{xx}^{-1}au - L_{xx}^{-1}(1+a)u^2 + L_{xx}^{-1}u^3 - L_{xx}^{-1}L_t^{-1}u + L_{xx}^{-1}L_t u \tag{32}$$

Define $u = \sum_{n=0}^{\infty} u_n$ and represent the nonlinear term $u^2 = \sum_{n=0}^{\infty} A_n\{u^2\}$ and $u^3 = \sum_{n=0}^{\infty} A_n\{u^3\}$ where the $A_n\{u^2\}$ are the Adomian polynomials for u^2 and $A_n\{u^3\}$ represents the A_n generated for u^3 as shown in (31). L_{xx}^{-1} is a two-fold (indefinite) integration; however, it is convenient to use L_t^{-1} as a definite integral from 0 to t [1-3]. Each u_n is computed in terms of the preceding component. The A_n $(u_0, \ldots, u_n)$ are calculated [1,3] by $A_0 = f(u_0)$ and for $n \geq 0$ $A_n = \sum_{n=0}^{\infty} c(\nu,n)f^{(\nu)}(u_0)$ where $f^{(\nu)}(u_0)$ is the νth derivative of f(u) at u_0 and $c(\nu,n)$ is the sum of possible combinations of ν components of u whose subscripts add to n, with the stipulation that we divide by the factorial of the number of repetitions of subscripts. Thus $c(1,3)=u_3$ and $c(3,3)=(1/3!)\,u_1^3$ while $c(2.6) = u_1u_5 + u_2u_4 + (1/2!)u_3^2$.

In (31) we identify $u_0 = u(x,0)$. Then for $n \geq 0$,

$$u_{n+1} = -L_t^{-1}au_n + L_t^{-1}(1+a)\,A_n\{u^2\} - L_t^{-1}A_n\{u^3\} + L_t^{-1}L_t^{-1}u_n + L_t^{-1}L_{xx}u_n.$$

In (32) identify $u_0 = A + Bx$. Then for $n \geq 0$,

$$u_{n+1} = L_{xx}^{-1}au_n - L_{xx}^{-1}(1+a)\,A_n\{u^2\} + L_{xx}^{-1}\,A_n\{u^3\} - L_{xx}^{-1}\,L_t^{-1}u_n + L_{xx}^{-1}\,L_t u_n$$

with which all components of u can be determined. In earlier work, u was computed using both equations, adding, and dividing by two; however, Adomian and Rach [32] have shown that each equation provides the complete solutions. Depending on given conditions, u_0 may be zero in one equation in which case, that equation makes no contribution. This can occur when we have u = 0 at the end of an interval of interest. If we have only initial conditions, it is still valid to add both equations and divide by two. For general boundary conditions in (32) we compute n terms for some desired n forming an approximation $\phi_n = \sum_{i=0}^{n-1} u_i$ which approaches $u = \sum_{n=0}^{\infty} u_n$ as $n \rightarrow \infty$ leaving "constants" of integration in u_0. Then ϕ_n is treated as u to satisfy the boundary conditions. This procedure is not necessary if only initial conditions are given.

An interesting case is that of u(x,0)=0 and u(0,t)=P(t) which corresponds to a resting nerve with a time-dependent forcing. Thus, $u_o=0$ in equation (31) and we must use equation (32) for a solution. The condition $u(0,t) = P(t)$ means that $A = P(t)$. A further condition on u_t, for example, would evaluate B and a complete solution follows. Let's suppose $u_t=0$ then $B = 0$ and consequently $u_o = P(t)$. We represent the nonlinearities u^2 by $\sum_{n=o}^{\infty} A_n\{u^2\}$, i.e., the A_n generated for u^2, and represent u^3 by $\sum_{n=o}^{\infty} A_n\{u^3\}$. Then all of the components can be determined through

$$u_n = L_{xx}^{-1}au_{n-1} - L_{xx}^{-1}(1+a)A_{n-1}\{u^2\} + L_{xx}^{-1}A_{n-1}\{u^3\} - L_{xx}^{-1} L_t^{-1}u_{n-1}$$

$$+ L_{xx}^{-1}L_t^{-1}(n \geq 1) \tag{33}$$

5.0 Closing Remarks

We have seen that the method is applicable also to the Hodgkin-Huxley system including cases in more than one space dimension as well as time, as in some Hodgkin-Huxley equations for heart tissue. Related work is found in references [7-10].

Since this method makes linearization, perturbation, or discretization and massive computation unnecessary, providing a physically realistic solution (providing, of course, that the model is physically realistic), it offers a possibly valuable potential for neural models as well as other physiological problems such as compartment models or blood flow.

REFERENCES

[1] G.ADOMIAN, Nonlinear Stochastic Operator Equations, Academic Press, New York, 1986.

[2] G.ADOMIAN, Stochastic Systems, Academic Press, NY, 1983.

[3] G.ADOMIAN, Applications of Nonlinear Stochastic Systems Theory to physics, Kluwer, Amsterdam, 1988.

[4] M.V.MASCAGNI, Numerical methods for neuronal modeling, (in) Methods in Neuronal Modeling: From Synapses To Networks, (eds.) C.Koch and I. Segev MIT Press, Cambridge, MA 1989.

[5] H.TUCKWELL, Stochastic processes in the neurosciences, Lecture Notes for NSF-CBMS Regional Conference, North Carolina State University, Raleigh NC 23-27 June 1986.

[6] R.RACH, A Convenient computational form for the Adomian polynomials, J.Math Anal. and Applic., 102 (1984) 415-419.

[7] G.ADOMIAN, Solution of nonlinear stochastic physical problems, Rendiconti del Seminario Mathematico, Stochastic Problems in Mechanics, Torino, Italy, 1982.

[8] G.ADOMIAN AND G.E.ADOMIAN, Cellular systems and aging models, Advances in Hyperbolic Partial Differential Equations-2, Pergamon Press, NY, 1985.

[9] G. ADOMIAN, G.E. ADOMIAN, AND R.E. BELLMAN, Biological system interactions, Proc. Nat. Acad. Sci., 81, 1984.

[10] G.ADOMIAN AND G.E.ADOMIAN, Solution of the Marchuk model of infectious disease and immune response, Advances in Mathematics and Computers in Medicine-2, Pergamon Press, NY, 1986.

[11] C.KOCH AND I.SEGEV (EDS.), Methods In Neuronal Modeling, MIT Press, Cambridge, MA, 1989.

[12] P.D. WASSERMAN, Neural Computing: Theory And Practice, Van Nostrand, New York, 1989.

[13] J.CRONIN, Mathematical Aspects Of Hodgkin-Huxley Theory, Cambridge University Press, Cambridge, England, 1987.

[14] G.F.ROACH (ED.), Mathematics In Medicine And Biomechanics, Birkhauser-Boston, Cambridge, MA, 1984.

[15] B.BUNOW, I.SEGEV, AND J.W.FLESHMAN, Modeling the electrical properties of anatomically complex neurons using a network analysis program : Excitable membrane, Biol. Cyber., 53 (1985) 41-56.

[16] N.T. CARNEVALE AND F.J. LEBEDA, Numerical analysis of electrotonus in multicompartmental neuron models, J. Neurosci. Meth., 19 (1987) 69-87.

[17] P.S. CHURCHLAND, C.KOCH, AND T.J.SEJNOWSKI, What is computational neuroscience?, (in) Computational Neuroscience (ed.) E.Schwartz, MIT Press, Cambridge, MA, 1989.

[18] J.W.COOLEY AND F.A.DODGE, Digital computer simulation of excitation and propagation of the nerve impulse, Biophys, J., 6(1966) 583-599.

[19] E.DESCHUTTER, Alternative equations for molluscanion currents described by Connor and Stevens, Brain Res.,382 (1986) 134-138.

[20] D.DURAND, The somatic shunt cable model for neurons, Biophys, J., 46 (1984) 645-653.

[21] J.EVANS AND N.SHENK, Solutions to axon equations, Biophysi, J.,6 (1970) 583-599.

[22] A.L. HODGKIN AND A.F. HUXLEY, A quantitative description of membrane current and its application to conduction and excitation in nerve, J. Physiol. (London), 117 (1952) 500-544.

[23] C.KOCH AND P.R. ADAMS, Computer simulation of bullfrog sympathetic ganglion cell excitability, Soc. Neurosci.Abst.,11 (1984) 48.7.

[24] C.KOCH AND T. POGGIO, A simple algorithm for solving the cable equation in dendritic trees of arbitrary geometry, J. Neurosci. Meth., 12 (1985) 303-315.

[25] C.KOCH AND T.POGGIO, Computations in the vertebrate retina :motion discrimination, gain enhancement and differentiation, Trends Neurosci., 9 (1986) 204-211.

[26] I.PARNAS AND I. SEGEV, A mathematical model for the conduction of action potentials along bifurcating axons, J. Physiol. (London), 295 (1979) 323-343.

[27] R.R. POZNANSKI, Techniques for obtaining analytical solutions for the somatic shunt cable, Math. Biosci., 85 (1987) 13-35.

[28] J. RINZEL, Repetitive nerve impulse propagation: Numerical results and methods, (in) Research Notes In Mathematics - Nonlinear Diffusion (eds.) W.E. Fitzgibbon and H.F. Walker, Pitman Publishing Ltd., London, 1977.

[29] J. RINZEL, Excitation dynamics: In sights from simplified membrane models, Fed. Proc., 44 (1985) 2944-2946.

[30] J.RINZEL AND Y.S.LEE, Dissection of a model for neuronal parabolic bursting, J. Math. Biol., 25 (1987) 653-675.

[31] J.H. VAN HATEREN, An efficient algorithm for cable theory, applied to blowfly photoreceptor cells and LMCs, Biol. Cyber., 54 (1986) 301-311.

[32] G.ADOMIAN AND R.RACH, Equality of partial solutions in the decomposition method for linear or nonlinear partial differential equations, Int. J. Comput. and Math. with Appl., 19 # 12 (1990) 9-12.

[33] G.ADOMIAN, Decomposition solution of nonlinear hyperbolic equations, seventh Int. Conf. on Mathematical Modeling, Chicago, IL 2-5 August 1989 and Int. J. Comput. and Math. with Appl.

[34] G.ADOMIAN, A review of the decomposition method and some recent results for nonlinear equations, Int. J.Comput. and Math. with Appl.

[35] G.ADOMIAN, On the solution of complex dynamical systems I and II, Simulation, 54 # 5 (1990) 245-252.

[36] G.ADOMIAN AND R.RACH, Analytic solution of nonlinear boundary-value problems in several space dimensions,submitted for publication.

[37] G.ADOMIAN AND R.RACH, An extension of the decomposition method for boundary-value problems, submitted for publication.

[38] M.WITTEN, On stochasticity in the McKendrick/Von Foerster hyperbolic partial differential equation system, Int. J. Computers and Math. with Appl., 9 (1983) 447-458.

Neuronal variability - Stochasticity or Chaos?

A. V. Holden and M. A. Muhamad*

Centre for Nonlinear Studies,
The University,
Leeds LS2 9JT,
United Kingdom

*Permanent address: *Jabatan Fisiologi, Fakulti Perubatan, Universiti Malaya, 59100 Kuala Lumpur, Malaysia.*

ABSTRACT

Irregularity in neuronal activity can be characterised in terms of either stochastic theory and dynamical system theory. The choice of approach is arbitrary but it may be possible to fuse these approaches in the near future.

INTRODUCTION

Sherrington [1] described the pattern of activity in the waking brain as "the head-mass becomes an enchanted loom where millions of flashing shuttles weave a dissolving pattern, always a meaningful pattern though never an abiding one; a shifting harmony of sub-patterns." This description encompasses the ideas of maintained, non periodic patterning and self-similarity, and emphasises the key fact that the activity of the nervous system is irregular.

It is impossible to observe the activity of all cells in the nervous system simultaneously as there are 10^{12} cells in the human brain. However, extracellular recordings from single cells in freely behaving animals, and intracellular recordings from single cells in restrained or anaesthesised animals are readily obtained.

Electrical recordings from larger neurones that have long axons are a sequence of stereotyped, all-or-none nerve impulses, or action potentials. In the analysis of experimental recordings, these action potentials are treated as events and so the output from these neurones can be considered as a point process. Smaller neurones, and neurones that do not have long axons, do not necessarily generate action potentials, and their activity is irregular, simple and complicated oscillations and so is best considered as a continuous process with a band limited spectral density [2].

There are two main problems concerning the irregular activity of neurones: how to *analyse* experimental recordings, and how to *model* the irregularity. In this paper we consider two different approaches, one

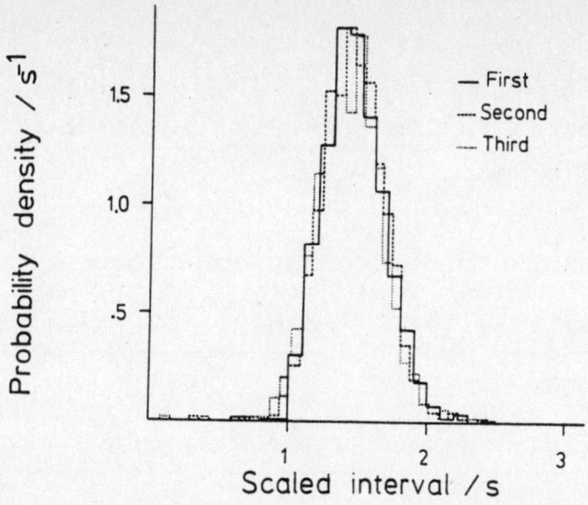

Figure 1. *Normalised first, second- and third-order interspike interval histograms for endogeneous discharge of an identified molluscan neurone.* Holden and Ramadan [6]. *Since they superimpose they represent stable densities.*

based on the theory of stochastic processes, and one on the qualitative theory of dynamical systems. The ideal approach would use elements of both.

STOCHASTIC MODELS OF NEURONAL ACTIVITY

Extracellular recordings of spike trains from single neurones provided an extensive test bed of data, for the analysis of multivariate point processes, and for developing models. Early analysis of spike train data was by hand, or using special purpose hardware. Using such hardware Gerstein and Mandelbrot [3] noted that the shapes of first-order, second- and higher order interspike interval histograms - estimators of interval probability density functions - of cochlear nucleus neurones of anaesthesized cats were similar. Since for a renewal process the n-th order interval probability density function is the n-fold self convolution of the first order interval probability density function this observation suggested that the interval probability density function had a form which is invariant under self-convolution i.e it was an example of a stable (or infinitely divisible) interval probability density function. Stable densities arise under a variety of circumstances: in the context of neuronal modelling a natural mechanism is the first passage time density of a random walk process, or when taken to appropriate limits, a diffusion process.

If the subthreshold membrane potential $\{V(t)\}$ is a stationary, continuous stochastic process defined on the real line with a transition probability density function $f(y, t|x)$, then if $\{V(t)\}$ is assumed to be a Markov process the Smoluchowski equation holds:

$$f(y, t|x) = \int_{-\infty}^{\infty} f(y, \tau|z)\, f(z, t-\tau|y)\, dz, \qquad (1)$$
$$0 \le \tau \le t$$

From (1) the forward and backward Kolmogorov equations

$$\frac{\delta f(y, t|x)}{\delta t} = \frac{1}{2} \frac{\delta^2\{\alpha(y)\, f(y, t|x)\}}{\delta y^2} - \frac{\delta\{\beta(y)\, f(y, t|x)\}}{\delta y} \qquad (2)$$

$$\frac{-\delta f(y, t|x)}{\delta t} = \frac{1}{2} \frac{\alpha(x)^2\{\alpha(y)\, f(y, t|x)\}}{\delta x^2} + \frac{\beta(x)\, \delta f(y, t|x)}{\delta x} \qquad (3)$$

which define a *diffusion process*, with infinitesimal moments $\alpha(x)$, $\beta(x)$, can be obtained [4, 5]. When $\alpha(x) = 1$, $\beta(x) = 0$, a Weiner process is obtained; when $\alpha(x) = \alpha$ and $\beta(x) = \beta$ an Ornstein-Uhlenbeck process.

For a neurone, if the membrane potential V(t) reaches a threshold value K in the spike initiation zone (the initial segment for neurones with myelinated axons) an action potential is generated; after an action potential the membrane potential returns to its resting value. Thus in a model where V(t) is instantaneously reset to its resting value $V(0) = x_o$ the distribution of first passage times from x_o to the threshold K ($K > x_o$) will be the interspike interval distribution.

If $\{V(t)\}$ has a transition probability density $f(y, t|x)$ that satisfies (1-3) then the probability density function $\rho(K, t|x_o)$ of the first passage time $T(K|x_o)$ satisfies

$$f(y, t|x_o) = \int_0^t \rho(K, t-\tau|x_o)\, f(y, \tau|K)\, d\tau \qquad (4)$$

This can be subsituted into (3) to give

$$-\rho(K,0|x_o)\, f(y, t|x_o) = \int_0^t \left(\frac{\delta}{\delta t} - \beta(x) \frac{\delta}{\delta x} - \frac{1}{2} \alpha(x) \frac{\delta^2}{\delta x^2} \right)$$
$$\cdot\ \rho(k, t-\tau|x_o)\, f(y, \tau|K)\, \delta\tau$$

and since, by definition $\rho(k, 0|x_o) = 0$ for $x_o \ne K$, a backward Kolmogorov equation with initial condition $\rho(K, 0|x_o) = \delta(K-x_o)$ and boundary conditions $\rho(k, t|K) = \delta t$ and $\lim_{x \to -\infty} \rho(x, t|x_o) = 0$ results:

$$\frac{\delta\rho(k, t|x_o)}{\delta t} = \beta(x) \frac{\delta\rho(k, t|x_o)}{\delta x_o} + \frac{1}{2} \alpha(x) \frac{\delta^2\rho(k, t|x_o)}{\delta x_o^2} \qquad (5)$$

It is possible to obtain the first passage time density for a Wiener process in closed form; for an Orstein-Uhlenbeck process (the leaky integrator model) the renewal equation (4) does not have a closed form solution.

The first passage densities of the leaky integrator, and more complicated models (where the threshold varies with time, and where there is an afterhyperpolarisation) can be obtained numerically. From the viewpoint of physiology, the irregularity arises as a result of a large number of independent inputs, giving rise to a random walk or diffusion model [4,5].

CHAOS

Deterministic chaos is unpredictable or complicated asymptotic behaviour generated by a deterministic nonlinear dynamical system. The generation of this irregular motion is highly sensitive to initial conditions: this means that any small perturbation of the initial condition will grow exponentially with time. In general a dynamical system can be considered as a series or collection of states which evolve with time, and the time variable can be either continuous or discrete.

A discrete dynamical system is characterised by the iteration of a function (or map)

$$x_{t+1} = f(x_t) \tag{6}$$

$f : I \to I$, where I is an interval.

A map f is said to have a *forward periodic orbit* through x_o if there exists a positive integer N such that $f^m(f^n(x_o)) = f^n(x_o)$ for all $m \in \mathbb{Z}$, the integers. The logistic map which is a 1-parameter family of

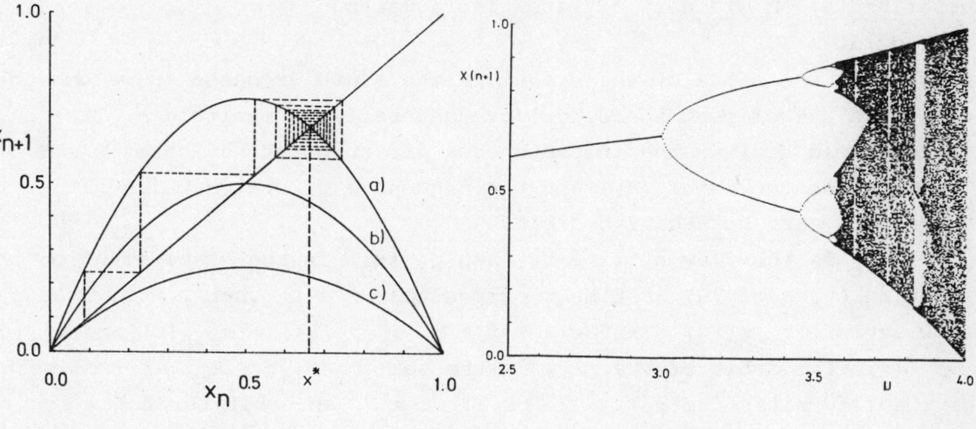

Figure 2. *The Logistic map and its bifurcation diagram.*

smooth maps $f_\rho : I \to I$, $I = [0,1]$, which depends on the parameter $\rho \in (0,4]$, is a simple example that exhibit complicated behaviour.

$$f_\rho(x) = \rho x(1-x) \tag{7}$$

The graph of the map is simply an inverted parabola within the square $I \times I$. The behaviour of the map f_ρ become increasingly complicated as the parameter ρ increases. The bifurcation values of (7) are all those values of ρ where the map f_ρ undergoes topological changes in its orbit structure.

It is clear that for all $\rho \in (0,1]$, there is exactly one fixed point at the origin. There is certainly no sensitive dependence on initial conditions for every orbit with initial points $x_o \in (0,1)$, and they approaches the attracting fixed point at $x = x^*$. As ρ increases a bifurcation occurs. The fixed point here has changed stability and a period-2 orbit is created. Further increase of ρ will lead another bifurcation to a period-4 orbit, and successively a cascade of bifurcations will produce periodic points of period-2^n, $n \in \mathbb{Z}$. This phenomenon is known as the *period doubling route* to chaos. This cascade of period doubling has universal features. If the sequence of bifurcations occur at a parameter values $\{\rho_n\}_{n=1}^{\infty}$ then the values of ρ accumulate at a parameter ρ^* and the sequence

$$\mu_n = (\rho_{n+1} - \rho_n) / (\rho_{n+2} - \rho_{n+1}) \tag{8}$$

has a limiting value 4.66920... known as the *Feigenbaum number* which is characteristic of a large class of maps which undergo period doubling cascades to chaos [7,8].

The behaviour of a continuous dynamical system can simply be described by an ordinary differential equation

$$dx/dt = X(x) \tag{9}$$

where x is the state of the system in the state or phase space embedded either in a Euclidean space, or more generally, a manifold M. The vector field X gives the instantaneous direction of motion at $x \in M$. The solution curves of (9), and the sequences of states which satisfy the map (6) are both called *orbits*.

If φ_t is the *flow* at time t, then $\varphi_t\,(x_o)$ is the state point on the solution curve of (9) at time t, given that $x = x_o$ when $t = 0$. Thus a flow satisfies, $\varphi_0(x) = x$ for all $x \in M$, $\varphi_t(\varphi_s(x)) = \varphi_{s+t}(x)$ and $\varphi_{-t}(x) = \varphi^{-1}(x)$. The orbit of the flow is the set $\{\varphi_t(x) | t \in \mathbb{R}\}$. A fixed point of the flow satisfies $\varphi_t(x) \equiv x$ for all $t \in \mathbb{R}$. An orbit through x is periodic if there exist a positive real τ such that $\varphi_{t+\tau}(x) = \varphi_\tau(x) = x$ and $\varphi_t(x) \neq x$ for all $t \in (0,\tau)$.

In simple cases, the long term behaviour of a continuous dynamical system are confined to well-defined regions of the manifold which show little change in character when small changes to the initial condition are made. In the ω-limit these solutions are on attractors such as a limit point or a periodic attractor. The linear systems (10) on $\mathbb{R}^n$ has fixed point solution at the origin and stability is determined by the eigenvalues of the matrix **A**.

$$dx/dt = \mathbf{A}\, x \qquad (10)$$

where $x \in \mathbb{R}^n$, and **A** is n x n matrix. If **A** has no eigenvalues with zero real part, then (10) is hyperbolic and the system is topologically equivalent to the system

$$dx_s/dt = -x_s, \qquad dx_u/dt = x_u \qquad (11)$$

where $x_s \in \mathbb{R}^s$, $x_u \in \mathbb{R}^u$. The subspaces $\mathbb{R}^s$ and $\mathbb{R}^u$ of (11) are called the unstable and stable manifolds associated with the fixed point at the origin. If both u, $s \neq 0$ then the fixed point is of *saddle* type.

In some instances the evolution of the system depends so sensitively on initial conditions that it appears to have random behaviour, although mathematically it is completely deterministic. The orbits in these persistent modes of chaotic behaviour form an attractor with a complicated structure, which on fine examination often exhibits self-similarity implying an underlying fractal structure. The *strange* attractor lies on a complex hypersurface in the phase space which attracts nearby trajectories but within which neighbouring trajectories diverge exponentially [9,10]. The hyperbolic nature of the strange attractor means that the solutions from neighbouring initial points diverge very rapidly, and any small uncertainty in initial condition will soon grow rapidly as the solution evolves. The concept of a strange attractor has its origins in the pioneering work of Birkhoff on planar diffeomorphisms with attracting sets other than fixed points or invariant circles [11].

Let **f** be a diffeomorphism on a manifold M. A closed invariant set $\mathcal{A}$ of **f** is said to be an attracting set for **f** if there exist a closed neighbourhood $\mathcal{N}$ of $\mathcal{A}$ such that $\mathbf{f}(\mathcal{N}) \subset \text{interior}(\mathcal{N})$ and $\mathcal{A} = \cap_{n=0}^{\infty} \mathbf{f}^n(\mathcal{N})$. $\mathcal{N}$ is the trapping region for $\mathcal{A}$. The attracting set is said to be strange if $\mathcal{A}$ contains a transverse homoclinic point of **f**. The strange attracting set is a *strange attractor* if $\mathbf{f}|\mathcal{A}$ has a dense orbit and a dense set of periodic orbits.

In elementary analysis there has been a strong belief that the solution to a problem consisted of finding functions, no matter how esoteric, which fitted the dynamic behaviour and initial conditions of the problem. An altenative approach is to describe and understand the qualitative feature of the system rather than paying attention to the quantitative details. Various techniques has been used to extract the qualitative dynamics in experiments on real dynamical system.

The graphical reconstruction of an attractor, and display of bifurcation pattern is often used to extract qualitative features of a dynamical system from a time series record. The phase [12] and time-delayed [13] portraits and Poincaré maps feature prominently in this approach. The phase portrait of the system is constructed by plotting derivatives of the measured variables against itself. A time delayed portrait is simply obtained by plotting a variable $x(t)$ against $x(t+\tau)$, i.e itself after a time delay τ, where τ is (in practice) about 1/3 the period of the predominant oscillations. The singular value decomposition analysis as developed by Broomhead and King [14] is a valuable improvement to this technique of attractor reconstruction.

In the construction of Poincaré maps, consider the system given in (9), and suppose the solution curve $x(t)$ is periodic, i.e there exist a minimum positive τ such that $x(t+\tau) = x(t)$, for all $t \in \mathbb{R}$. Then a periodic orbit is a closed loop in state space. If γ denote the periodic orbit so formed and $x_o \in \gamma$, then a *local section* (n-1) dimensional hyperplane Σ will contain x_o which is transverse to the orbit γ. Successive orbits will return to the section, giving rise to a map $f : \Sigma \to \Sigma$ defined on the neighbourhood of x_o. The periodic orbit γ on Σ is replaced by a fixed point x_o of f. Thus the dynamics of the vector field X can be now interpreted by discrete dynamics of the Poincaré map f on Σ.

The sensitivity to initial condition can be quantified by means of a spectrum of Lyapunov exponents, or by estimates of the *attractor* dimension. The Lyapunov exponents, λ_i are defined by

$$\lambda_i = \lim_{n\to\infty} \frac{1}{t} \log_2 \left[P_i(t) \ / \ P_i(0) \right] \qquad (12)$$

where an infinitesimal *n*-sphere in state space is deformed by the dynamics into an *n*-ellipsoid with the principal axes $P_i(t)$.

The sum of exponents measure the average logarithmic rate of growth of infinitesimal volume elements as they are carried along by the flow in the phase space. The exponents thus quantify the rate of trajectories divergence in the strange attractor and degree of chaos. A positive Lyapunov exponent indicates exponential divergence of trajectories within the attractor, a known characteristic of a strange attractor, and is taken as sufficient evidence of chaos.

The folding and stretching processes responsible for deterministic chaotic dynamics often form an attractor with extremely complicated structure. Although state space volumes are contracting in dissipative systems, length is not neccessarily shrinking. In order for the attractor to remain bounded, stretching in some directions must be compensated for by folding, producing after many stretches and folds, a multi-layered self similar structure. Eventually this produces an attractor with non-integer dimension, resembling a Cantor set. Estimated low and non-integer dimensions of the reconstructed attractor may indicates fractal geometry, another signature of a chaotic attractor. There are a number of definations of dimensions and methods of estimating these dimensions that can be applied to reconstructed attractors [15].

The spectrum of Rényi dimensions provides a more complete description of the metric properties of an attractor. If one set is the image under a diffeomorphism of another, then they have identical Rényi spectra and thus its difference may be use to measure the "distance" between two attractors [16].

IDENTIFICATION OF CHAOS IN NEURAL ACTIVITY

The nervous system and its components are highly nonlinear and may oscillate periodically as well as displaying irregular behaviour under certain conditions. It is therefore possible that the cause of irregular behaviour observed in neural systems might be attributed to chaos rather than as a consequence of the influence of random noise.

A neurone as a threshold device may be modelled by a simple, piece-wise linear interval map introduced by Labos [17].

$$\begin{aligned} x_{t+1} &= m\,x_t, & x_t &\le a \\ x_{t+1} &= u\,x_t - u, & x_t &> a \\ m &= 1 + (1-a)^2 \\ u &= 1\,/\,(a-1) \end{aligned} \tag{13}$$

where x_t represents the membrane potential and a, the threshold. This primitive recursion is illustrated in Figure 3: iteration generates a series of "action potential-like" events. These are chaotic if the starting value x_o is irrational.

Biophysical excitation equations for neurones are derived from voltage clamp measurements of ionic currents, and are generally high order, nonlinear systems of ordinary differential equations [18]. A simple, third order biophysical excitation system [19] shows patterned, bursting and chaotic solutions (see Figure 4).

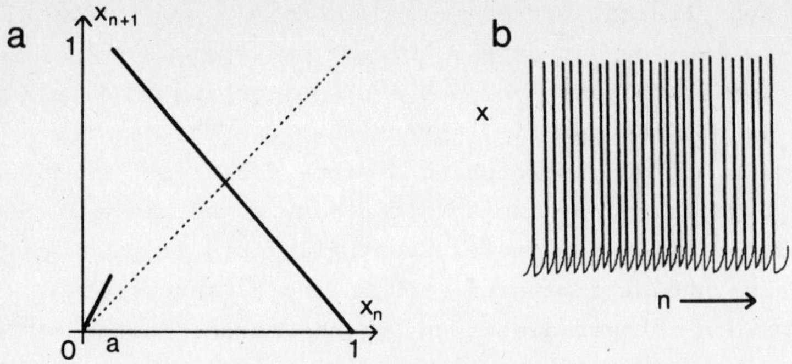

Figure 3. *The neuron as a threshold device* (Labos [17]).

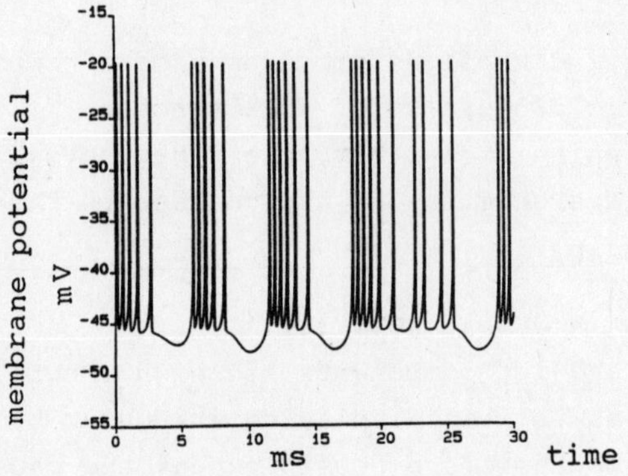

Figure 4. *Chaotic bursting in the Chay system of equations* (Chay [19]).

The best studied example of chaotic behaviour in single neurones is the sinusoidally forced membrane of the squid giant axon, which exhibit periodic, quasiperiodic and chaotic responses. The fractal characteristics of a reconstructed attractor for the sinusoidally forced Hodgkin-Huxley equations [20] is illustrated in Figure 5.

Techniques for identifying chaos in recordings from single neurones are discussed in [20]. The responses of a spontaneously active neurone to a range of sinusoidal input was recorded by Hayashi in giant neurones of *Onchidium* [21] and in the squid giant axon by Matsumoto et al [22]. The periodic and nonperiodic spike trains were analysed with stroboscopic plots, one- dimensional transfer functions and power spectrum analysis.

The power spectra of the recorded time series is frequently used as a means to identify transitions between periodic activities and from periodic to irregular activity. Periodic motion contributes to a sharp peak at its fundamental frequency to the power spectrum while irregular behaviour is characterised by a broad band with no discernable structure. The power spectrum by itself cannot be used to distinguish between stochasticity and chaos [23]. However, the path from oscillatory motion to aperiodic behaviour may be used in comparison with known transition routes to chaos.

Everson [16] applied the singular value decomposition method to the experimental records of Aihara et al [24] and estimated the Lyapunov exponents for a range of input regimes. The oscillations exhibited by the squid giant axon during sinusoidal current stimulation agrees with the result of numerical computations by Jensen et al [25] and Holden and Muhamad [20] on the Hodgkin-Huxley system of equations.

Chaos has been identified in numerous physical, chemical and biological systems and are reviewed in [26, 27, 28].

PROBLEMS OF NOISY CHAOS

Real neurones are nonlinear, and can, in certain circumstances, generate a chaotic output in the absence of any inputs. In the nervous system most nerve cells are subjected to a large (approximately 10^4) number of excitatory and inhibitory inputs, and so activity will have a stochastic component. Irregularity can be a mixture of stochasticity (noise) and deterministic chaos.

This may be modelled by adding a Langevan (stochastic) term to the

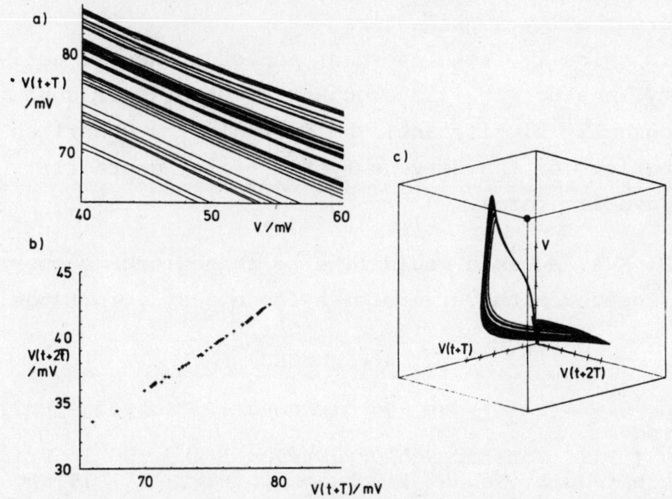

Figure 5. (a) *Self-similarity of reconstructed attractor of sinusoidally forced 4-dimensional Hodgkin-Huxley system* [20]. (b) *A cut across the reconstructed attractor obtained by plotting* $V(t+2\tau)$ *against* $V(t+\tau)$ (c) *3-dimensional representation of reconstructed attractor.*

deterministic equations, but the main problem is how to analyse an irregular signal that is mixed with stochasticity.

One approach is to view the stochasticity as a contaminating noise, which may be removed to expose the underlying chaos. This is implicit in the singular value decomposition method of Broomhead and King [14] developed as a basis for selecting rational coordinates and embedding dimension. The trajectory matrix

$$X = \hat{X} + E \tag{14}$$

$\hat{X}$ is the deterministic component of the signal and E the noise. The singular value decomposition of X is given by

$$X = U \Sigma C^T \tag{15}$$

where U and C are the orthogonal matrices and Σ is an N x d diagonal matrix with non negative elements and $\Sigma_{ij} = \sigma_i$ and $\Sigma_{ij} = 0$, $i \neq j$ and σ_i are the singular values of X.

However, treating stochasticity as noise ignores important phenomena like noise induced order and the biological significance of the high degree of synaptic convergence onto a single neuron.

Another approach is to generalise the idea of an attractor of a deterministic dynamical systems to stochastic dynamical systems. Demongeot et al (see [29]) have generalised the idea of attractor to that of a *confiner* to deal with noisy nonlinear oscillations, as observed in experimental observations, or generated by a nonlinear oscillations perturbed by a noise term.

However, biologically, the important aspect of neural activity is its irregularity: whether this is stochastic, chaotic or some mixture is not physiologically significant. The irregularity underlies the interesting behaviour of the nervous system - its adaptability, creativity and dynamic forms.

ACKNOWLEDGEMENT: M.A. Muhamad would like to acknowledge support by the Association of Commonwealth Universities for his stay in Leeds.

REFERENCES

[1] Sherrington, C.S. (1946) *Man and his nature.* Cambridge University Press.United Kingdom.

[2] J.H. Martin (1981) *Somatic Sensory System I and II.* In Principles of Neural Science (eds E. Kandel & Schwartz). Part III, **15** and **16**, 158 -83.Edward Arnold.

[3] Gerstein, G.L. and B. Mandelbrot (1964) *Random walk models for the spike activity of a single neuron.* Biophysical J. **4**, 41-67.

[4] A.V. Holden (1976) *Models of the Stochastic Activity of Neurones.* Lecture Notes in Biomathematics. **12**, 1-368. Springer Verlag: Berlin.

[5] G. Sampath and S.K. Srinivasan (1977) *Stochastic Models for Spike Trains of Single Neurons.* Lecture Notes in Biomathematics. **16**, 1-188.

Springer Verlag: Berlin.
[6] Holden, A.V. and Ramadan, S.M. (1980) *Identification of endogeneous and exogeneous activity in a molluscan neurone by spike train analysis.* Biol. Cybern. **37**, 107-14.
[7] Eckmann, J.P. (1981) *Roads to turbulence in dissipative dynamical system.* Rev. Modn. Phys. **53**, 543-654.
[8] Ott, E. (1981) *Strange attractors and chaotic motions of dynamical system.* Rev. Modn. Phys. **53**, 665-71.
[9] Ruelle, D. and Takens, F. (1971) *On the nature of turbulence.* Comm. Math. Phys. **20**, 167-92.
[10] Ruelle, D. (1980) *Strange attractors.* Math. Intelligenser **3**, 126-37
[11] R. Abraham and C.D. Shaw (1986) *Dynamics: Geometry of behaviour.* Part I-IV. Aerial press. Santa Cruz. USA.
[12] Holden, A.V., Winlow, W. and Haydon, P.G. (1982) *The induction of periodic and chaotic activity in molluscan neurone.* Biol. Cybern. **43**, 169-73.
[13] Packard, N.H., Crutchfield, J.P., Farmer, J.P. and Shaw, R.S. (1980) *Geometry from a time-series.* Phys. Rev. Letts. **45**, 712-6.
[14] D. Broomhead and G. King (1986) *Extracting qualitative dynamics from experimental data.* Physica **20D**, 217.
[15] Hertschel, H.G.E. and Procaccia, I. (1983) *The infinite number of generalised dimensions of fractals and strange attractors.* Physica **8D**, 435-44.
[16] Everson, R. (1988) *Quantification of chaos from periodically forced squid axons.* In Chaos in Biological systems (eds. Degn, H, A.V. Holden and L.F. Olsen).NATO ASI series. **138**, 133-42. Plenum. New York.
[17] E. Labos (1984) *Periodic and nonperiodic motions in different classes of formal neuronal network and chaotic spike generators.* In Cybernetic and System Research 2 (ed. R. Trappl) 237-43. Elsevier, Amsterdam.
[18] Hodgkin, A.L. and Huxley, A.F. (1952) *A quantitative description of membrane current and its application to conduction and excitation in nerve.* J. Physiol. **117**, 500-44.
[19] T.R. Chay (1985) *Chaos in three-variable model of excitable cell.* Physica **16D**, 233.
[20] A.V. Holden and M.A. Muhamad (1984) *The identification of deterministic chaos in the activity of single neurones.* J. Electrophysiol. Techniques. **11**, 135-47.
[21] Hayashi, H., Ishizuku, S., Ohta, M. and Hirakawa, K. (1982) *Chaotic behaviour in Onchidium giant neuron under sinusoidal stimulation.* Phys. Lett. **88A**, 435-8.
[22] Matsumoto, G., Aihara, K., Ichikawa, M and Tasaki, A. (1984) *Periodic and nonperiodic responses of membrane potentials in squid giant axons during sinusoidal current stimulation.* J. Theoret. Neurobiol. **3**, 1-14.
[23] Farmer, J.D., Crutchfield, J., Packard, N.H., Shaw, R.S. (1980) *Power spectra and mixing properties of strange attractors.* Ann. of N.Y. Acad. Sci. **357**, 453-72.
[24] Aihara, K. and Matsumoto, G. (1987) *Forced oscillations and routes to chaos.* In Chaos in biological system (Eds. Degn, H., Holden, A.V. and Olsen, L.F.). NATO ASI series. **138**, 121-32. Plenum. New York.
[25] Jensen, J.H., Christiansen, P.L., Scott, A.C. and Skovgaard, O. (1983) *Chaos in nerve.* Proc.Iasted Symp. A.C.I. **2**, 15/6-15/9. Copenhagen.
[26] *Universality in Chaos* (1984) P. Cvitanovic (ed), Adam Hilger, United Kingdom.
[27] *Chaos* (1986) A.V. Holden (ed) Manchester University Press, United Kingdom. Princeton University Press, USA.
[28] *Dynamical chaos* (1987) Berry, M.V., Percival, I.C. and Weiss, N.O

(eds) Proc. of the Royal Society, Royal Soc. United Kingdom.
[29] J. Demongeot, C. Jacob and P. Cinquin (1988) *Periodicity and chaos in biological system: New tools for the study of attractors.* In Chaos in biological system (Eds. Degn, H., Holden, A.V. and Olsen, L.F.). NATO ASI series. 138, 255-66. Plenum. New York.

A LIMIT THEOREM AND ASYMPTOTICAL STATISTICAL CHARACTERISTICS FOR A SELECTIVE INTERACTION MODEL OF A SINGLE NEURON

A. Rangan
Department of Mathematics
Indian Institute of Technology
Madras-600 036, INDIA.

ABSTRACT

We consider in this article, interaction of two types of inputs, one excitatory and the other inhibitory arriving according to a stationary renewal point process, resulting in a renewal point process of events consisting of neural discharge (response yielding events). Apart from obtaining the explicit expressions for the asymptotical statistical characteristics of time for the first response yielding event, a limit theorem is proved.

1. INTRODUCTION

In recent years, several renewal process models have been proposed for the spontaneous single neuron discharge (e.g. see for reference S.E. Fienberg [1974], A.V. Holden [1976], Srinivasan and Sampath [1977], J.W. De Kwaadsteniet [1982], P. Lansky and C.E. Smith [1989]). Basically, these models describe the interspike interval distribution of the neuron as a function of excitation, inhibition, temporal summation of excitation and decay. Such studies deal with two types of events (excitation and inhibitory), each of which is assumed to arrive according to an independent stationary recurrent point process. The two events are assumed to interact in a variety of ways giving rise to response yielding events which correspond to the neuronal discharge.

The statistical characteristics of the response yielding events which form a point process, form the core of such studies, as they can be made to correspond to the experimentally observable interspike interval histograms. However, it is well known that a neuron receives synapses from neighbouring neurons and the synaptic connections can be both excitatory and inhibitory [Eccles [1977]]. Thus the synaptic transmission to the neuron in question due to an electrical impulse can be an Excitatory post synaptic potential or an Inhibitory post synaptic potential due to the various synaptic connections. Also in an inhibitory interneuron connected through a Renshaw cell, the action of the Renshaw cell on the inhibitory interneuron is to effect an inhibition of its inhibitory action, that is to diminish its inhibitory action which is equivalent to excitation. This process is called disinhibition (Eccles [1977], Jewett and Rayner [1984]). The above view points clearly bring to the fore the necessity of considering correlated pair of events.

We analyse a stochastic model for the selective interaction of two types of events arriving according to a recurrent point process, with the two events identified according to a point binomial distribution. The purpose of this paper is to study the mean and variance of the time to a response yielding event S_z, and their asymptotic behaviour as the threshold value z tends to infinity and to develop a limit theorem for S_z, which can be visualised as a random sum of independent random variables.

2. THE MODEL

We assume that events occur according to a renewal process characterized by the interval density f(.). An event, on occurrence is classified either as an e-event with probability p or an i-event with probability q=1-p. The quantal changes effected by the two types of events on the potential level X(t) and their interaction scheme are as follows:

(i) An e-event increases X(t) by a random amount, successive such increases being i.i.d. with p.d.f. $\chi(.)$

(ii) An i-event on arrival sets X(t) to the rest level, assumed to be zero.

(iii) When X(t) crosses a pre-specified threshold level 'Z', a response yielding event (r-event) occurs and instantaneously X(t) drops down to zero and the process continues with further e and i-events.

It is worth observing that the response yielding events form a renewal process and hence is characterised by the interval density between two such events. Let $f_r(t)dt$ be the interval density between successive r-events; an explicit expression for $f_r(.)$ is obtained by arguing that either the first e-event or the accumulation of a certain number of e-events yield the r-event and further dividing the classification so that they are either intercepted by an i-event or not. Thus we obtain

$$f_r(t) = pf(t)\int_z^\infty \chi(x)dx + \sum_{n=1}^{\infty} p^{n+1}\int_0^t f^{(n)}(u)f(t-u)du\int_0^z \chi^{(n)}(x)dx\int_{z-x}^{\infty} \chi(y)dy$$
$$+ \sum_{n=1}^{\infty} p\int_0^t \Pi_n(u)f(t-u)du\int_z^\infty \chi(x)dx + \sum_{n=1}^{\infty}\sum_{m=1}^{\infty} p^{m+1}\int_0^t \Pi_n(u)\int_u^t f^{m}(v)f(t-v)dudv$$
$$\times \int_0^z \chi^{(m)}(x)dx\int_{z-x}^{\infty} \chi(y)dy \qquad (2.1)$$

where $f^{(n)}(.)$ and $\chi^{(n)}(.)$ are the n-fold convolutions of f(.) and $\chi(.)$ respectively and $\Pi_n(t)dt$ is the probability that the nth renewal event occurs in (t,t+dt), starting initially with a renewal event at t=0. We see that $\Pi_n(t)$ satisfies the recursive equation

$$\Pi_n(t) = \int_0^t \Pi_{n-1}(u)\Pi_1(t-u)du \quad , \; n > 1 \tag{2.2}$$

$$\text{and } \Pi_1(t) = qf(t)+q\sum_{n=1}^{\infty} p^n \int_0^t f^n(u)f(t-u)du \int_0^z \chi^{(n-1)}(x)\chi(z-x)dx \tag{2.3}$$

2.1 SPECIAL CASES

(i) When $f(t)=\lambda e^{-\lambda t}$ and $\chi(x)=\alpha e^{-ax}$, equation (2.1) reduces to

$$f_r(t) = p\lambda e^{-(\alpha z+p\lambda t)} I_0(2\sqrt{\alpha p\lambda zt}) \tag{2.4}$$

where $I_0(.)$ is the modified Bessel function of order zero.

(ii) When $\chi(x)=\alpha^2 xe^{-\alpha x}$, the transformed equation of (2.1) is given by

$$f_r^*(s) = g^*(s)\; e^{-\alpha z}(\sinh(g^*(s)\alpha z)\; g^*\cosh(g^*(s)\alpha z)) \tag{2.5}$$

where

$$g^*(s) = \left[\frac{pf^*(s)}{1-qf^*(s)}\right]^{1/2}$$

3. ASYMPTOTIC STATISTICAL CHARACTERISTICS

In the last section, while we have derived the pdf of the time S_z to an r-event, z>0 being the threshold value, in this section we study the mean and variance of S_z and their asymptotic behaviour as the threshold value $z\to\infty$. We also develop a central limit theorem for S_z, which can be visualised as the random sum of i.i.d. random variables for a particular case.

Since the z-crossings form a renewal sequence, let

$$\Pr\{S_z \le t\} = \Pr\{X(t)>z\} = P(t,z) \text{ (say)} \tag{3.1}$$

It is to be noted that while P(t,z) is the distribution function of the density function $f_r(.)$ given in (2.1); we have introduced z above as parameter entering in P(t,z) for further analysis. Denoting the double Laplace transform of P(t,z) by $\hat{p}^*(s,\omega)$ so that

$$\hat{p}^*(s,\omega) = \int_0^{\infty}\int_0^{\infty} e^{-(st+\omega x)} p(t,x)dtdx \tag{3.2}$$

we obtain from (2.1)

$$\hat{p}^*(s,\omega) = \frac{pf^*(s)[1-\hat{\chi}(\omega)]}{\omega[1-f^*(s)(q+p\hat{\chi}(\omega))]} \tag{3.3}$$

where

$$f^*(s) = \int_0^{\infty} e^{-st}f(t)dt \qquad \text{and} \qquad \hat{\chi}(\omega) = \int_0^{\infty} e^{-\omega x}\,\hat{\chi}(x)dx$$

Using the identity

$$\hat{\bar{p}}^*(s,\omega) = \frac{1}{s}\left[\frac{1}{\omega} - \hat{p}^*(s,\omega)\right]$$

where

$$\hat{\bar{p}}^*(s,\omega) = \int_0^\infty\int_0^\infty e^{-(st+\omega x)}\left\{\int_t^\infty p(u,x)du\right\}dtdx$$

we obtain

$$\hat{\bar{p}}^*(s,\omega) = \frac{\bar{F}^*(s)}{\omega[1-f^*(s)(q+p\hat{\chi}(\omega))]} \tag{3.4}$$

LEMMA 3.1:

If $\mu_x = \int_0^\infty x\,\chi(x)dx$ and $\mu_y = \int_0^\infty yf(y)dy$ and $\sigma_x^2 = \int_0^\infty (x-\mu_x)^2\chi(x)dx$ then

(i) $$E[S_z] = \frac{\mu_y}{p}[1 + H_y(z)] \tag{3.5}$$

(ii) $$E[S_z] = \frac{\mu_y}{p\mu_x}z + \frac{\mu_y}{2p}\left[1 + \left(\frac{\sigma_x}{\mu_y}\right)^2\right] + O(1), \text{ as } z\to\infty \tag{3.6}$$

where $H_x(.)$ is the renewal function corresponding to the interval density $\chi(.)$.

PROOF: Observe that

$$E[S_z] = \int_0^\infty Pr(S_z>t)dt$$

and

$$\hat{\bar{p}}^*(0,\omega) = \frac{\mu_y}{\omega p(1-\hat{\chi}(\omega))}$$

$$= \frac{\mu_y}{\omega p}\left[1 + \frac{\hat{\chi}(\omega)}{1-\hat{\chi}(\omega)}\right]$$

$$= \frac{\mu_y}{p}\left[\frac{1}{\omega} + \frac{\hat{h}_x(\omega)}{\omega}\right]$$

where

$$\hat{h}_x(\omega) = \frac{\hat{\chi}(\omega)}{1-\hat{\chi}(\omega)}$$

is the transform of the renewal density $h_x(.)$ corresponding to the interval density $\chi(.)$. Taking the inverse Laplace transform with respect to ω

$$\bar{p}^*(0,z) = \frac{\mu_y}{p}[1 + H_x(z)] = E[S_z]$$

Thus statement (i) is proved. Statement (ii) follows from statement (i) by using the asymptotic expansion for the renewal function

$$H_x(z) = \frac{z}{\mu_x} + \frac{\sigma_x^2 - \mu_x^2}{2\mu_x^2} + O(1) \text{ as } z\to\infty \tag{3.7}$$

Using (3.7) in (3.5), we obtain (3.6).

LEMMA 3.2:

If $\sigma_y^2 = \int_0^\infty (t-\mu_y)^2 f(t)dt$ then

$$\text{(i)}\quad E[S_z^2] = \left[\frac{\sigma_y^2-\mu_y^2}{p}\right]\left[1+H_x(z)\right] + \frac{2\mu_y^2}{p^2}\left[1+2H_x(z)+\int_0^z H_x(u)h_x(z-u)du\right] \tag{3.8}$$

$$\text{(ii)}\quad E[S_z^2] = \left(\frac{\sigma_y^2-\mu_y^2}{p}\right)\left[1+\frac{z}{\mu_x}+\frac{\sigma_x^2-\mu_x^2}{2\mu_x^2}\right] + \frac{\mu_y^2}{p^2}\left[\frac{z^2}{\mu_x^2}+2\frac{\sigma_x^2}{\mu_x^2}+\frac{z}{\mu_x}\left[4+\frac{\sigma_x^2-\mu_x^2}{\mu_x^2}\right]\right]+O(z)$$

as $z \to \infty$ (3.9)

PROOF: The second moment of S_z is given by

$$\frac{1}{2}\int_0^\infty e^{-\omega z}E[S_z^2]dz = -\left.\frac{\partial \hat{p}^*(s,\omega)}{\partial s}\right|_{s=0} \tag{3.10}$$

Using equation (3.4) in (3.10), we obtain

$$\int_0^\infty e^{-\omega z}E[S_z^2]dz = -\frac{1}{\omega}\left[\left[\frac{\sigma_y^2-\mu_y^2}{p}\right](1+\hat{h}_x(\omega))+\frac{2\mu_y^2}{p}\left[1+2\hat{h}_x(\omega)+[\hat{h}_x(\omega)]^2\right]\right] \tag{3.11}$$

Statement (i) follows by inverting equation (3.11). To prove statement (ii), we use the asymptotic expansion of $H_x(.)$ and $h_x(.)$ in equation (3.8).

COROLLARY 3.3: Let

$$C^2 = \frac{\sigma_y^2}{p\mu_x} + \frac{\mu_y^2}{p^2\mu_x}\left[(1+q) + (1-\mu_x)(1+\left[\frac{\sigma_x}{\mu_x}\right]^2)\right]$$

Then
$$\lim_{z\to\infty}\frac{\text{Var } S_z}{z} = C^2 \tag{3.12}$$

PROOF: From Lemmas 3.1 and 3.2, we have

$$\text{Var } S_z = \frac{\sigma_y^2-\mu_y^2}{p}\left[\frac{z}{\mu_x}+\frac{1}{2}\left(1+\left(\frac{\sigma_x}{\mu_y}\right)^2\right)\right] + \frac{\mu_y^2}{p^2}\left[\frac{3}{2}\left(\frac{\sigma_x}{\mu_y}\right)^2+\frac{2z}{\mu_x}+z\left(1+\left(\frac{\sigma_x}{\mu_x}\right)^2\right)\left(\frac{1}{\mu_x}-1\right)-\frac{1}{4}-\left(\frac{\sigma_x}{\mu_x}\right)^2\right] + O(z)$$

Hence, it follows that $\frac{\text{Var } S_z}{z} = C^2$ as $z \to \infty$

Let us now restrict our attention to the case where the inputs to X(t) are only positive so that the effects of i-events are neglected. Thus p is taken to be unity. Now the time to the z-crossing, S_z, could be expressed by using the counting process $\{M(z); z\geq 0\}$ associated with the renewal process of jump magnitudes. Since the number of events before a z-crossing is given by M(z)+1, we observe that $S_z \overset{d}{=} \sum_{i=1}^{M(z)+1} Y_i$, where Y_i's are the time between the arrivals of two e-events, which are assumed to be i.i.d. (Shanthikumar and Sumita [1983]).

LEMMA 3.4:

$$\frac{S_z}{z} \to \frac{\mu_y}{\mu_x} \text{ a.s. as } z\to\infty$$

PROOF: Since $\frac{S_z}{z} = \frac{S_z}{M(z)+1} \times \frac{M(z)+1}{z}$, the lemma follows from the observation that

$$S_z \overset{d}{=} \sum_{i=1}^{M(z)+1} Y_i$$

the strong law of large numbers and the almost sure convergence of $\frac{M(z)}{z}$ to $\frac{1}{\mu_x}$.

Since we have already established the sufficient condition under which the central limit theorem for S_z holds, namely $\lim_{z\to\infty}\frac{\text{Var } S_z}{z} = C^2$, and since S_z is the random sum of random (i.i.d.) variables, Anscombe's theorem (Anscombe [1952]) stated below, is directly applicable to S_z.

Let $\{\nu_n; n\geq 1\}$ be a sequence of strictly positive integer random variables such that $\frac{\nu_n}{n}\to\alpha$ in probability as $n\to\infty$, where α is a positive constant. For the renewal process $\{Z_j; j\geq 1\}$ with $E[Z_j] = \mu_z$ and Var $[Z_j] = \sigma_z^2<\infty$, let $S_n = \sum_{j=1}^{n} Z_j$ and consider $S_{\nu_n} = \sum_{j=1}^{\nu_n} Z_j$, where $\{\nu_n\}$ and $\{Z_j\}$ may be correlated. Then

$$\frac{S_{\nu_n} - \mu_z \nu_n}{\sqrt{\nu_n}\,\sigma_z} \overset{d}{\to} N(0,1) \text{ as } n\to\infty.$$

THEOREM 3.5:

$$\frac{S_z - [M(z)+1]\,\mu_y}{\sigma_y \sqrt{[M(z)+1]}} \overset{d}{\to} N(0,1) \text{ as } z\to\infty$$

PROOF: The theorem follows from Lemma 3.4 and Anscombe's theorem.

REFERENCES

1. ANSCOMBE, F.J. [1952], Large sample theory of sequential estimation. *Proc. Camb. Phil. Soc.* **48**, 600-607.
2. ECCLES, J.C. [1977], The understanding of the brain. *Second edition, McGraw-Hill*, New York.
3. FIENBERG, S.E. [1974], Stochastic models for single neuron firing trains, a survey, *Biometrics*, **30**, 399-427.
4. HOLDEN, A.V. [1976], Models of stochastic activity of Neurons. *Lecture notes in Bio Mathematics, Springer,* **Vol. 12**.
5. JEWETT, D.C. and RAYNER, M.D. [1984], Basic concepts of neuronal function. *Acad. Press*, New York.
6. DE KWAADSTENIET, J.W. [1982], Statistical analysis and stochastic modelling of neuronal spike train activity, *Math. Bio Sciences,* **60**, 17-71.

7. LANSKY, P. and SMITH, C.E. [1989], Effect of a random initial value in neural first passage time models. *Math. Bio Sciences,* **93**, 191-215.

8. SAMPATH, G. and SRINIVASAN, S.K. [1977], *Lecture notes in Bio Mathematics.* **16,** *Springer Verlag*, Berlin.

9. SHANTHIKUMAR, J.G. and SUMITA, U. [1983], General shock models associated with correlated renewal sequences, *J. Appl. Prob.*, **20**, 600-614.

PHASE DEPENDENT POPULATION GROWTH MODELS

C. R. RANGANATHAN
DEPARTMENT OF MATHEMATICS
TAMIL NADU AGRICULTURAL UNIVERSITY
COIMBATORE - 641 003
INDIA

ABSTRACT

Growth of biological populations is phase dependent; individuals can give birth to offspring only after reaching a 'maturity age'. Assuming that the life time of an individual consists of two phases, we study the implication of phase dependence in deterministic and stochastic population growth models. By taking the first phase to be of constant duration in the deterministic model an explicit expression for the size of the population is derived. Developing a birth and death process in which the birth and death rates are age and phase dependent, an explicit expression for the mean number of individuals has been obtained in the case when the death rate is constant. Two particular cases of the model are also discussed.

INTRODUCTION

The literature on models of population growth is vast. These models, whether deterministic or stochastic generally deal with the size and age structure of populations (see for example Pollard (1) for deterministic models, and Iosifescu and Tauto (2) for stochastic models). An important aspect of the evolution of biological populations is the constraint on the age of the individuals for reproduction; they begin giving birth to offspring only after reaching a 'maturity age'. In otherwords the life time of each member can be assumed to consist of two phases. During the first phase individuals are inactive producing no offspring at all. Those who complete the first phase enter into the second phase in which period they give birth to new members who inturn behave in a like fashion.

Van Der Plank (3), a plant epidemiologist introduced the notion of phases in deterministic population growth models. He was concerned with the estimation of 'infection rate' of a plant epidemic. In plant and animal disease epidemics (Bailey (4) newly infected individual becomes infectious only after passing through a 'latent period' In these cases the population consists of the infected individuals and the conversion of a healthy plant (animal) into an infected one can be termed as 'birth' and the latent period corresponds to phase 1. Parthasarathy (5) studied a modified Markov branching process (Harris (6)) by assuming that each individual surviving atleast upto a constant time duration T produces offspring at the time of its death according to a known probability distribution. More recently a method of phases has been introduced by Srinivasan (7) to study a population point process model of cavity radiation.

The object of the present paper is to study the implication of phase in some well known deterministic and stochastic population growth models. Explicit expressions have been obtained for the size of the population for the deterministic model and mean number of individuals in a stochastic model when the birth rate is age and phase dependent and death rate is a constant.

NOTATIONS

$N(t)$	:	Size of the population at time t
R	:	'Specific growth rate' or 'Malthusian parameter'
p	:	Duration of the phase in which individuals produce no offspring
$\lambda(.)$	:	Birth rate of individuals
$\mu(.)$	:	Death rate of individuals
$f_1(x\|x_o;t)$	:	First order product density corresponding to the distribution of the age of an individual existing at time t.
$f_2(x,y\|x_o;t)$	:	Second order produce density corresponding to the distribution of the ages of two individuals existing at time t.
$m(t;x_o)$	:	Mean number of individuals at time t due to an individual of age x_o at time t=0.
$m^*(s;x_o)$	:	Laplace transform of $m(t;x_o)$ with respect to t.
$m^{**}(s_1,s_2)$	:	Double Laplace transform of $m(t;x_o)$.

ANALYSIS

(i) A Deterministic Model

Consider a population developing continuously under abundant supply of resources. If N(t) denotes the size of the population at time t then the growth of the population can be described by the equation

$$\frac{dN(t)}{dt} = RN(t).$$

The quantity R which may be either a constant or a function of time is known as 'specific growth rate' or 'Malthusian parameter'. The above equation can be readily solved to yield

$$N(t) = N_o \exp\left(\int_0^t R(u)du \right)$$

where N_o is the size of the initial population; all individuals areof the same age.

We now assume that the life span of every individual consists of two phases. The first phase is assumed to be of constant duration denoted by p. During this phase the individuals produce no offspring at all; they can give birth only during the second phase. If $\Delta N(t)$ denotes the change in the population size in the time interval $(t,t+\Delta)$ then it is reasonable to assume that $\Delta N(t)$ is proportional to the size at time t-p, that is, N(t-p). Hence $\Delta N \propto N(t-p)\Delta t$ which leads to the equation

$$\frac{dN(t)}{dt} = RN(t-p) \tag{1}$$

Although in the most general form the proportionality constant R may depend on both time t and the duration of the first phase p, we shall assume that R is a function of t only. Thus (1) becomes

$$\frac{dN(t)}{dt} = R(t)N(t-p) \tag{2}$$

The solution of (2) depends on the value of N(t) in the interval (O,P). We shall solve (2) by assuming that at t=0 there are N_o newly born individuals. Then it is clear that

$$N(t) \equiv N_o, \quad 0 \le t \le p \tag{3}$$

Consider the interval $p \le t \le 2p$. In view of (3)

$$N(t-p) \equiv N_o.$$

So (2) becomes

$$\frac{dN(t)}{dt} = R(t)N_o$$

Integrating and using (3) we get

$$N(t) = N_o[1+ \int_p^t R(u)du] \tag{4}$$

Let $R_1(t) = \int_p^t R(u)du$. Then (4) becomes

$$N(t) = N_o(1+R_1(t)), \quad p \le t \le 2p \tag{5}$$

Solution (5) can be used in (2) to solve it when $2p \le t \le 3p$. A recursive procedure can be followed to obtain the solution of (2) when $np \le t \le (n+1)p$, n =2,3,... By mathematical induction it can be shown that the solution of (2) is

$$N(t) = N_o \sum_{o}^{n} R_k(t), \quad np \leq t \leq (n+1)p$$

or

$$N(t) = N_o \sum_{o}^{n} R_k(t)H(t-kp) \tag{6}$$

where

$$R_o(t) \equiv 1;$$

$$R_k(t) = \int_{kp}^{t} R(u)R_{k-1}(u-p)du, \quad k = 1,2,... \tag{7}$$

and H(.) is the Heaviside unit function defined by

$$H(t) = 1 \quad \text{if} \quad t \geq 0$$

$$= 0 \quad \text{if} \quad t < 0$$

The functions $R_k(t)$ can be obtained recursively using (7) and hence solution (6) gives N(t) explicitly.

A special case of the model discussed above arises when $R(t)= r$, a constant. In this case it is easily seen that

$$R_k(t) = \frac{r^k(t-kp)^k}{k!}, \quad k = 1,2,... \tag{8}$$

and

$$N(t) = N_o \sum_{k=o}^{\infty} \frac{r^k(t-kp)^k}{k!} H(t-kp) \tag{9}$$

This result is in agreement with that obtained by Van Der Plank (3).

In the model discussed above it was assumed that all the individuals at t=0 were newly born. This need not be true always. Since the subpopulation generated by coexisting individuals can be assumed to develop in complete independence of one another, it is sufficient to consider a single individual of age x_o at time t=0. Two cases arise.

Case 1: $x_o < p$

By a procedure exactly similar to that used to arrive at equation (6) it can be shown that

$$N(t) = \sum_{n=o}^{\infty} R_k(t)\ H(t-kp)$$

where

$$R_o(t) = 1,$$

$$R_k(t) = \int_{kp-x_o}^{t} R(u)R_{k-1}(u-p)du, \ k = 1,2,... \qquad (10)$$

Case 2: $x_o \geq p$

For this case

$$N(t) = \sum_{k=-1}^{\infty} R_k(t)H(t-kp)$$

where

$$R_{-1}(t) \equiv 1,$$

and

$$R_k(t) = \int_{kp}^{t} R(u)R_{k-1}(u-p)du, \ k = 1,2,... \qquad (11)$$

(ii) A Stochastic Model

The population is assumed to develop according to an age dependent birth and death process. There is a primary of age x_o at time $t = 0$. It and its secondaries generate a population in accordance with the following assumptions.

(i) The subpopulation generated by two coexisting individuals develop in complete independence of one another.

(ii) An individual of age x existing at time t has a probability $\lambda(x)H(x-p)dt+o(dt)$ of producing another individual of age 0 in the time interval $(t,t+dt)$ where H(.) is the Heaviside unit function.

(iii) An individual of age x existing at time t has a probability $\mu(x)dt+o(dt)$ of death in the interval $(t,t+dt)$

(iv) The birth and death rates $\lambda(x)$ and $\mu(x)$ depend only on the age x of the individual and not on t, the time of its existence.

The model described above is identical with Kendall's age dependent birth and death process (Kendall (8)) except for the assumption (ii). This assumption in turn implies that every newly born individual has a constant phase 1 of duration p. Individuals who complete this phase give birth to new offspring according to the age dependent function $\lambda(.)$.

The model is best studied by the techniques of product densities introduced by Ramakrishnan (9). Let $f_1(x|x_o;t)$ and $f_2(x,y|x_o;t)$ respectively denotes the product densities of degrees 1 and 2 so that $f_1(x|x_o;t)dx$ denotes the probability that there exists an individual of age between x and x+dx at time t. Similarly $f_2(x,y|x_o;t)dxdy$ is the probability that there are two individuals one of age between x and x+dx and the other of age between y and y+dy at time t. If N(t) and $m(t;x_o)$ respectively denote the size of the population and its mean value at time t, then it is clear that

$$m(t;x_o) = E[N(t)] = \int^t f_1(x|x_o;t)dx$$

and

$$E[N^2(t)] = \int_o^t f_1(x|x_o;t)dx + 2\int_o^t \int_o^y f_2(x,y|x_o;t)dx\, dy \tag{12}$$

The primary object of this section is to obtain an expression for m(t;x). To obtain the equation satisfied by the function $f_1(x|x_o;t)$ we use the invariant imbedding technique of Bellman et al (10). We imbed the process corresponding to the interval (0,t) into a class of processes corresponding to the interval (Δ,t). By simple probabilistic arguments we get

$$\begin{aligned} f_1(x|x_o;t) = {} & [1-\{\lambda(x_o)H(x_o-p) + \mu(x_o)\}\,\Delta]f_1(x|x_o+\Delta;\ t-\Delta) \\ & + \lambda(x_o)H(x_o-p)\ \Delta[f_1(x|x_o;t)+f_1(x|x_o;t)] \end{aligned} \tag{13}$$

The first term of the right hand side of the above equation refers to the situation that the individual at t=0 neither gives birth nor dies in the interval (0,Δ). If the individual gives birth to a new offspring then there are two individuals one of age x_o and another of age 0 at time t=0. This leads to second term of (13). Proceeding to the limit as Δ→0, equation (13), leads to the partial differential equation

$$\frac{\partial f_1(x|x_o;t)}{\partial t} - \frac{\partial f_1(x|x_o;t)}{\partial x_o} = -\mu(x_o)f_1(x|x_o;t)+\lambda(x_o)\ H(x_o-p)f_1(x|0\ ;t) \tag{14}$$

By assumption there is only one individual of age x_o at time t = 0. So the initial condition is

$$f_1(x|x_o;0) = \delta(x-x_o) \tag{15}$$

To solve (14) subject to (15) we proceed along the characteristic curve $\psi = x_o+t$ and obtain

$$\frac{\partial f_1(x|\psi-t;t)}{\partial t} + \mu(\psi-t)f_1(x|\psi-t;t) = \lambda(\psi-t)\ H(\psi-t-p)f_1(x|o;t).$$

This is only an ordinary differential equation and its solution is

$$\begin{aligned} f_1(x|\psi-t;t) = {} & \exp\left(-\int_o^t \mu(\psi-t')dt'\right)\ \left[\int_o^t \exp\left(\int_o^{t'} \mu(\psi-t'')dt''\right)\right. \\ & \left. \times\ \lambda(\psi-t')\ H(\psi-t'-p)f_1(x|o;t')dt' + F(\psi)\right] \end{aligned}$$

Using the initial condition (15) we get finally

$$f_1(x|x_0;t) = \exp(-\int_0^t \mu(x_0+t-t')dt')\ [\int_0^t \exp(\int_0^{t'} \mu(x_0+t-t'')dt'')$$

$$\times\ \lambda(x_0+t-t')\ H(x_0+t-t'-p)f_1(x|o;\ t')dt' + \delta(x_0+t-x)] \tag{16}$$

Equation (16) is a Volterra's integral equation of the third kind. Its analytic solution is difficult to obtain when $\mu(.)$ is a general function. Hence we shall choose $\mu(.)$ to be a constant say μ. Then equation (16) becomes

$$f_1(x|x_0;t) = e^{-\mu t}[\int_0^t e^{\mu t'} \lambda(x_0+ t-t')\ H(x_0+t-t'-p)f_1(x|0;t')dt'$$

$$+ \delta(x_0+t-x)] \tag{17}$$

Let

$$F_1(x|x_0;t) = f_1(x|x_0;t)e^{\mu t} \tag{18}$$

Then equation (17) simplifies to

$$F_1(x|x_0;t) = \int_0^t \lambda(x_0+t-t')\ H(x_0+t-t'-p)\ F_1(x|o;t')dt' + \delta(x_0+t-x) \tag{19}$$

To solve the equation (19) we define the following Laplace transforms

$$L(s_1,s_2) = \int_0^\infty \int_0^\infty e^{-s_1x-s_2t}\ f_1(x|0;t)\ dxdt \tag{20}$$

$$L(s_1,s_2,s_3) = \int_0^\infty \int_0^\infty \int_0^\infty e^{-s_1x-s_2t-s_3x_0}\ f_1(x|x_0;t)\ dxdtdx_0 \tag{21}$$

$$\bar{L}(s_1,s_2) = \int_0^\infty \int_0^\infty e^{-s_1x-s_2t}\ F_1(x|0;t)\ dxdt \tag{22}$$

$$\bar{L}(s_1,s_2,s_3) = \int_0^\infty \int_0^\infty \int_0^\infty e^{-s_1x-s_2t-s_3x_0}\ F_1(x|x_0;t)\ dxdtdx_0 \tag{23}$$

In view of (18) the following relations can be easily verified

$$L(s_1,s_2) = \bar{L}(s_1,s_2+\mu) \tag{24}$$

$$L(s_1,s_2,s_3) = \bar{L}(s_1,s_2+\mu,s_3) \tag{25}$$

Hence in the discussions to follow we shall derive expressions for $\bar{L}(s_1,s_2)$ and $\bar{L}(s_1,s_2,s_3)$ only. Substituting (19) in (23) we get

$$\bar{L}(s_1,s_2,s_3) = I_1+I_2 \tag{26}$$

where

$$I_1 = \int_0^\infty \int_0^\infty \int_0^\infty e^{-s_1 x - s_2 t - s_3 x_0} \left\{ \int_0^t \lambda(x_0+t-t')\, H(x_0+t-t'-p)\, F_1(x|0;t') \right\} dx\,dt\,dx_0$$

and

$$I_2 = \int_0^\infty \int_0^\infty \int_0^\infty e^{-s_1 x - s_2 t - s_3 x_0}\, \delta(x_0+t-x)\, dx\,dt\,dx_0$$

On interchanging the order of integration and after some calculations it can be shown that

$$I_1 = \bar{L}(s_1,s_2)\, \frac{[\lambda^*(s_2) - \Lambda(s_2,p) - \lambda^*(s_3) + \Lambda(s_3,p)]}{s_3 - s_2} \tag{27}$$

where $\lambda^*(s)$ is the Laplace transform of $\lambda(.)$ and

$$\Lambda(s,p) = \int_0^p e^{-sx} \lambda(x)\, dx$$

Also it is easy to see that

$$I_2 = \frac{1}{(s_1+s_3)(s_1+s_2)} \tag{28}$$

Substituting (27) and (28) in (26) we get

$$\bar{L}(s_1,s_2,s_3) = \bar{L}(s_1,s_2)\, \frac{[\lambda^*(s_2) - \Lambda(s_2,p) - \lambda^*(s_3) + \Lambda(s_3,p)]}{s_3 - s_2} + \frac{1}{(s_1+s_3)(s_1+s_2)} \tag{29}$$

From (22) and (23) it can be seen that $\bar{L}(s_1,s_2)$ = Inverse Laplace Transform of $\bar{L}(s_1,s_2,s_3)$ with respect to s_3 at $x_0 = 0$. By a well-known Tauberian theorem

$$\bar{L}(s_1,s_2) = \underset{s_3 \to \infty}{\mathrm{Lt}}\ s_3\, \bar{L}(s_1,s_2,s_3)$$

Hence from (29) we have

$$\bar{L}(s_1,s_2) = \bar{L}(s_1,s_2)\, [\lambda^*(s_2) - \Lambda(s_2,p)] + \frac{1}{(s_1+s_2)}$$

from which we obtain

$$\bar{L}(s_1,s_2) = \frac{1}{(s_1+s_2)[1 - \lambda^*(s_2) + \Lambda(s_2,p)]} \tag{30}$$

Substituting (30) in (29) we get

$$\bar{L}(s_1,s_2,s_3) = \frac{\lambda^*(s_2) - \Lambda(s_2,p) + \Lambda(s_3,p) - \lambda^*(s_3)}{(s_1+s_2)(s_3-s_2)[1-\lambda^*(s_2)+\Lambda(s_2,p)]} + \frac{1}{(s_1+s_3)(s_1+s_2)} \tag{31}$$

Finally using (30) and (31) in (24) and (25) we get

$$L(s_1,s_2) = \frac{1}{(s_1+s_2+\mu)[1-\lambda^*(s_2+\mu)+\Lambda(s_2+\mu,p)]} \tag{32}$$

$$L(s_1,s_2,s_3) = \frac{\lambda^*(s_2+\mu) - \Lambda(s_2+\mu,p) + \Lambda(s_3,p) - \lambda^*(s_3)}{(s_1+s_2+\mu)(s_3-s_2-\mu)\,[1-\lambda^*(s_2+\mu)+\Lambda(s_2+\mu,p)]}$$

$$+ \frac{1}{(s_1+s_3)(s_1+s_2+\mu)} \tag{33}$$

Expressions (32) and (33) can be inverted to obtain $f_1(x|o;t)$ and $f_1(x|x_o;t)$ respectively. But as stated earlier our goal is to obtain an expression for $m(t;x_o)$. To this end equations (32) and (33) can be used as follows.

It is easy to see from (12), (20) and (21) that

$$m^*(s_2;\, o) = \text{Laplace Transform of } m(t;o)$$

$$= L(0,s_2)$$

and

$$m^{**}(s_2,s_3) = \text{Double Laplace Transform of } m(t;x_o)$$

$$= L(0,s_2,s_3).$$

Putting $s_1 = 0$ in (32) and (33) we get

$$m^*(s_2;0) = \frac{1}{(s_2+\mu)\,[1-\lambda^*(s_2+\mu)+\Lambda(s_2+\mu,p)]} \tag{34}$$

$$m^{**}(s_2,s_3) = \frac{\lambda^*(s_2+\mu) - \Lambda(s_2+\mu,p) + \Lambda(s_3,p) - \lambda^*(s_3)}{(s_2+\mu)(s_3-s_2-\mu)\,[1-\lambda^*(s_2+\mu)+\Lambda(s_2+\mu,p)]} + \frac{1}{s_3(s_2+\mu)} \tag{35}$$

For a given function $\lambda(.)$, (34) and (35) can be inverted to obtain $m(t;0)$ and $m(t;x_o)$ respectively.

To illustrate our method we consider two special forms of $\lambda(.)$and obtain $m(t;o)$. Let $\lambda(.)=\lambda$, a constant. Then from (34) it can be shown that

$$m^*(s_2;o) = \frac{1}{s_2+\mu} \sum_{k=o}^{\infty} \frac{\lambda^k\, e^{-k(s_2+\mu)p}}{(s_2+\mu)^k}$$

So

$$m(t;o) = \sum_{k=0}^{\infty} \lambda^k L^{-1} \left[\frac{e^{-k(s_2+\mu)p}}{(s_2+\mu)^{k+1}}\right]$$

$$= e^{-\mu t} \sum_{k=0}^{\infty} \frac{\lambda^k (t-kp)^k H(t-kp)}{k!} \tag{36}$$

Result (36) is in agreement with its deterministic counterpart (equation (9)) when $\mu = 0$ and $N_o = 1$. We now let

$$\lambda(x) = 0, \quad \text{if } x < p$$

$$= \alpha(x-p)\, e^{-\lambda(x-p)}, \quad \text{if } x \geq p$$

This choice of the form of $\lambda(.)$ implies that the birth rate is zero upto time p and afterwards increases with age, reaches a maximum and then decreases. Such a form for the birth rate was observed in the egg production process in poultry (McNally (11)) and in the division process in yeast cells (Mortimer and Johnston (12)). For this choice of $\lambda(.)$ it can be shown after some algebra that (34) becomes

$$m^*(s_2;o) = \frac{1}{(s_2+\mu)[1-(\alpha(e^{-(s_2+\mu)p}/(s_2+\lambda+\mu)^2))]}$$

The inverse Laplace Transform of this expression can be shown to be

$$m(t;o) = e^{-\mu t}[1 + \sum_{k=1}^{\infty} \frac{\alpha^k}{(2k-1)} \int_0^t e^{-\lambda(u-pk)} (u-pk)^{2k-1} H(u-pk)du] \tag{37}$$

If $f_2(x,y|x_o;t)$ is known then the variance of the process can be obtained by employing equation (12). To this end, by a procedure exactly similar to the one used for obtaining (17) we arrive at the following integral equation.

$$f_2(x,y|x_o;t) = e^{-\mu t} [\int_0^t e^{\mu t'} \lambda(x_o+t-t')\, H(x_o+t-t'-p) \times$$

$$\left\{f_2(x,y)|o;t) + f_1(x|x_o;t)\, f_1(y|o;t) + f_1(x|o;t)\, f_1(y|x_o;t)\right\} dt'] \tag{38}$$

However in view of the tedious calculations involved in solving (38) (see Srinivasan and Koteswara Rao (13)) for the solution when p =0) we shall not attempt to solve it here.

REFERENCES

1. J.H.Pollard, (1973), Mathematical models of human populations, Cambridge University press, London.

2. M.Iosifescu and P.Tauto, (1973), Stochastic processes and applications in biology and medicine, Vol.2, Springer-Verlag, Berlin.

3. J.E. Van Der Plank, (1963), Plant Diseases: Epidemics and Control, Academic Press, New York.

4. N.T.J.Bailey, (1957), The mathematical theory of epidemics, Griffin, London.

5. P.R. Parthasarathy, (1979), On a modified Markov branching process, J. Math. Biol., 7: 95-97.

6. T.E.Harris,. (1963), The theory of branching processes, Springer-Verlag, Berlin.

7. S.K. Srinivasan, (1986), An age-dependent model of cavity radiation and its detection, Pramana - J. Phys., 27: 19-31.

8. D.G. Kendall, (1949), Stochastic processes and population growth, J.R. Statist. Soc., Bll; 230-264.

9. A. Ramakrishnan, (1950), Stochastic processes relating to particles distributed in a continuously infinity of states, proc.Camb. Phil. Soc. 46: 595-602.

10. R. Bellman, R.E. Kalaba and G.M. Wing, (1960), Invariant imbedding and mathematical physics-I: Particle processes, J. Math.Phys., 1: 280-308.

11. D.H. McNally, (1971), Mathematical model for poultry egg production, Biometrics, 27: 735-737.

12. R.K. Mortimer and J.R. Johnston, (1959), Life span of individual yeast cells, Nature, 183: 1751-1752.

13. S.K. Srinivasan and N.V. Koteswara Rao, (1968), Invariant imbedding technique and age dependent birth and death process, J. Math. Anal.Appl., 21: 43-52.

THE OPTIMAL INVESTMENT PROCESS IN GERMAN INDUSTRY

Horst Albach
Bonn University
Adenauerallee 24 - 42, 5300 Bonn, Germany

Abstract

This paper provides an explanation of investment behaviour of German industrial firms. Regarding the actual development of business investment, there are two main problems of interest:

(1) its decline between 1970 - 1976,

(2) its recovery, first slow between 1976 - 1984, then sudden after 1984.

The theoretical framework of our analysis is provided by the Bonn Model of Firm Development, which represents an econometric model of firm growth consisting of 27 equations dependent on each other. Firms are divided into functional sections which act independent from each other and therefore come to globally suboptimal decisions.
Data material of annual statements of accounts of 134 German industrial firms yield empirical support of our theory.

Two models of business investment are motivated, discussed and tested:
In the first model, the cost of the production process are minimized subject to a Cobb-Douglas production function with Hicks-neutral technical progress. Factor inputs react highly elastic to changes in relative factor prices. Factor stocks adjust to their optimal level with high speed. As a consequence, investment behaviour is explained by the investment rate of the previous period, weighted averages of growth rate of sales and weighted averages of growth rate of relative prices for capital goods.

Whereas user cost of capital remained relatively stable overtime, wage rates and raw material prices rose significantly. For the reason of revaluation of the deutschmark, relative prices remained rather stable, so that no major investments were undertaken. After 1984, the price indices of factor inputs developed in opposite directions in favour of capital goods prices. As a consequence, investment increased substantially.

In the second model, the intertemporal allocation of resources is optimized simultaneously. Capital and labour are treated as quasi-fix factors. The speed of adaptation is considered rather low since reaching the optimal level of factor input stocks is costly. Therefore our cost function is extended to an adaptation cost term. First, this model reveals a complementary relationship between labour and capital. Secondly, the amount of adaptation cost is crucial for the level of employment. In order to achieve full employment, conditions for equity financing have to be improved. To stress the importance of this "delicious circle" is one fundamental purpose of this paper.

A. Dedication

This paper is dedicated to Professor S.K. Srinivasan on the occasion of his 70th birthday. Professor Srinivasan has made significant contributions to the theory of optimal investment decisions, including the theory of maintenance and replacement. He has inspired many outstanding young mathematicians and mathematical economists to work in this interesting and complex field. His work and the work of his group at the Indian Institute of Technology have provided new insights in optimal replacement strategies and optimal stand-by policies. These results have stimulated research in the field of industrial investment processes in many parts of the world. The following paper is the outcome of research in the optimal investment behaviour of firms that started with an analysis of optimal investment behaviour under uncertainty, with optimal capital budgeting under uncertain cash constraints, and was greatly influenced by ideas of investing in stand-by facilities as an insurance against risk. In this paper we try to model the investment process over time with the cost of adaptation being influenced also by the institutional framework that the economy provides for the investing firm.

I gratefully acknowledge the inspiration that Professor Srinivasan has given me directly and indirectly during many years. Even though this paper may seem more like an econometric analysis than an operations research paper, the operations researcher will easily accept my thesis that his thinking permeates all analyses of optimal economic behaviour.

B. The Problem

Business investment is crucial for employment. It is therefore important to explain investment behaviour by firms in an economy. Now there are two puzzling problems in explaining business investment by German firms:

1. How could we explain the significant decline in business investment between 1970 and 1976?
2. Why was recovery of business investment so slow between 1976 and 1983?

If we can answer these two questions, we may be in position to explain why investment in plant and equipment has been bullish since 1984.

C. The Empirical Data:

The Bonn Data Bank

The empirical data used for attacking these questions are data of 295 German industrial firms in the Bonn data bank. These are all the industrial corporations in Germany that are quoted on a German stock exchange. This data bank was developed over the years with the financial assistance of the Deutsche Forschungsgemeinschaft, and this financial support is gratefully acknowledged. Out of these 295 corporations 161 had to be eliminated for various reasons, incompleteness of published data being one reason and character of a holding company with only small production activity being another. Figure 1 shows total real investment of the firms in the Bonn data bank. The decline in business investment between 1970 and 1976 is quite evident. It appears obvious also that pick-up of investment activity after 1976 was rather slow until 1984.

One may question whether 134 industrial corporations are representative in their investment behaviour of the economy as a whole. From figures 2-5 we notice that the Bonn data bank represents roughly 15% of total investment in the producing sector. The share of the producing sector in total investment was roughly 21% until 1972, declined to about 19% in the period afterwards and has never again reached the former level. It is the service sector that has increased its share of total investment in the economy. However, this has not been sufficient to match the drop of investment in the producing sector: While total investment was 24% of total gross national product until 1972, it dropped to about 21% in 1976 and has remained at this low level since.

Figure 6 shows that the Bonn data bank is fairly representative of business behaviour in the producing sector.

Figure 1: Real investment in plant and equipment - Bonn data bank; 134 German industrial firms. Prices of 1970

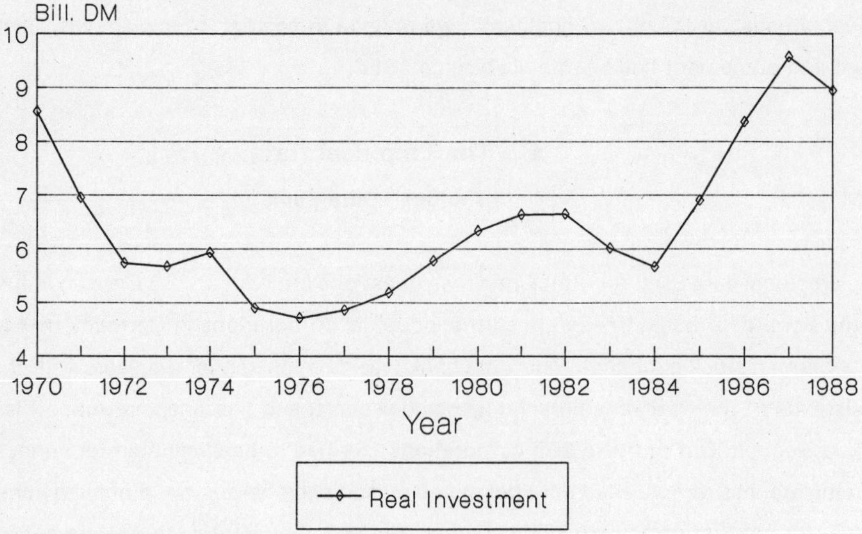

Source: Bonn data bank

Figure 2: Real Investment in housing, plant and equipment in prices of 1980 - Total economy - [Bill. DM]

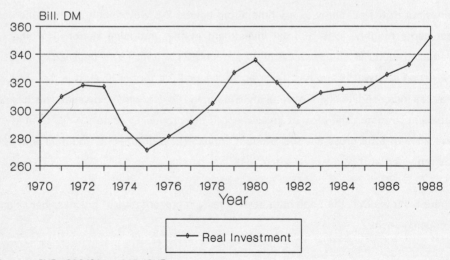

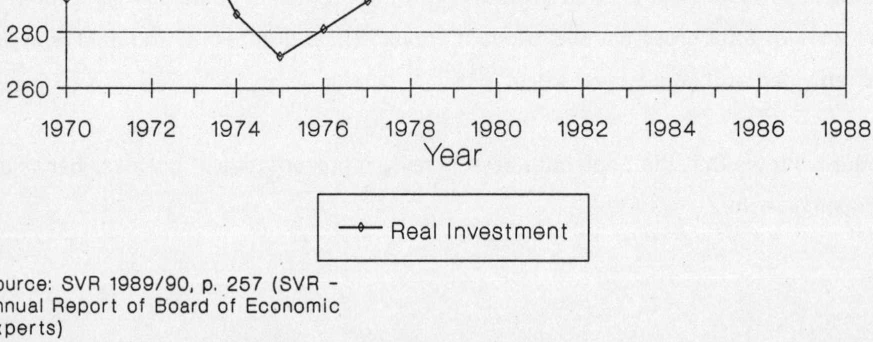

Source: SVR 1989/90, p. 257 (SVR - Annual Report of Board of Economic Experts)

Figure 3: Real Investment in housing, plant and equipment in prices of 1980 - Share of real investment to G.N.P. -

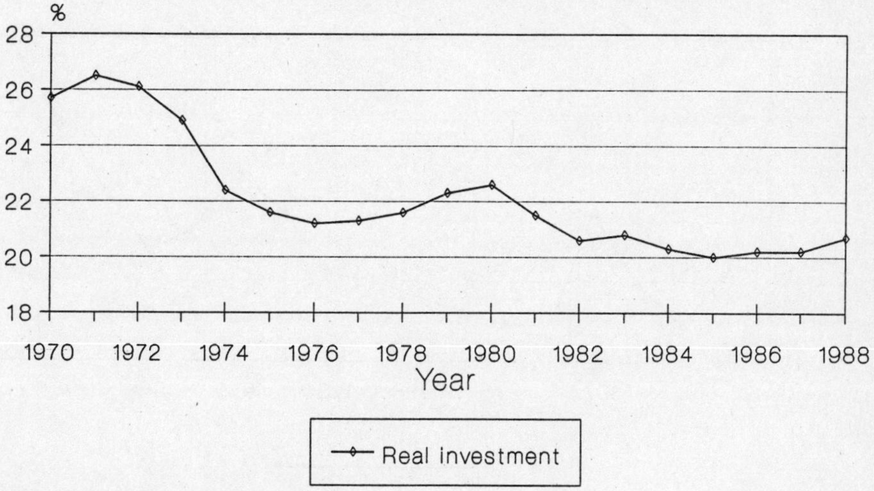

Source: SVR 1989/90, p. 257 (SVR - Annual Report of Board of Economic Experts)

Figure 4: Share of investment in the productive sector to total investment - nominal investment - in %

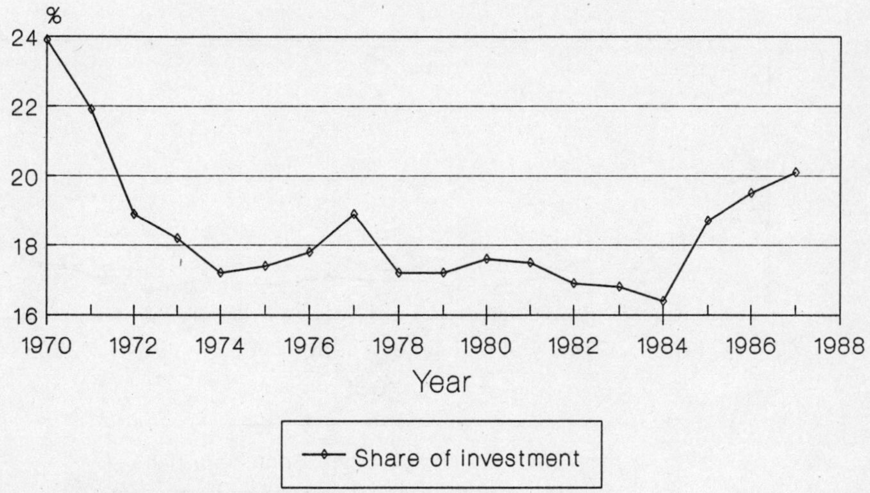

Source: SVR 1989/90, p. 260, col. 1 and 5 (SVR - Annual Report of Board of Economic Experts)

Figure 5: Share of investment of 134 industrial corporations to total producing industry in %

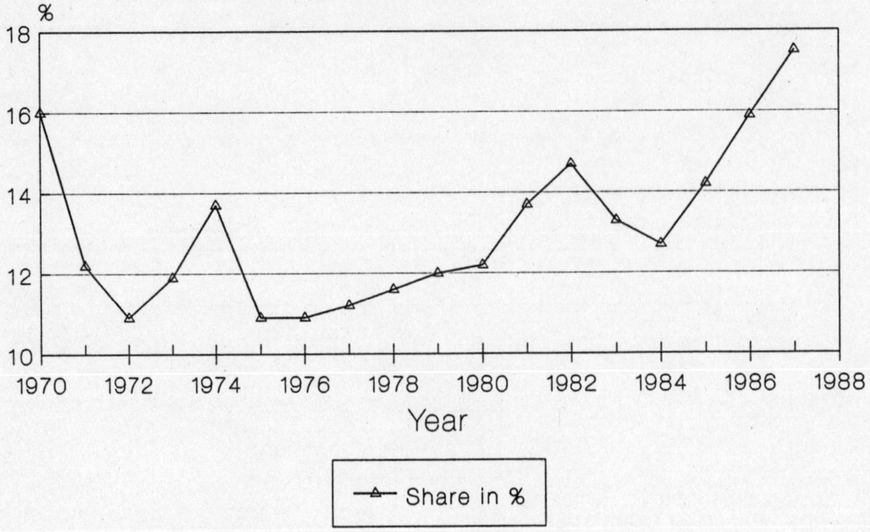

Source: Bonn data bank; SVR

Figure 6: Investment-Sales Ratio, producing sector - nominal values - Total economy versus Bonn data bank

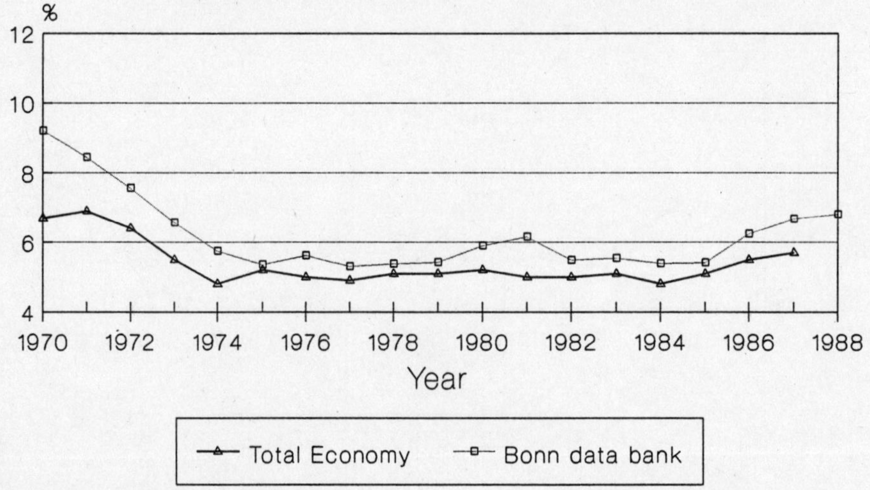

Source: SVR 1989/90, p. 260, p. 286, (SVR - Annual Report of Board of Economic Experts); Bonn data bank

D. The Business Investment Function

1. A Model of Business Investment with Separate Optimization and Adaptation

In order to explain business investment behaviour in German industry within the Bonn model, a neoclassical framework is used to determine optimal investment in plant and equipment. Various publications report on this approach and its empirical results for different branches of industry in detail (1).

If we assume that the firms react fast and efficiently to changes in relative prices, we can deduce that in selecting a production process they minimize the total cost of the productive factors needed for production subject to a level of production determined by demand. The relative factor inputs in the production process are highly elastic to changes in relative prices. Such a productive situation is described by a Cobb-Douglas production function.

Therefore we minimize

$$C = \int_0^{\infty} (qI + wL + mM)\ e^{-rt}dt \tag{1}$$

subject to

$$X = a_0 K^{a_1} \cdot L^{a_2} \cdot M^{a_3} \cdot e^{pt} \tag{2}$$

where X is planned output, K, L, and M inputs of capital, labour, and raw materials respectively, q, w, and m the unit prices of the respective inputs, C total discounted cost, p denotes Hicks-neutral technological progress, and r is the discount rate. a_1, a_2, and a_3 are parameters which are not restricted to sum to unity.

Solving for K we get

$$K^{*} = \left(\frac{1}{a_0}\right)^{\frac{1}{s}} \cdot \left(\frac{a_1}{a_2}\right)^{\frac{a_2}{s}} \cdot \left(\frac{a_1}{a_3}\right)^{\frac{a_3}{s}} \cdot X^{\frac{1}{s}} \cdot \left(\frac{w}{c}\right)^{\frac{a_2}{s}} \cdot \left(\frac{m}{c}\right)^{\frac{a_3}{s}} \cdot e^{-\frac{pt}{s}} \tag{3}$$

K^{*} is the optimal stock of capital given the relative prices for the productive factors in period t. Now if actual stock of capital is K_{t-1}, then the optimal volume of investment is $K^{*} - K_{t-1}$. The optimal stock of capital cannot be realized immediately. It takes time to plan the

investments, to order the equipment and to set it up. All of this is time-consuming and costly. In a separate optimization procedure we weigh profits lost from less than immediate realization of the optimal stock of capital against the cost of adjusting actual stock of capital to the optimal level. Profits foregone are balanced against cost of adaptation.
Assuming quadratic adaptation cost we get

$$K_t = (K_t^*)^z \cdot K_{t-1}^{1-z} \qquad 0<z<1 \tag{4}$$

with z denoting the speed of adaptation and $\frac{(1-z)}{z}$ the period of adaptation.

Since all the parameters in the investment function (3) are expected values, we have to show how these expectations are derived from observable values. We assume that expected values are the weighted averages of past and present values of the respective variables. Then we have for the rate of capacity expansion

$$\begin{aligned} \frac{I}{K} = -zb_0 &+ zb_1 \sum_{i=0}^{n_1} g_i^1 \frac{X_{-i}-X_{-i-1}}{X_{-i-1}} \\ &+ zb_2 \sum_{i=0}^{n_2} g_i^2 \frac{k_{-i}-k_{-i-1}}{k_{-i-1}} \\ &+ (1-z)\cdot \frac{I_{-1}}{K_{-1}} \end{aligned} \tag{5}$$

where

$$\frac{\dot{k}}{k} = \frac{\dot{w}}{w}\left(\frac{a_2}{a_2+a_3}\right) + \frac{\dot{m}}{m}\left(\frac{a_3}{a_2+a_3}\right) - \frac{\dot{c}}{c} \tag{6}$$

and where the user cost of capital is defined by

$$c = y \cdot q(\bar{r}+d) \tag{7}$$

$\bar{r}$ is the rate of interest on long-term debt, d the real depreciation rate, q is the price for investment goods, and y is the tax factor with y = 1 indicating neutrality of the tax system.

In order to test the investment function (5) against the empirical data of investment in plant and equipment by German industrial firms, we rewrite (3) with (4) to obtain

$$lnK_t - lnK_{t-1} = z \cdot (lnK_t^* - lnK_{t-1}) \tag{8}$$

with

$$K_t^* = h + f\left(\frac{w_t}{c_t}, \frac{m_t}{c_t}, X_t\right) \tag{9}$$

The empirical analysis yields results that are given in figure 7. They are derived from a pooled regression estimation using the data of the 134 firms for 19 years. There are therefore 2546 different observations. Despite the fact that actual and estimated figures for 1976 and 1987 point in different directions we can use the estimated investment function in order to answer the two questions.

The neoclassical explanation would argue with relative prices. Since user cost of capital and wage rates are endogenous factors of the Bonn model and since raw material prices are used on an industry basis in the analysis, an interpretation of the development of the cost components will be given first. Relative prices are dealt with subsequently.

Figure 8 shows the development of the indices of wages, raw materials prices and user cost of capital between 1970 and 1988. Wages continued to rise significantly after 1973. Raw materials prices show significant increases between 1972 and 1974 and again between 1978 and 1980 as a consequence of the oil price hikes, but their rates of increase are modest in between. User cost of capital remains stable between 1974 and 1978, obviously due to the revaluation of the deutschmark, but rises rather significantly between 1978 and 1981.

Figure 9 gives relative prices. Here we have the combined effects of wage increases after the 1973/74 oil shock and of the drop in interest rates due to the revaluation of the deutschmark. Relative wage costs would have called for heavy investment between 1974 and 1978. But since the relative prices of raw materials remained substantially unchanged for the reason of revaluation, the combined effect on k is such that no pressure on investment

Figure 7: Real investment in plant and equipment - Bonn data bank and Bonn model of firm development (BMFE)

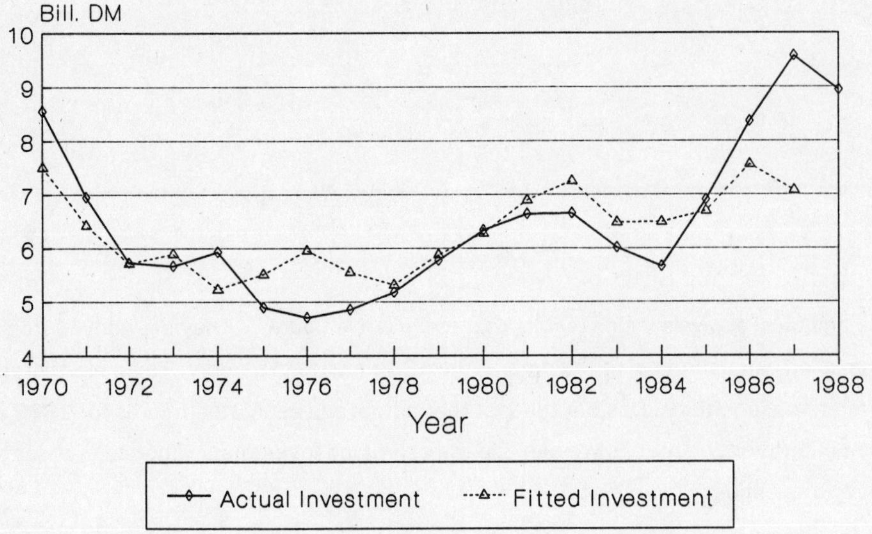

Source: Bonn data bank

Figure 8: Indices of wages, raw material prices and user cost of capital (1970 = 1)

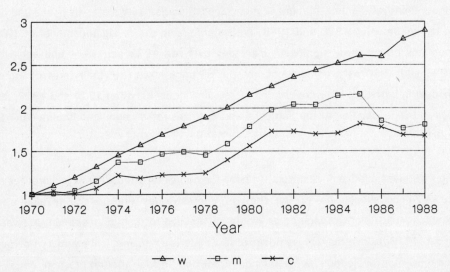

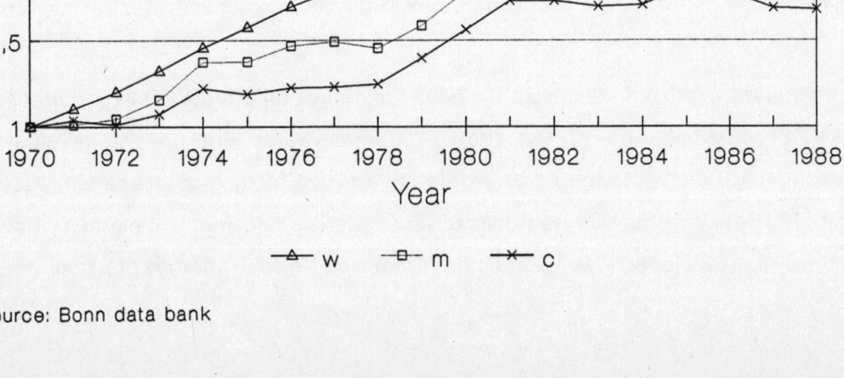

Source: Bonn data bank

Figure 9: Relative prices

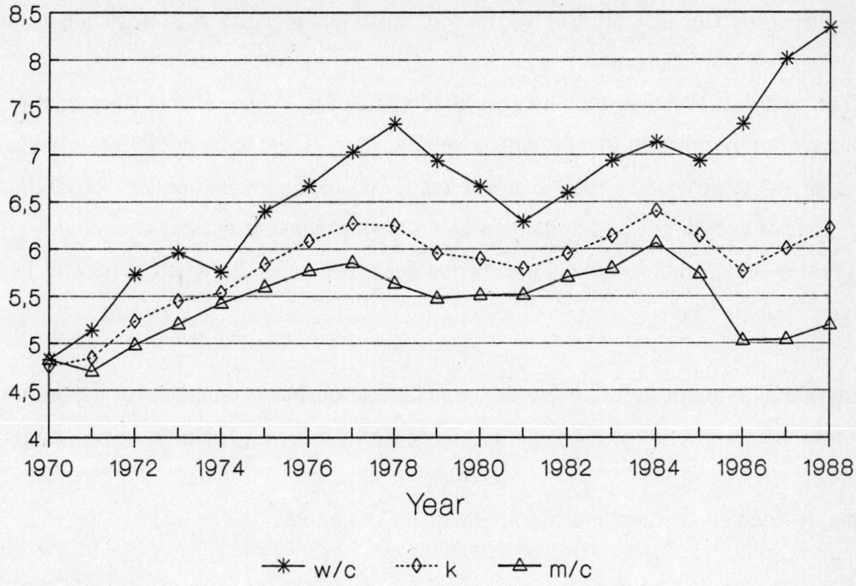

Source: Bonn data bank

was exerted. Revaluation of the deutschmark retarded investment which would have been necessary to restructure our industry, and then the user cost of capital rose so much that investment became less profitable. The turn in k came in 1981 with declining interest rates.

The second question may thus be answered as follows: While relative labor cost would have called for an increase in investment, investing abroad and thus substituting investment by material inputs provided negative incentives for domestic investment. After 1978 increasing user cost of capital put additional brakes on business investment in Germany.
The first question is better explained by decreasing demand. Between 1969 and 1975 output dropped substantially.

From the investment function we derive price and output elasticities for capital which offer further insights into investment behaviour of German firms. Table 1 shows results.

Table 1: Price elasticities and output elasticity of capital

	short-term	long-term
user cost of capital	- .108	- .175
wage rate	.043	.070
raw materials prices	.065	.105
output	.266	.431

The elasticities are all in the expected direction but very small indeed. Even in the long run (after the adaptation period has elapsed) elasticities are rather low. Since long-term elasticities are short-term elasticities divided by speed of adaptation, the speed of adaptation is seen to be rather high: between 7 and 8 months.

Such a speed of adaptation cannot explain convincingly why investment has not reacted faster to the improved conditions for business investment, in particular to the improvement in relative prices after 1981. Therefore another explanation of the adaptation process may be helpful.

2. A Model of Business Investment with Simultaneous Optimization and Adaptation

The model of optimal intertemporal allocation of resources is more suitable to explain adaptation behaviour. Results of intertemporal duality theory with two quasi-fix production factors (2) yield the following model.

We minimize

$$C = \int_0^{\infty} (mM+cK+wL+u(I^K, I^L, K, L))\, e^{-rt}dt \tag{10}$$

subject to a production technology which is described by very general properties. u is the adaptation cost function.

Raw material is a variable factor of production, and capital and labor are quasi-fix factors. Therefore adaptation of material inputs does not require time and can be made at zero cost. It is costly, however, to adapt the stocks of capital and personnel to the optimal levels. Thus from (10)

$$\min_{I>0} C = \int_0^{\infty} (v(X, K, L, I^K, I^L) + (cK+wL))\, e^{-rt}dt \tag{11}$$

The first part in the sum represents variable cost of material inputs and adaptation. The second part is the cost of the two quasi-fix factors.

Minimizing (11) yields

$$\begin{pmatrix} \dot{K} \\ \dot{L} \end{pmatrix} = \begin{pmatrix} z_{11} & z_{12} \\ z_{21} & z_{22} \end{pmatrix} \cdot \begin{pmatrix} K - K^* \\ L - L^* \end{pmatrix} \tag{12}$$

where K^* is defined as above and L^* is the optimal number of persons in the work force. It is obvious that if we treat labor as a variable factor we have z_{12}, z_{21} and z_{22} equal to zero, which yields (8) above.

The matrix of adaptation coefficients for the firms in the analysis is computed as follows

$$z = \begin{pmatrix} -.213 & +.470 \\ +.038 & -.672 \end{pmatrix}$$

This indicates that in a growth process ($K < K^*$) there is a direct positive effect on investment through z_{11} and an indirect negative effect through z_{12}. Since it takes time to increase the work force, investment is lower than would be the case without adaptation costs of personnel. In fact, the period of adaptation for capital to the desired optimal level is much longer now than was computed from the model of separate optimization and adaptation. It is 3.7 years. Conversely, in a period of down-turn ($K > K^*$) the direct effect on investment is negative but slowed down rather significantly by the high adaptation cost of the work force.

The analysis also shows that labor is complementary to capital in the long run. The net effect of innovative investment and rationalization investment on the work force has over the past 18 years been positive. Therefore it is conducive for full employment not only to improve relative prices of the input fators but also to reduce the costs of adaptation.
Organization theories are still in the beginning of attempts to specify adaptation cost. One hypothesis is that adaptation costs depend on the agency costs combined with the financial structure of the firm. The lower the coverage ratio of fixed assets, the higher the adaptation costs. This hypothesis yields the following investment function

$$\begin{pmatrix} \dot{K} \\ \dot{L} \end{pmatrix} = \begin{pmatrix} z_{11}+\beta_1 D & z_{12} \\ z_{21} & z_{22}+\beta_2 D \end{pmatrix} \begin{pmatrix} K-K^* \\ L-L^* \end{pmatrix} \tag{13}$$

with D the coverage ratio (equity divided by fixed assets).
Estimating (13) we get

$$z=\begin{pmatrix} -.221 - .067D & .650 \\ -.018 & -.539-.025D \end{pmatrix}$$

This is an interesting result. The higher the coverage ratio, the larger the volume of investment and the greater the number of additions to the work force. Our analysis thus asserts the thesis that it is mandatory to improve conditions for equity financing in order to increase investment and employment.

We can now answer the second question in a more comprehensive way. It is not only relative prices that have kept investment down during the period from 1976 to 1980 but it is also the high cost of adaptation to the optimal level of capital and to the optimal number of persons in the work force that retarded recovery in Germany after 1981. It was not until 1984 that German investment recovered from an extended period of adverse conditions and from the impact of misguided economic policy.

E. The Investment Forecast

Over the past seven years conditions for investment have improved. The relative prices are conducive to investment, and due to improved profits the equity ratio has risen. One may therefore well forecast high rates of investment in industry during the next few years. The investment function is therefore used to make an ex post forecast for investment in 1988. The scenario used for determining the exogenous factors is given in table 2.

Table 2: Scenario of the German Economy

	1988	1989
Real Growth Rate	4.0	3.5
Wage rate increase	2.5	2.5
Increase, materials prices	1.0	1.5
Increase, prices of investment goods	2.5	2.7
Interest rate	8.0	9.0
Increase, stock prices	3.0	5.0

The result is given in figure 10. The growth rate of real investment in plant and equipment is estimated at 18,2%.

The model seems slow to capture the impact of improved investment conditions after 1984. The government's deregulation policy has certainly helped reduce the period of adaptation, and a forecast which takes the whole period as a basis for estimating the matrix Z does not reflect fully the impact of the structural change in regulation policy after 1981. So the rather poor forecast for 1988 may also be considered as the necessary result of a learning process among politicians that was brought about by the neoclassical approach to understanding the difficult period from 1970 to 1981.

Industrial corporations invest in capital under favorable economic conditions. By investing they contribute to improving economic conditions. Politicians who understand this "delicious circle" will improve the wealth of their nation and that of their neighbors. Mathematical economists have contributed significantly to bringing about this understanding. But it takes time for politicians as well to adapt to better theories. A wise man like Professor Srinivasan will not expect anything but gradual progress. What matters, is progress, and he has certainly contributed to it.

Figure 10: Real investment in plant and equipment - Bonn data bank and Bonn model of firm development (BMFE)

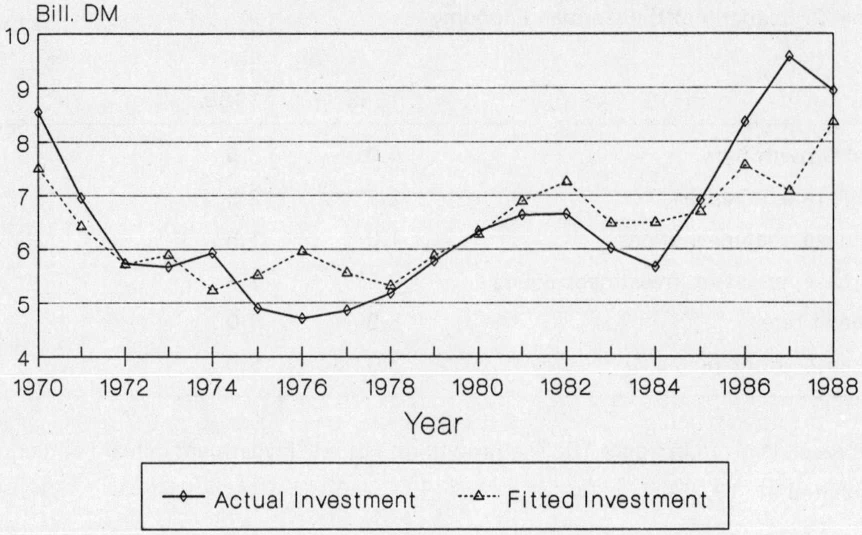

Source: Bonn data bank

References

(1) Albach, H. (1983): Investment Forecasts for German Industrial Corporations, in: Beckmann, M.J. (ed.): Mathematische Systeme in der Ökonomie, Königstein, p. 27-40.

Albach, H. (1984): The Rate of Return in German Manufacturing Industry: Measurement and Policy Implications, in: Holland, D.M.(ed.): Measuring Profitability and Capital Costs, Lexington (Mass.), p. 273-311.

Werhahn, M. (1978): Kapazität und Arbeitsplätze, Bonner Betriebswirtschaftliche Schriften, vol. 2, Bonn.

Maltzan, B.v. (1978): 'Average'- Produktionsfunktionen und Effizienzmessung über 'Frontier-Production Functions', Bonner Betriebswirtschaftliche Schriften, vol. 3, Bonn.

Geisen, B. (1979): Das Finanzierungsverhalten deutscher Industrieaktiengesellschaften, Bonner Betriebswirtschaftliche Schriften, vol. 5, Bonn.

Holzberg, B. (1980): Das Lagerverhalten industrieller Unternehmen, Bonner Betriebswirtschaftliche Schriften, vol. 6, Bonn.

Bruse, H. (1980): Der Absatzbereich von Unternehmen - Theorie, Methoden, Empirische Untersuchung, Bonner Betriebswirtschaftliche Schriften, vol. 8, Bonn.

Wanik, B. (1981): Die Lohnentwicklung deutscher Industrieaktiengesellschaften, Bonner Betriebswirtschaftliche Schriften, vol. 10, Bonn.

Schleiter, M. (1982): Steuersystem und Unternehmenspolitik - Theorie und empirische Ergebnisse zum Einfluß des Steuersystems auf die Investitionsentscheidungen deutscher Industrieaktiengesellschaften, Dissertation, Bonn.

Fischer, K.H. (1984): Die Messung von totaler Faktorproduktivität, Effizienz und technischem Fortschritt, Bonner Betriebswirtschaftliche Schriften, vol.16, Bonn.

Sieger, G. (1984): Die Finanzierung der deutschen Industrieaktiengesellschaften bei vollkommenem und unvollkommenem Kapitalmarkt, Bonner Betriebswirtschaftliche Schriften, vol. 17, Bonn.

(2) Epstein, L.G. (1981): Duality Theory and Functional Forms for Dynamic Factor Demands, Review of Economic Studies, vol. 48, p. 81-95.

Epstein, L.G. (1982): Comparative Dynamics in the Adjustment-cost Model of the Firm, Journal of Economic Theory, vol. 27, p. 77-100.

Mc Laren, K.R., Cooper, R. (1980): Intertemporal Duality: Application to the Theory of the Firm, Econometrica, vol. 48, p. 1755-1776.

Claßen, K. (1987): Determinanten der Investitionstätigkeit deutscher Unternehmen, Bonner Betriebswirtschaftliche Schriften, vol. 23, Bonn.

Fischer, K.-H. (1984): Die Messung von totaler Faktorproduktivität, Effizienz und technischem Fortschritt, Bonner Betriebswirtschaftliche Schriften, vol. 16, Bonn.

(3) See for a full description Burg, K.-H.(1987): Das Bonner Modell der Firmenentwicklung, Bonner Betriebswirtschaftliche Schriften, vol. 24, Bonn.

Albach, H., Burg, K.-H. (1986): Empirische Untersuchungen der Firmenentwicklung, in: Krelle, W. (Hrsg.), Ökonomische Prognose-, Entscheidungs- und Gleichgewichtsmodelle, Weinheim.

INCENTIVES AND REGULATION IN QUEUES

Kashi R. Balachandran
Stern School of Business
New York University
40 West 4th Street
New York, NY 10003
U.S.A.

ABSTRACT

It is known that customers seeking service in a queuing system tend to overcrowd the facility when making their individual decisions based on a consideration of the benefits they derive from the service and the cost due to waiting at the system. To obtain an optimal utilization it is necessary to restrict entry by pricing. In this paper, the effect of applying the operating cost of the service center to its users is analyzed.

In a single class customers case it is shown that when service center cost is applied, the optimal arrival rate may be higher than the individual equilibrium rate and thus a subsidy to increase the arrival rate to the system may be required. For the multi-class case the class dominance property states that for optimal utilization, the system should be dominated by a single class of customers. It is shown here that when service center costs are absorbed by the users, there exists conditions for which the class 1 users may restrict their own arrival rate in order to allow class 2 customers to utilize the facility. The conditions for such are derived for the FIFO and nonpreemptive priority rules.

1. INTRODUCTION

Firms operate service centers for the benefit of users within the firm. The users such as departments or product lines are evaluated based on their profitability. The costs of operating the service center should also be allocated to the product lines or departments in order to measure their profitability accurately. The objective of the firm is typically to maximize their total profits. The work to date on consideration of costs due to congestion at service facilities have assumed a social service center where the costs of operating the facilities are ignored. For a social service center, it has been established that customers tend to overcrowd the facility if they are allowed to make individually optimal decisions to the disregard of public welfare. The earliest work establishing this fact for a single class case was by Naor [9].

For a multi-class scenario with Poisson arrivals and exponential service, Balachandran and Schaefer [1] established pricing policies to induce optimal behavior on the part of the individual customers using a public facility. Balachandran and Schaefer [2, 3] further extended these results to queues with general service time distributions and established the class dominance properties which state that a single class will dominate usage of the service facility if all users are allowed to optimize their individual objectives. For social optimality, the facility should again be utilized by a single class only; however, the class chosen for social optimum may not be the same class that would dominate under individual optimization. Further work has been done by several authors. (For example, see Hassin [5], Lippman and Stidham [6], Mendelson [8], Mendelson and Yechiali [7] and Whang [11].)

This paper deals with a firm that operates a service center and requires that the operating costs of the center be allocated to its users. Explicit solutions for the individually optimal and firmwide optimal arrival rates are obtained for the single class case in Section 3. Section 4 deals with a multi-class case where conditions to allow for diversification among user classes for optimality are obtained.

2. NOTATION AND ASSUMPTIONS

Assume a single server queuing model with Poisson arrivals, general independent service times with finite second moment. For customers in the ith class, let λ_i be the continuously variable arrival rate to the facility, g_i be the gross reward or the value of the service to be obtained at the center, and h_i be the cost of waiting in queue for one unit of time. Let s_i be the expected time spent in service at the facility and D_i be the expected waiting time in queue prior to the start of service. The second moment of the service time of a class i customer will be denoted by $E(S_i^2)$.

Let c represent the cost of running the service facility. It does not depend on the number of customers served or their service times. In fact, we do not lose any generality with this assumption, since all directly variable costs can be included in the calculation of the benefit g_i.

For notational convenience, denote $V_i = g_i - h_i s_i$. We assume $V_i \geq 0$.

3. INDIVIDUAL EQUILIBRIUM AND FIRM OPTIMAL ARRIVAL RATES IN A SINGLE CLASS CASE

We shall consider a single class of users in this section. We omit all subscripts in the notation that refer to a class designation. Given the arrival rate is λ, the service center cost allocated to each individual user is $\frac{c}{\lambda}$. This allocation is determined at the beginning of the period and hence the diviser is λ and not the actual number of users which is not known at the beginning.

An individual customer will enter the center for service whenever there is any positive benefit to be obtained. The equilibrium arrival rate, denoted λ_e is a solution to the equation:

$$V - hD - \frac{c}{\lambda_e} = 0, \tag{1}$$

where $D = \dfrac{\lambda_e \quad E(S^2)}{2\ (1 - \lambda_e s)}$.

(Note that the required condition $\lambda_e s < 1$ is implied by (1).)

Substituting for D in (1),

$$V - \frac{h\,\lambda_e \quad E(S^2)}{2\ (1 - \lambda_e s)} - \frac{c}{\lambda_e} = 0.$$

After some easy algebraic manipulations, we get

$$\lambda_e = \frac{(V + cs) \pm [\ (V - cs)^2 - 2\ ch\ E(S^2)\ ]^{\frac{1}{2}}}{[h\ E(S^2) + 2\ Vs]} \tag{2}$$

We can show that the positive root should be chosen in (2).

In fact, consider $hD + \frac{c}{\lambda}$.

Note that $\frac{dD}{d\lambda} > 0$ *and* $\frac{d^2D}{d\lambda^2} > 0$.

The graph looks as follows:

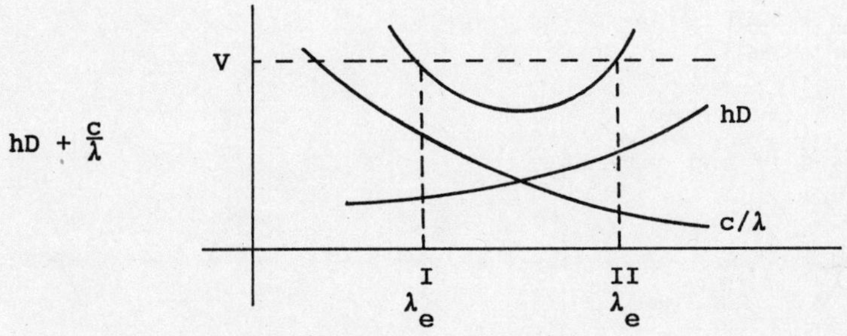

At the negative root λ_e^I, the cost portion $hD + \frac{c}{\lambda}$ is decreasing with λ and hence the expected net benefit $V - hD - \frac{c}{\lambda}$ is increasing. Hence λ_e cannot be at the negative root. At the positive root λ_e^{II}, the cost portion is increasing with λ.

Hence,

$$\lambda_e = \frac{(V + cs) + [\ (V - cs)^2 - 2\ ch\ E(S^2)\ ]^{\frac{1}{2}}}{[h\ E(S^2) + 2\ Vs]} \tag{3}$$

When the service center cost is ignored, set $c = 0$, to obtain

$$\lambda_e(0) = \lambda_e\ (c = 0) = \frac{2\ V}{[h\ E(S^2) + 2\ Vs]} . \tag{4}$$

We state Theorem 1 to describe the properties of λ_e.

<u>Theorem 1</u>: (i) The equilibrium arrival rate λ_e decreases with the service center cost c, if and only if $\lambda_e hD > c(1 - \lambda_e s)$.

(ii) For any c, $\lambda_e < \lambda_e(0)$.

<u>Proof</u>: (i) Take derivative of (1) implicitly with respect to c.

$$-h \frac{dD}{d\lambda_e} \frac{d\lambda_e}{dc} - \frac{1}{\lambda_e} + \frac{c}{\lambda_e^2} \frac{d\lambda_e}{dc} = 0.$$

$$\frac{d\lambda_e}{dc} = [-h\, \lambda_e \frac{dD}{d\lambda_e} + \frac{c}{\lambda_e}]^{-1} .$$

It follows that $\frac{d\lambda_e}{dc} < 0$ if and only if

$$\lambda_e^2 \quad \frac{dD}{d\lambda_e} > \frac{c}{h}$$

Substituting $\lambda_e^2 \frac{dD}{d\lambda_e} = \frac{\lambda_e^2 \; E(S^2)}{2(1 - \lambda_e s)^2} = \frac{\lambda_e D}{(1 - \lambda_e s)}$ and

rearranging the terms the result follows.

(ii) From (3) and (4)

$$\lambda_e = \lambda_e(0) \frac{[\,(V + cs) + [\,(V - cs)^2 - 2\,ch\,E(S^2)\,]^{\frac{1}{2}}\,]}{2V} \tag{5}$$

To prove the result, we need to have

$$(V + cs) + [\,(V - cs)^2 - 2\,ch\,E(S^2)\,]^{\frac{1}{2}} < 2\,V$$

i.e., $[\,(V - cs)^2 - 2\,ch\,E(S^2)\,]^{\frac{1}{2}} < (V - cs)$.

Squaring both sides, we need

$2\,ch\,E(S^2) > 0$, which is true.

Q.E.D.

The part (i) of Theorem 1 has a good interpretation. The left side is the total expected cost due to waiting per unit time and the right side is the expected cost of maintaining the service center while it is idle. The equilibrium arrival rate decreases with increase in c if and only if it is cheaper to keep the center idle as compared to the cost due to waiting of the customers.

We now turn to the determination of the firm optimal arrival rate, denoted by λ^*.

The sum of individual customer benefits equal the benefit to the firm of operating the service center.

The expected benefit to the firm per unit time is:

$$\pi = \lambda \left[V - hD - \frac{c}{\lambda} \right]$$

$$= \lambda \left[V - hD \right] - c.$$

For optimal λ, setting $\frac{d\pi}{d\lambda} = 0$ and substituting for D and $\frac{dD}{d\lambda}$, we get,

$$V - \frac{h \lambda^* E(S^2)}{2(1 - \lambda^* s)} - \frac{h \lambda^* E(S^2)}{2(1 - \lambda^* s)^2} = 0. \qquad (6)$$

This yields a quadratic equation in λ^*.

Noting that $\lambda^* < \frac{1}{s}$, we should take the negative root.

After some simplifications, we get,

$$\lambda^* = \frac{1}{s} - \frac{1}{s} \left[\frac{h E(S^2)}{2 Vs + h E(S^2)} \right]^{\frac{1}{2}} \qquad (7)$$

The optimal λ^* does not depend on the service center cost c. Hence the optimal arrival rate for society and firm are the same. We state the following Theorem 2 relating the equilibrium arrival rate for c = 0 and the optimal arrival rate.

<u>Theorem 2</u>:

The equilibrium arrival rate for c = 0 and the optimal arrival rate are related as follows:

$$(1 - \lambda_e (0) s) = (1 - \lambda^* s)^2. \qquad (8)$$

Further, the equilibrium arrival rate for $c = 0$ is greater than the optimal arrival rate, i.e., $\lambda_e(0) > \lambda^*$.

Proof:

(8) is easily proven by substituting from (4) and (7) for $\lambda_e(0)$ and λ^* respectively. Since $1 - \lambda^* s < 1$, it is clear that

$$(1 - \lambda_e(0)s) < (1 - \lambda^* s).$$

Hence, $\lambda_e(0) > \lambda^*$.

Q.E.D.

Theorem 2 establishes the well known result that $\lambda_e(0) > \lambda^*$ but goes further in establishing (8). (For further analysis of the relationship between $\lambda_e(0)$ and λ^* and an application of these concepts to cost control, see Balachandran and Srinidhi [4].) From Theorem 1, $\lambda_e < \lambda_e(0)$ for $c > 0$. It is possible for λ_e to be less than λ^*. We wish to obtain the value of c for which the equilibrium and optimal arrival rates are equal.

That is, to find c such that $\lambda^* = \lambda_e = \hat{\lambda}$.

The $\hat{\lambda}$ is obtained by equating the left side expressions in (1) and (6),

$$\text{i.e.,} \quad V - \frac{h\,\hat{\lambda}\,E(S^2)}{2(1 - \hat{\lambda}\,s)} - \frac{h\,\hat{\lambda}\,E(S^2)}{2(1 - \hat{\lambda}\,s)^2}$$

$$= V - \frac{h\,\hat{\lambda}\,E(S^2)}{2(1 - \hat{\lambda}\,s)} - \frac{c}{\hat{\lambda}}$$

$$\Longrightarrow \quad c = \frac{h\,\hat{\lambda}^2\,E(S^2)}{2(1 - \hat{\lambda}\,s)^2} \quad . \tag{9}$$

We can provide a scenario for which the cost of running the service center is given by (9). Suppose, the planned cost of running the center per unit time is given by:

$$c\,(\hat{\lambda}, s) = k(\hat{\lambda})\,e^{-\hat{\lambda} s}$$

where k is a function of $\hat{\lambda}$ and independent of s and $\hat{\lambda}$ is planned arrival rate. The expected service time, s is a decision variable that is chosen to minimize the total expected costs. Note that $c(\hat{\lambda}, s)$ is convex in s.

Let $c(\hat{\lambda}, s)$ be an increasing function of $\hat{\lambda}$. The costs consist of $c(\hat{\lambda}, s)$ and the expected waiting time costs.

The expected cost per unit time is:

$$F(\hat{\lambda}, s) = c(\hat{\lambda}, s) + h\,\hat{\lambda}\,D(\hat{\lambda}, s)$$

$$= k\,e^{-\hat{\lambda} s} + h\,\frac{\hat{\lambda}^2\,E(S^2)}{2(1 - \hat{\lambda} s)}$$

Setting $\frac{dF(\hat{\lambda}, s)}{ds} = 0$ for optimal s^*,

$$\frac{dF(\hat{\lambda}, s)}{ds} = -k\,\hat{\lambda}\,e^{-\hat{\lambda}s} + h\,\frac{\hat{\lambda}^3\,E(S^2)}{2(1 - \hat{\lambda}s)^2}$$

$$= -\hat{\lambda}\,c(\hat{\lambda}, s^*) + h\,\frac{\hat{\lambda}^3\,E(S^2)}{2(1 - \hat{\lambda}s^*)^2} = 0.$$

We get, $c(\hat{\lambda}, s^*) = \frac{h\,\hat{\lambda}^2\,E(S^2)}{2(1 - \hat{\lambda}\,s^*)^2}$ as given in (9).

We state the result of this development as Theorem 3.

<u>Theorem 3</u>:

Given that the planned arrival rate to the service center is $\hat{\lambda}$ and the service center operating cost is given by the function $k\,e^{-\hat{\lambda}s}$ where k is any positive constant, a determination of the expected service time that maximizes the sum of expected operating and waiting time costs and an application of the resultant costs to the users will result in optimal utilization of the service center. In particular the planned arrival rate, equilibrium arrival rate and the optimal arrival rates coincide.

In the next section, we shall analyze the case of multiple classes of arrivals.

4. CLASS BEHAVIOR IN A MULTI-CLASS CASE

In this section, we assume two classes of customers. Balachandran and Schaefer [3] established the class dominance properties in a social service center setting. According to these properties if the classes are ranked in the decreasing order of the ratio $\frac{V_i}{h_i}$, then, in the individually determined equilibrium without social interference, the facility will be dominated by class 1 to the exclusion of other classes. In the socially optimal policy, again, the facility will be dominated by a single class. However, the dominating class may not be identical to the individually dominating class. Thus to effect social optimum, it may be necessary to bar the dominating class under individual equilibrium.

The class dominance properties do not carry over to the service center setting within a firm where the cost of running the center needs to be absorbed by its users. We show that a class may not desire to dominate the facility. The class may restrict its own entry to allow customers from other classes to obtain service. We obtain conditions for this phenomena for two cases: one, in which the customers are served on a first-come-first-served basis and second, where the class 1 customers are given a nonpreemptive priority over the class 2 customers. For firm optimality, we prove that the policy calls for dominance of the center by a single class.

Assume FIFO order of service. The expected delay in queue is given by:

$$D = \frac{\lambda_1 E(S_1^2) + \lambda_2 E(S_2^2)}{2(1 - \lambda_1 s_1 - \lambda_2 s_2)} .$$

The expected return per unit time to class 1 customers from obtaining service is:

$$R_1 = \lambda_1 \; [V_1 - h_1 D)] - \frac{c\lambda_1}{\lambda_1 + \lambda_2} . \tag{10}$$

where $\frac{c\lambda_1}{\lambda_1 + \lambda_2}$ is the absorption of the center's cost by class 1.

Taking the derivative of (10) and setting it to 0, the optimal λ_1^{**}, should satisfy:

$$(V_1 - h_1 D) - h_1 \frac{dD}{d\lambda_1} - \frac{(\lambda_1^{**} + \lambda_2) c - c\lambda_1^{**}}{(\lambda_1^{**} + \lambda_2)^2} = 0. \qquad (11)$$

The last term in (11) is zero if and only if $\lambda_2 = 0$. Comparing (11) and (6), it is clear that $\lambda_1^{**} < \lambda^*$ provided $\lambda_2 > 0$. We wish to show that the to show that the optimal value for λ_2 as desired by class 1 is indeed positive under a certain condition. We can prove this by showing that:

$$\frac{dR_1}{d\lambda_2} > 0 \quad \text{evaluated at } \lambda_1 = \lambda_1^{**} \text{ and } \lambda_2 = 0$$

$$\frac{dR_1}{d\lambda_2} > 0 \quad \Longleftrightarrow \quad \frac{c\lambda_1}{(\lambda_1 + \lambda_2)^2} > \lambda_1 h_1 \frac{dD}{d\lambda_2}$$

$$\Longleftrightarrow \quad \frac{c}{h_1 (\lambda_1 + \lambda_2)^2} > \frac{E(S_2^2) - \lambda_1 E(S_2^2) s_1 + \lambda_1 s_2 E(S_1^2)}{2 (1 - \lambda_1 s_1 - \lambda_2 s_2)^2}$$

$$\Longleftrightarrow \quad \frac{c}{h_1 (\lambda_1 + \lambda_2)^2} > \frac{s_2}{(1 - \lambda_1 S_1 - \lambda_2 s_2)} [s_{2_e} + D]$$

where $s_{2_e} = \dfrac{E(S_2^2)}{2 s_2}$, the equilibrium excess.

λ_1^{**} evaluated at $\lambda_2 = 0$ is equal to the expression given in (7), i.e.,

$$\lambda_1^* = \lambda_1^{**} \Big|_{\lambda_2 = 0} = \frac{1}{s_1} - \frac{1}{s_1} \left[\frac{h_1 E(S_1^2)}{2V_1 s_1 + h_1 E(S_1)^2} \right]^{\frac{1}{2}} \qquad (12)$$

$$\left.\frac{dR_1}{d\lambda_2}\right|_{\lambda_2 = 0} > 0$$

$$\Longleftrightarrow \frac{c}{\lambda_1^*} > h_1 \lambda_1^* s_2 \left[\frac{s_2 e + D}{1 - \lambda_1^* s_1} \right].$$

The result we have proven is stated as the following theorem.

Theorem 4:

With first-in-first-out order of service, the existence of a second class of customers will make the class 1 reduce its optimal arrival rate (when there is no second class of customers) to a new optimal arrival rate that is lower and allows class 2 customers to obtain service provided the following condition is met:

$$\frac{c}{\lambda_1^*} > h_1 \left[\lambda_1^* s_2 \frac{s_2 e}{(1 - \lambda_1^* s_1)} + \lambda_1^* D \frac{s_2}{(1 - \lambda_1^* s_1)} \right]. \qquad (13)$$

To interpret (13) note that in the right side the first term is the probability of a class 1 customer arriving during the service of a class 2 customer multiplied by the remaining service time and the ensuing busy period consisting of all class 1 customers. The second term multiplies the probability of class 1 arrival during the waiting time of a class 2 customer already in queue multiplied by the expected busy period caused by this additional class 2 customer. Multiplying this by h_1 gives the cost to class 1 customer caused by allowing a class 2 customer to enter. On the left side is the service center cost application per class 1 customer when there are no class 2 customers. As long as this is larger than the right side, the class 1 customers would allow class 2 customers to obtain service.

We have established that given the existence of multiple classes, class 1 will allow class 2 to obtain service provided (13) holds. Will class 2 seek the service under these circumstances allowing the arrival rate of class 1 to be λ_1^{**}? Given class 1 is using the facility at a rate of λ_1^{**}, class 2 will be willing to enter provided their expected net benefit evaluated at λ_1^{**} and $\lambda_2 = 0$ is positive.

$$\text{i.e.,} \quad V_2 - h_2 \frac{\lambda_1^{**} E(S_1^2)}{2(1 - \lambda_1^{**} s_1)} - \frac{c}{\lambda_1^{**}} > 0. \tag{14}$$

where we have set $\lambda_2 = 0$ at $\lambda_1^{**} = \lambda_1^{*}$ given by (7).

If class 1 is given a nonpreemptive priority over class 2, we state the Theorem 5 that gives the necessary condition for class 1 to allow class 2 to obtain service.

<u>Theorem 5</u>:

With a nonpreemptive priority scheme where class 1 has a higher order of service, class 1 will reduce its optimal rate (when there is no second class of customers) to a new optimal rate that is lower and allows class 2 to obtain service provided the following condition is met:

$$\frac{c}{\lambda_1^{*}} > h_1 \lambda_1^{*} s_2 \frac{s_{2e}}{(1 - \lambda_1^{*} s_1)} \tag{15}$$

<u>Proof</u>: Note that

$$R_1 = \lambda_1 [V_1 - h_1 D_1]$$

where

$$D_1 = \frac{\lambda_1 E(S_1^2) + \lambda_2 E(S_2^2)}{2(1 - \lambda_1 s_1)} .$$

Again the optimal arrival rate λ_1^{**} evaluated at $\lambda_2 = 0$ is λ_1^{*} given in (7).

$$\left.\frac{dR_1}{d\lambda_2}\right|_{\lambda_2 = 0} > 0$$

$$\Longleftrightarrow \quad \frac{c}{\lambda^*} > h_1 \lambda_1^* s_2 \frac{s_{2e}}{(1 - \lambda_1 s_1)}$$

follows easily. Q.E.D.

The interpretation of (15) is very similar to (13).

We now turn to the determination of firm optimal arrival rates.

For the FIFO case, the expected net return per unit time to the firm is given by the following:

$$\pi_F = \sum_{i=1}^{2} \lambda_i (V_i - h_i D) - c$$

If optimal λ_1 and λ_2 are chosen after the service center is designed (i.e., after the cost c is determined) then c is not a function of λ_1 and λ_2.

In this case, the problem reduces to the one analyzed by Balachandran and Schaefer [3] and the optimal solution calls for domination by one class excluding the other class altogether.

If class 1 is the optimally dominating class, using a non-preemptive priority for class 1 is superior to excluding the class 2 from service altogether. This requires a condition on the parameters to be satisfied but is independent of c.

To prove this assertion, note that the expected return to the firm using the non-preemptive priority discipline is:

$$\pi = \lambda_1 \left[V_1 - h_1 \frac{\lambda_1 E(S_1^2) + \lambda_2 E(S_2^2)}{2(1 - \lambda_1 s_1)} \right]$$

$$+ \lambda_2 \left[V_2 - h_2 \frac{\lambda_1 E(S_1^2) + \lambda_2 E(S_2^2)}{2(1 - \lambda_1 s_1)(1 - \lambda_1 s_1 - \lambda_2 s_2)} \right] - c.$$

Taking the derivative of π with respect to λ_1, setting it to zero and evaluating at $\lambda_2 = 0$, the value of λ_1 can be seen to be identical to the optimal λ_1 under the FIFO discipline. (This is also the same expression as given in (7).)

Our task is to obtain conditions under which the optimal λ_2 will not be zero. This is shown by proving that the derivative of Đ with respect to λ_2 evaluated at $\lambda_2 = 0$ is positive.

First observe that

$$\frac{d\pi}{d\lambda_1}\bigg|_{\lambda_2 = 0} = V_1 - \frac{h_1\ \lambda_1\ E(S_1^2)}{2(1 - \lambda_1\ s_1)} - \frac{h_1\ \lambda_1\ E(S_1^2)}{2(1 - \lambda_1\ s_1)^2} \tag{16}$$

setting this to 0 and solving we obtain λ_1 to be equal to (7).

Now,

$$\frac{d\pi}{d\lambda_1}\bigg|_{\lambda_2 = 0} = V_2 - \frac{h_2\ \lambda_1\ E(S_2^1)}{2(1 - \lambda_1\ s_1)^2} - \frac{h_1\ \lambda_1\ E(S_2^2)}{2(1 - \lambda_1\ s_1)}$$

$$= V_2 - \frac{h_2}{h_1}\left[V_1 - \frac{h_1\ \lambda_1\ E(S_1^2)}{2(1 - \lambda_1\ s_1)} - \frac{h_1 \lambda_1\ E(S_2^2)}{2(1 - \lambda_1\ s_1)} \right.$$

using (16)

It follows that

$$\frac{d\pi}{d\lambda_1}\bigg|_{\lambda_2 = 0} > 0 \textit{ if and only if}$$

$$\left[V_2 - \frac{h_1\ \lambda_1\ E(S_2^2)}{2(1 - \lambda_1\ s_1)} \right] / h_2 > \left[V_1 - \frac{\lambda_1\ h_1\ E(S_1^2)}{2(1 - \lambda_1\ s_1)} \right] / h_1 .$$

The left side is the ratio of the net of value of serving a class 2 customer paying for the delay to class 1 of experiencing an additional busy period caused by a class 2 customer in service to the delay cost of a class 2 customer. This should be greater than the ratio of the net value to class 1 being served alone to its delay cost. (Radhakrishnan [10] analyzed this assertion for the special case where the classes have the same service time distribution.)

If c is dependent on the arrival rates (i.e., these firm-wide optimal rates are determined while designing the service system), the diversification is more apparent.

REFERENCES

1. Balachandran, K. R. and M. E. Schaefer, "Regulation by Price of Arrivals to a Congested Facility", cahiers du C.E.R.O., Vol. 21, No. 2, pp. 149-154, 1979.

2. Balachandran, K. R. and M. E. Schaefer, "Public and Private Optimization at a Service Facility with Approximate Information on Congestion", European Journal of Operational Research, Vol. 4, pp. 195-202, 1980.

3. Balachandran, K. R. and M. E. Schaefer, "Class Dominance Characteristics at a Service Facility", Econometrica, Vol. 47, No. 2, pp. 515-519, 1979.

4. Balachandran, K. R. and B. N. Srinidhi, "A Rationale for Fixed Charge Application", Journal of Accounting, Auditing and Finance, Vol. 2, no. 2, pp. 151-183, 1987.

5. Hassin, R., "Consumer Information in Markets with Random Product Quality: The Case of Queues and Balking", Econometrica, Vol. 54, No. 5, pp. 1185-1195, 1986.

6. Lippman, S. and S. Stidham, "Individual Versus Social Optimization in Exponential Congestion Systems", Operations Research, Vol. 25, pp. 233-247, 1977.

7. Mendelson, H. and U. Yechiali, "Controlling the GI/M/1 Queue by Conditional Acceptance of Customers", European Journal of Operational Research, Vol. 7, pp. 77-85, 1981.

8. Mendelson, H., "Pricing Computer Services: Queuing Effects", Communications of the ACM, Vol. 28, No. 3, pp. 312-321, 1985.

9. Naor, P., "On the Regulation of Queue Size by Levying Tolls", Econometrica, Vol. 37, No. 1, pp. 15-24, 1969.

10. Radhakrishnan, S., "The Relevance of Cost Application in a Common Facility", Ph.D. Thesis, New York University, 1990.

11. Whang, S., "The Value of System-State Information in Computer Systems", Unpublished Paper, University of Rochester, 1986.

TWO MODELS OF BRAND SWITCHING

Martin J. Beckmann
Department of Economics
Brown University
Providence, RI 02912
U.S.A.

ABSTRACT

It is well known that brand choice can be described by a Markov chain. In order to put some structure into the transition probabilities we model brand choice as a two-stage decision process: (1) whether to continue or whether to reconsider the last choice, (2) in the latter case which brand to choose. A distinction must then be made as to whether the last brand is ruled out (hypothesis II) or not (hypothesis I) giving rise to two different probability models. In the case of only two brands, the transition probabilities can always be modelled in either way.

The following problems are considered. How to test the existence of choice probabilities? How in the ergodic case the state probabilities, i.e., the long-run market shares are determined? What are the implications of zero-one probabilities of retention or choice? Under what conditions are market shares equalized? It is also suggested that the retention probabilities depend on product attributes while the choice probabilities respond to advertising.

INTRODUCTION

Brand switching has been described traditionally in a straightforward way by Markov chains (Bass 1961, Montgomery and Urban 1969, Schmalensee 1972).

In order to put some structure into the transition probabilities it has been suggested (Schmalensee 1972, p. 106, equation (4.7)) that brand loyalty or switching involves two-stage decision process: 1) whether to stick to the same brand, and if not, 2) which new brand to choose? There are two basic possibilities here, whether the old brand will be ruled out or not. While Schmalensee assumed the probabilities of brand loyalty to be the same for all brands, we allow it to depend on the brand, since it is after all an indication of consumer satisfaction with the product.

When the last brand may be chosen again, we have Hypothesis I:

$$p_{ii} = r_i + (1 - r_i)p_i \tag{1}$$

$$p_{ij} = (1 - r_i)p_j, \qquad j \neq i \tag{2}$$

Here p_{ij} denotes the probability of buying brand j in the next period when brand i is purchased in this period. r_i is the probability that no new choice is made. Notice the fundamental assumption that the probability p_j of selecting brand j is independent of the last brand used.

When the old brand is ruled out we have Hypothesis II:

$$p_{ii} = r_i \tag{3}$$

$$p_{ij} = (1 - r_i)\frac{p_j}{1 - p_i}, \qquad i \neq j, \qquad p_i < 1 \tag{4}$$

TESTS

While under Hypothesis II the probabilities of choosing j depend on i, the proportions

$$\frac{p_{ij}}{p_{ik}}$$

do not. A test for the existence of probabilities p_j in (2) or (4) is in fact

$$\frac{p_{ij}}{p_{ik}} = \frac{p_j}{p_k} \text{ for all } i \neq j, k \tag{5}$$

In particular, let

$$\frac{p_{ij}}{p_{i1}} = \frac{p_j}{p_1} = \lambda_j \qquad \text{(say)} \tag{6}$$

implying

$$\frac{p_{ij}}{p_{ik}} = \frac{\lambda_j}{\lambda_k} \tag{7}$$

The least squares estimate of λ_j is in fact

$$\lambda_j = \frac{1}{n-2} \sum_{i \neq 1} \frac{p_{ij}}{p_{i1}} \tag{8}$$

Another test for assumption (2) is

$$\frac{p_{ik}}{p_{jk}} = \frac{1 - r_i}{1 - r_j} \qquad \text{for all } k \neq i, j \tag{9}$$

In particular

$$\frac{p_{ik}}{p_{1k}} = \frac{1 - r_i}{1 - r_1} = \mu_i \quad \text{(say)} \tag{10}$$

implying

$$\frac{p_{ik}}{p_{jk}} = \frac{\mu_i}{\mu_j} \tag{11}$$

For assumption (4) the test is

$$\frac{p_{ik}}{p_{jk}} = \frac{(1 - r_i)(1 - p_j)}{(1 - p_i)(1 - r_j)} \tag{12}$$

In particular, let

$$\frac{p_{ik}}{p_{1k}} = v_i \tag{13}$$

then

$$\frac{p_{ik}}{p_{jk}} = \frac{v_i}{v_j} \tag{14}$$

Comparing (11) and (14) and recalling (7) it is seen that the tests for the two hypotheses are the same. To distinguish between them we must use estimates of r_i and p_j. This will not be pursued here.

TWO AND THREE BRANDS

For n = 2, hypothesis II, i.e., equation (4) is always satisfied since

$$p_{ij} = (1 - r_i)\frac{p_j}{1 - p_i} = 1 - r_i \; , \; j \neq i \quad \text{(say)} \tag{15}$$

In fact, no probabilities p_j are needed, only the probabilities r_i of retaining a brand.

For hypothesis I, i.e., equation (2) three parameters r_1, r_2, p_1 are available to satisfy two equations

$$p_{12} = (1 - r_1)\, p_2 = (1 - r_1)(1 - p_1) \tag{16}$$

$$p_{21} = (1 - r_2)\, p_1 \tag{17}$$

It is, therefore, also satisfied. Now Occam's razor suggests Hypothesis II which uses only two numbers r_1 , r_2.

For n = 3, we have five parameters r_1, r_2, r_3, p_1, p_2 to satisfy six independent equations for p_{ij}, i = 1, 2, 3; j = 1, 2. This means that one can expect to get a reasonably good fit almost always.

Our interpretation of the p_j is that they depend on advertising whereas the r_i should depend on consumer's satisfaction with the product, i.e., on product quality. $p_j \gtrless r_j$ then means that advertising is $\left\{\begin{matrix}\text{more}\\ \text{less}\end{matrix}\right\}$ persuasive than product quality turns out to be.

STEADY STATE

Suppose that all r_i are in the range $0<r_i<1$ and that all p_j are positive: $p_j > 0$. Then the transition probabilities (1), (2), (3), (4) are all positive. They define an ergodic Markov chain whose steady-state probabilities are all positive and independent of initial states. Hypothesis I implies

$$\pi_j = \pi_j r_j + \sum_i \pi_i (1 - r_i) p_j \tag{18}$$

Letting

$$\sum_i \pi_i (1 - r_i) = \lambda$$

we have

$$\pi_j = \frac{\lambda p_j}{1 - r_j}$$

or

$$\pi_i = \frac{p_i/(1 - r_i)}{\sum_j p_j/(1-r_j)} \tag{19}$$

The state probabilities are proportional to the choice probabilities p_i and inversely proportional to the disloyalty probabilities $1- r_i$.

Hypothesis II implies

$$\pi_j = \pi_j r_j + \sum_{i \neq j} \frac{\pi_i (1 - r_i) p_j}{1 - p_j}$$

$$= \pi_j r_j - \pi_j \frac{(1 - r_j) p_j}{1 - p_j} + \sum_{i=1}^{n} \frac{\pi_i (1 - r_i) p_j}{1 - p_i} \tag{20}$$

Write

$$\mu = \sum_{i=1}^{n} \frac{\pi_i (1-r_i)}{1 - p_i}$$

$$\pi_j = \frac{\mu p_j}{1-r_j + \dfrac{1-r_j}{1-p_j} p_j} = \mu \frac{p_j (1 - p_j)}{1 - r_j}$$

$$\pi_i = \frac{\frac{p_i(1-p_i)}{1-r_i}}{\sum_{j=1}^{n} \frac{p_j(1-p_j)}{1-r_j}} \tag{21}$$

Once more, the state probabilities are inversely proportional to the disloyalty probabilities $1-r_j$, but their dependence on the choice probabilities p_i is quadratic rather than proportional.

From equations (19) and (21) it is seen that the ergodic assumption requires merely

$$r_i<1, \quad i=1,\dots n$$

with $r_i=0$ admitted. In the latter case under hypothesis II

$$p_{ii} = 0$$

$$p_{ii}^{(2)} = \sum_{j\neq i} p_{ij}p_{ji} = \sum_{j\neq i} p_j(1-r_j)p_i>0$$

and under hypothesis I

$$p_{ii}= p_i>0$$

As equation (21) shows, hypothesis II gives rise to a parradoxical result. When $p_i> \frac{1}{2}$, π_i is a decreasing function of p_i. Of course, when the number of brands is large, $p_j> \frac{1}{2}$ is unlikely for any j. Still it is curious that a choice probability $p_j= \frac{1}{2}$ should maximize that brand's share. The reason is that a large choice probability p_i raises the transition probabilities p_{ij}, $j\neq i$ by decreasing the denominator in (4).

UNIFORM BRAND LOYALTY

We now consider various simplifying assumptions. Suppose first that $r_i \equiv r$ as assumed by Schmalensee. Then for Hypothesis I

$$p_{ii}= r + (1 - r)\, p_i \tag{22}$$

$$p_{ij}= (1 - r)\, p_j\, , \qquad i\neq j \tag{23}$$

It follows that

$$\frac{p_{ii}}{p_{ij}} > \frac{p_i}{p_j}$$

while

$$\frac{p_{ki}}{p_{kj}} = \frac{p_i}{p_j} \quad i \neq j, \quad j \neq k, \quad k \neq i$$

For Hypothesis II

$$p_{ii} = r \tag{24}$$

$$p_{ij} = \frac{(1 - r)p_j}{1 - p_i}, \qquad i \neq j \tag{25}$$

Under hypothesis I

$$\pi_i = p_i \tag{26}$$

while under hypothesis II

$$\pi_i = \frac{p_i(1 - p_i)}{\sum_{j=1}^{n} p_j(1 - p_j)} = \frac{p_i - p_i^2}{1 - \sum_{j=1}^{n} p_j^2} \tag{27}$$

Thus when $r_i \equiv r$, this probability of retention has no effect on the steady-state. Hypothesis I implies that the steady-state probabilities π_i equal the choice probabilities p_i, while Hypothesis II means that

$$\pi_i \sim p_i(1 - p_i)$$

ZERO-ONE CHOICE PROBABILITIES

Suppose that

$$p_1 = 1; \quad p_j = 0, \quad j = 2, \ldots, n \tag{28}$$

Then under hypothesis I

$$\pi_1 = r_1\pi_1 + \sum_{i=1}^{n} (1 - r_i)\pi_i = \pi_1 + \sum_{j=2}^{n} (1 - r_j)\pi_j$$

$$\pi_i = r_i\pi_i, \qquad i = 2, \ldots, n$$

whose unique solution is

$$\pi_1 = 1; \ \pi_i = 0, \ i = 2, \ldots, n \tag{29}$$

All brands but one are extinguished.

Consider hypothesis II. Let

$$p_1 = 1 - (n-1)\varepsilon, \qquad p_j = \varepsilon, \qquad j = 2, \ldots, n \tag{30}$$

Now

$$p_{11} = r_1 p_{1j} = \frac{(1 - r_1)\varepsilon}{(n - 1)\varepsilon} = \frac{1 - r_1}{n - 1}, \qquad j \neq 1$$

$$p_{ii} = r_i (p_{ij} = \frac{(1 - r_i)\varepsilon}{1 - \varepsilon}), \quad i, \quad j \neq 1; \qquad p_{i1} = 1 - r_i, \qquad i \neq 1$$

As $\varepsilon \to 0$

$$p_{ii} = r_i, \quad i = 1, \dots, n \tag{31}$$

$$p_{1j} = \frac{1 - r_1}{n - 1}, \qquad j \neq 1 \tag{32}$$

$$p_{ij} = 0, \ i \neq 1, \quad j \neq 1 \tag{33}$$

$$P = \begin{bmatrix} r_1 & \frac{1-r_i}{n-1} & \dots & \frac{1-r_i}{n-1} \\ 1-r_2 & r_2 & \dots & 0 \\ \vdots & 0 & \ddots & \vdots \\ 1-r_n & 0 & \dots & r_n \end{bmatrix} \tag{34}$$

From this

$$\pi_i = \pi_1 \frac{1 - r_1}{n - 1} + \pi_i r_i$$

It follows that

$$\pi_i = \frac{1 - r_1}{1 - r_i} \frac{\pi_1}{n - 1} > 0 \tag{35}$$

where π_1 is determined by the condition

$$1 = \pi_1 + \frac{\pi_1(1 - r_1)}{n - 1} \sum_{i=2}^{n} \frac{1}{1 - r_i}$$

$$\pi_1 = \frac{1}{1 + \frac{1-r_1}{n-1} \sum_{i=2}^{n} \frac{1}{1-r_i}} \tag{36}$$

In this case the shares of brands $i \neq 1$ do not fall to zero. Hypothesis II leads to results different from hypothesis I.

TOTAL BRAND LOYALTY

Suppose next that $p_j > 0$ for all j and that

$$r_1 = 1,\ r_i < 1,\ i = 2,\ldots,n\ . \tag{37}$$

Then under both hypotheses

$$P = \begin{bmatrix} 1 & 0 & \ldots & 0 \\ p_{21} & \ldots & & p_{2n} \\ \vdots & & & \vdots \\ p_{n1} & & & p_{nn} \end{bmatrix} \tag{38}$$

Let $\pi_i^{(n)}$ denote brand i's market share in the nth period. Assume that $p_{i1}>0$ for $i = 2,\ldots,n$, and $\pi_1^{(n)} < 1$ then

$$\pi_1^{(n+1)} = \pi_1^{(n)} + \sum_{i=2}^{n} \pi_i^{(n)} p_{i1} > \pi_1^{(n)} \tag{39}$$

so that the market share of brand 1 keeps increasing. The unique solution of

$$\pi' P = \pi'$$

is

$$\pi_1 = 1; \quad \pi_i = 0, \qquad i=2,\ldots,n$$

with brand 1 once more conquering the whole market in the steady-state.

Assume now that both $r_1=1$ and $r_2=1$. In a two-brand system the initial market shares are then preserved. When $n > 2$ only brands 1 and 2 will survive with shares π_i depending on the initial share distribution $\pi_i^{(o)}$.

EQUAL CHOICE PROBABILITIES

Consider next that all selection probabilities are the same $p_j = \frac{1}{n}$. Hypothesis I states

$$p_{ii} = r_i + (1 - r_i)\,\frac{1}{n}$$

$$p_{ij} = (1 - r_i)\,\frac{1}{n}, \qquad i \neq j$$

yielding

$$\pi_i = \frac{\dfrac{1}{1 - r_i}}{\sum_{j=1}^{n} \dfrac{1}{1 - r_j}} \tag{40}$$

The state probabilities are inversely proportional to the disloyalty probabilities.

Under hypothesis II

$$P_{ii} = r_i p_{ij} = \frac{(1 - r_i)}{n - 1}, \quad j \neq i$$

Once more (40) results so that in this case Hypotheses I and II are indistinguishable in the long run.

In addition, let now the probabilities of retention $r_i = r$ be the same for all i. Then (40) shows that all market shares must become equal.

$$\pi_i = \frac{1}{n} \tag{41}$$

PROPORTIONALITY OF CHOICE AND RETENTION

Next suppose that for all i, p_i and $1 - r_i$ are proportional.

$$p_i = \lambda\ (1 - r_i) \tag{42}$$

Under hypothesis I it follows that once more

$$\pi_i = \frac{1}{n}$$

and all brands have equal market shares. Under Hypothesis II

$$\pi_i = \frac{1 - p_i}{\sum_{j=1}^{n} (1 - p_j)} = \frac{1 - p_i}{n - 1} \tag{43}$$

Paradoxically, equilibrium markets share is now a decreasing function of selection probability p_i. But recall that this choice probability was proportional to the rejection probability.

Finally, assume that

$$1 - p_i = \lambda(1 - r_i) \tag{44}$$

The probability of not choosing brand i is proportional to the probability of dropping it which would be more realistic than assumption (42).

Hypothesis I implies

$$\pi_i = \frac{\dfrac{p_i}{1 - r_i}}{\sum_j \dfrac{p_j}{1 - r_j}} = \frac{\dfrac{1}{1 - r_i} - \lambda}{\sum_j \dfrac{1}{1 - r_j} - n\lambda}$$

Now λ is determined by

$$\sum_{j=1}^{n} (1 - p_j) = \lambda \sum_{j=1}^{n} (1 - r_j)$$

$$n - 1 = \lambda\,(n - R)$$

where

$$R = \sum_{j=1}^{n} r_j \gtreqless 1$$

$$\lambda = \frac{n - 1}{n - R}$$

Thus

$$\pi_i = \frac{\dfrac{1}{1 - r_i} - \dfrac{n - 1}{n - R}}{\sum_j \dfrac{1}{1 - r_j} - \dfrac{n(n-1)}{n - R}} \tag{45}$$

The equilibrium market shares are increasing functions of the retention probabilities r_i. Alternatively from (44)

$$\frac{1}{1 - r_i} = \frac{\lambda}{1 - p_i}$$

Hence

$$\pi_i = \frac{\dfrac{1}{1-p_i} - 1}{\sum_j \dfrac{1}{1 - p_j} - n} \tag{46}$$

showing that the equilibrium market shares are increasing functions of the selection probabilities p_i.

Under hypothesis II

$$\pi_i = \frac{\lambda p_i}{\lambda \sum_j p_j} = p_i \tag{47}$$

Equilibrium market shares now equal the selection probabilities, as under hypothesis I when $r_i \equiv r$.

CONCLUSION

Brand choice has been modelled as a two-stage decision process where the consumer first decides on retention or rejection of the last brand used , and in the case of rejection makes a new choice. Whether that choice may include the last brand or not does in general make a difference. In any case a rich set of possibilities is opened up demonstrating once more how probabilistic thinking can benefit marketing theory in particular and economic analysis in general. The potential of stochastic processes for economic analysis is in fact becoming more and more recognized, notably in the area of finance.

Economists are increasingly aware of their indebtedness to probability theorists who like S.K.Srinivasan have advanced the field of stochastic processes and have made it operationally useful to those of us engaged in economic applications. It is a pleasure to salute him on this auspicious occasion and to congratulate him on his many accomplishments.

REFERENCES

Bass, F.M., et al. (1961), Mathematical Models in Marketing Homewood, Illinois: R. D.Irwin.

Montgomery, D.B., and G.L. Urban (1969), Management Science in Marketing, Englewood Cliffs, New Jersey: Prentice-Hall.

Schmalensee, R. (1972), The of Advertising. Amsterdam: North-Holland Publishing Co.

Srinivasan, S.K. (1974) Stochastic Point Proccesses and Their Applications, New York: Hafner Press.

Stochastic Processes:
Use and Limitations in Reliability Theory

Prof. Dr. A. Birolini
Reliability Laboratory
Swiss Federal Institute of Technology (ETH)
Zurich/Switzerland

Abstract

Stochastic processes are powerful tools for the investigation of the reliability and availability of repairable equipment and systems. Because of the involved models and in order to be mathematically tractable, these processes are generally confined to the class of regenerative stochastic processes with a finite state space, to which belong renewal processes, Markov processes, semi-Markov processes, and more general regenerative processes with only few (in the limit case only one) regeneration states. This contribution introduce briefly these processes and uses them to solve some reliability problems encountered in pratical applications. Investigations deal with different kinds of reliabilities and availabilities for one item, series, parallel, and series/parallel structures. For series/parallel structures useful approximate expressions are developed.

1. Introduction

For complex equipment and systems, reliability analyses are generally performed at two different levels. At assembly level, the designer performs failure rate and failure mode analyses to check fulfilment of reliability requirements and to detect and eliminate reliability weaknesses as early as possible in the design phase. At equipment and system level, the reliability engineer also investigates time behaviour, taking into account reliability, maintainability, and logistical aspects. Depending upon the system complexity, upon the assumed distribution functions for failure-free and repair times, and with thought toward maintenance policy, investigations are performed either analytically, making use of stochastic processes, or numerically with the help of Monte Carlo simulations.

This contribution summarizes the main properties of the stochastic processes used in reliability analyses and investigates some problems encountered in practical applications. Besides Section 4, the models investigated here assume that systems have only one repair crew and that no further failure can occur at system down. Repair and failure rates are generalized step-by-step, up to the case in which the involved process is regenerative with only one (or a few) regeneration state(s).

2. Stochastic Processes Used in Modeling Reliability Problems

Stochastic processes used in the modeling of reliability problems include renewal and alternating renewal processes, Markov processes with a finite state space, semi-Markov processes, regenerative stochastic processes with only one (or a few) regeneration state(s), and some kinds of non-regenerative stochastic processes. This section briefly introduce these processes.

2.1 Renewal Processes

In reliability theory, renewal processes describe the basic model of an item in continuous operation which is replaced at each failure in a negligible amount of time by a new, statistically identical item. To define the renewal process, let $\tau_0, \tau_1, \dots$ be statistically independent, non-negative random variables (e.g. failure-free operating times) distributed according to

$$F_A(x) = \Pr\{\tau_0 \le x\} \quad \text{and} \quad F(x) = \Pr\{\tau_i \le x\}, \quad i = 1,2,\dots. \tag{1}$$

The random variables $S_n = \tau_0 + \dots + \tau_{n-1}$ $(n = 1, 2, \dots)$ or equivalently the sequence $\tau_0, \tau_1, \dots$ itself constitute a *renewal process*. The points $S_1, S_2, \dots$ are renewal points (regeneration points). Renewal processes are ordinary for $F_A(x) = F(x)$ or modified otherwise. To simplify analyses, let us assume in the following that

$$F_A(0) = F(0) = 0,$$

$$f_A(x) = \frac{dF_A(x)}{dx} \quad \text{and} \quad f(x) = \frac{dF(x)}{dx} \quad \text{exist, and}$$

$$T = E[\tau_i] = \int_0^\infty (1 - F(x))\,dx < \infty, \qquad i \ge 1. \tag{2}$$

For renewal process, the distribution function of the number of renewal points $\nu(t)$ in the time interval $(0, t]$ is given by

$$\Pr\{\nu(t) \le n-1\} = 1 - \Pr\{\tau_0 + \dots + \tau_{n-1} \le t\} = 1 - F_n(t), \qquad n = 1,2,\dots \tag{3}$$

with

$$F_1(t) = F_A(t) \quad \text{and} \quad F_{n+1}(t) = \int_0^t F_n(t-x) f(x)\,dx, \qquad n = 1,2,\dots. \tag{4}$$

From Eq. (3) follows

$$\Pr\{\nu(t) = n\} = F_n(t) - F_{n+1}(t), \qquad n = 1,2,\dots,$$

and thus, for the mean of $\nu(t)$,

$$E[\nu(t)] = \sum_{n=1}^{\infty} n(F_n(t) - F_{n+1}(t)) = \sum_{n=1}^{\infty} F_n(t) = H(t). \tag{5}$$

The function $H(t)$ defined by Eq. (5) is called *renewal function*. Due to $F(0) = 0$ it is $H(0) = 0$. The distribution functions $F_n(t)$ have densities

$$f_1(t) = f_A(t) \quad \text{and} \quad f_n(t) = \int_0^\infty f(x) f_{n-1}(t-x)\,dx, \qquad n = 2,3,\dots \tag{6}$$

and are thus the convolutions of $f(x)$ with $f_{n-1}(x)$. The function

$$h(t) = \frac{dH(t)}{dt} = \sum_{n=1}^{\infty} f_n(t) \tag{7}$$

is called *renewal density*. The renewal density has the following important meaning: Due to the assumption $F_A(0) = F(0) = 0$,

$$\lim_{\delta t \downarrow 0} \frac{1}{\delta t} \Pr\{\nu(t+\delta t) - \nu(t) > 1\} = 0$$

and thus, for $\delta t \downarrow 0$,

$$\Pr\{\text{exactly one renewal point } (S_1 \text{ or } S_2 \text{ or } \ldots) \text{ lies in } (t, t+\delta t]\} \approx h(t)\delta t. \tag{8}$$

This interpretation of the renewal density is useful for analyses. From Eq. (8) it follows also that the renewal density $h(t)$ basically differs from the failure rate $\lambda(t)$ defined by

$$\lambda(t) = \lim_{\delta t \downarrow 0} \frac{1}{\delta t} \Pr\{t < \tau_0 < t + \delta t \mid \tau_0 > t\} = \frac{f_A(t)}{1 - F_A(t)}. \tag{9}$$

However, there is the special case of the *Poisson process* for which one has $F_A(x) = F(x) = 1 - e^{-\lambda x}$ and thus $h(t) = \lambda(t) = \lambda$.

Important for the investigation of renewal processes are the *forward recurrence time* $\tau_R(t)$ and the *backward recurrence time* $\tau_S(t)$. $\tau_R(t)$ and $\tau_S(t)$ are the time intervals from an arbitrary t forward to the next renewal point and backward to the last renewal point (or to the time origin), respectively. It is not difficult to see that the distribution functions of $\tau_R(t)$ and $\tau_S(t)$ are given by

$$\Pr\{\tau_R(t) \le x\} = F_A(t+x) - \int_0^t h(y)(1 - F(t+x-y))\,dy \tag{10}$$

and

$$\Pr\{\tau_S(t) \le x\} = \int_{t-x}^{t} h(y)(1 - F(t-y))\,dy \quad \text{for } x < t \quad \text{and} \quad 1 \quad \text{for } x \ge t. \tag{11}$$

The asymptotic behavior of a renewal process (i.e. the behavior for $t \to \infty$) can be investigated using the *renewal density theorem*

$$\lim_{t \to \infty} h(t) = \frac{1}{T}, \tag{12}$$

with T as in Eq. (2), and the *key renewal theorem*

$$\lim_{t \to \infty} = \int_0^t U(t-y)h(y)\,dy = \frac{1}{T}\int_0^\infty U(z)\,dz. \tag{13}$$

Both theorem holds under very general assumptions on $F_A(t)$ and $F(t)$ [5, 9]. $U(z)$ is ≥ 0, non-increasing, bounded, and Riemann integrable over $(0, \infty)$. Using Eqs. (12) and (13) it follows that

$$\lim_{t \to \infty} \Pr\{\tau_R(t) \le x\} = \lim_{t \to \infty} \Pr\{\tau_S(t) \le x\} = \frac{1}{T}\int_0^x (1 - F(y))\,dy. \tag{14}$$

From the above result, the following definition can be given for a stationary renewal process: A renewal process is *stationary* (in steady-state) if for all $t > 0$ the distribution function of $\tau_R(t)$ does not depend on t. It is intuitively clear that such a situation can only occur if a particular relationship exists between $F_A(x)$ and $F(x)$ defined by Eq. (1). Assuming, according to Eq. (14),

$$F_A(x) = \frac{1}{T}\int_0^x (1 - F(y))\,dy \tag{15}$$

one obtains for all $t \ge 0$

$$h(t) = \frac{1}{T} \tag{16}$$

and, from Eq. (10),

$$\Pr\{\tau_R(t) \le x\} = \frac{1}{T}\int_0^x (1 - F(y))\,dy. \tag{17}$$

It is not difficult to see that the counting process $\nu(t)$ belonging to a stationary renewal process is time-homogeneous. Equation (15) is a necessary and sufficient condition for the stationarity of the renewal process with $\Pr\{\tau_i \le x\} = F(x)$, $i \ge 1$. For the applications, the following interpretation is valid: *A stationary renewal process can be regarded as a reneval process with arbitrary inital condition $F_A(x)$, which has been started at $t = -\infty$ and which will be considered only for $t \ge 0$ ($t = 0$ being an arbitrary time point).*

2.2 Alternating Renewal Processes

The generalization of the renewal process of Section 2.1 achieved by introducing a random replacement time, distributed according to $G(x)$, leads to the alternating renewal process. An alternating renewal process is a process with two states which alternate from one state to the other after a sojourn time distributed according to $F(x)$ and $G(x)$, respectively. Considering the reliability and avaiability analyses of a repairable item in Section 3, and in order to simplify the notation, these two states will be referred to as up state and down state, abbreviated as u and d, respectively. To define the *alternating renewal process*, consider two independent renewal processes $\{\tau_i\}$ and $\{\tau_i'\}$. For reliability applications, τ_i denotes the i-th failure-free operating time and τ_i' the i-th repair time. These random variables are distributed according to $F_A(x)$ for τ_0 and $F(x)$ for t_i, $i \ge 1$, and $G_A(x)$ for τ_0' and $G(x)$ for τ_i', $i \ge 1$, with densities $f_A(x)$, $f(x)$, $g_A(x)$, and $g(x)$, and with finite means

$$MTTF = E[\tau_i] = \int_0^\infty (1 - F(t))\,dt \quad \text{and} \quad MTTR = E[\tau_i'] = \int_0^\infty (1 - G(t))\,dt, \qquad i \ge 1, \tag{18}$$

where *MTTF* stands for Mean Time To Failure and *MTTR* for Mean Time To Repair. The sequences $\tau_0, \tau_1', \tau_1, \tau_2', \tau_2, \tau_3', \ldots$ and $\tau_0', \tau_1, \tau_1', \tau_2, \tau_2', \tau_3, \ldots$ form two modified alternating renewal processes, starting at $t = 0$ with τ_0 and τ_0', respectively.

Embedded in each of the above processes are two renewal processes with renewal points S_{udui} or S_{uddi} and S_{duui} or S_{dudi} (the index udu means a transition from up to down given up at $t = 0$). These four embedded renewal processes are statistically identical up to the time intervals starting at $t = 0$. The corresponding densities are

$$f_A(x),\ f_A(x) * g(x),\ g_A(x) * f(x),\ g_A(x)$$

for the time interval starting at $t = 0$, and

$$f(x) * g(x)$$

for all others (the symbol $*$ denoting convolution). The results of Section 2.1 can be used for investigation of the embedded renewal processes. Equation (7) yields the Laplace transform of renewal densities $h_{udu}(t)$, $h_{duu}(t)$, $h_{udd}(t)$, and $h_{dud}(t)$

$$\tilde{h}_{udu}(s) = \frac{\tilde{f}_A(s)}{1 - \tilde{f}(s)\tilde{g}(s)}, \qquad \tilde{h}_{duu}(s) = \frac{\tilde{f}_A(s)\tilde{g}(s)}{1 - \tilde{f}(s)\tilde{g}(s)},$$

$$\tilde{h}_{udd}(s) = \frac{\tilde{g}_A(s)\tilde{f}(s)}{1 - \tilde{f}(s)\tilde{g}(s)}, \qquad \tilde{h}_{dud}(s) = \frac{\tilde{g}_A(s)}{1 - \tilde{f}(s)\tilde{g}(s)}. \tag{19}$$

An alternating renewal process is thus completely caracterized by the distribution functions $F_A(x)$, $F(x)$, $G_A(x)$, $G(x)$, and by the probability p to be up at $t=0$

$$p = \Pr\{\text{item up at } t=0\}. \tag{20}$$

For

$$p = \frac{MTTF}{MTTF + MTTR}, \qquad F_A(x) = \frac{1}{MTTF}\int_0^x (1-F(y))\,dy, \qquad \text{and}$$

$$G_A(x) = \frac{1}{MTTR}\int_0^x (1-G(y))\,dy, \tag{21}$$

the alternating renewal process is stationary. It can be shown that with p, $F_A(x)$, and $G_A(x)$ as in Eq. (21), it holds

$$p\,h_{udu}(t) + (1-p)\,h_{udd}(t) = p\,h_{duu}(t) + (1-p)\,h_{dud}(t) = \frac{MTTF}{MTTF + MTTR},$$

and in particular

$$\Pr\{\text{item up at } t\} = PA(t) = \frac{MTTF}{MTTF + MTTR}, \qquad t \ge 0. \tag{22}$$

Furthermore, irrespective of its initial conditions p, $F_A(x)$, and $G_A(x)$, an alternating renewal process has for $t \to \infty$ an asymptotic behavior which is identical with the stationary state (steady-state). In other words: *A stationary alternating renewal process can be regarded as an alternating renewal process with arbitrary initial conditions p, $F_A(x)$, and $G_A(x)$, which has been started at $t = -\infty$ and which will be considered only for $t \ge 0$ ($t = 0$ being an arbitrary time point).*

2.3 Markov Processes with Finitely Many States

Markov processes are processes without memory. They are characterized by the property that *for any time point t (arbitrarily chosen) their evolution after t depends on t and the state occupied at t, but not on the process evolution up to the time t.* In the case of time-homogeneous Markov processes, dependence on t also disappears so that future evolution of the process depends only of the current state. These processes describe in reliability theory the behaviour of repairable systems with constant failure and repair rates for all elements.

A stochastic process $\xi(t)$ in continuous time with finitely many states $Z_1, \ldots, Z_m$ is a *Markov process* if for $n = 1, 2, \ldots$, arbitrary time points $t + a > t > t_n > \ldots > t_1$, and arbitrary $i, j, i_1, \ldots, i_n \in \{0, \ldots, m\}$,

$$\Pr\{\xi(t+a) = Z_j \mid (\xi(t) = Z_i \cap \xi(t_n) = Z_{i_n} \cap \ldots \cap \xi(t_1) = Z_{i_1})\}$$
$$= \Pr\{\xi(t+a) = Z_j \mid \xi(t) = Z_i\}. \tag{23}$$

The conditional state probabilities in Eq. (23) are called *transition probabilities* of the Markov process and are indicated by $P_{ij}(t, t+a)$

$$P_{ij}(t, t+a) = \Pr\{\xi(t+a) = Z_j \mid \xi(t) = Z_i\}. \tag{24}$$

The Markov process is time-homogeneous if the transition probabilities are independent of t

$$P_{ij}(t, t+a) = P_{ij}(a). \tag{25}$$

In the following only time-homogeneous Markov processes will be considered. The transition probabilities $P_{ij}(a)$ satisfy the conditions

$$P_{ij}(a) \geq 0 \quad \text{and} \quad \sum_{j=0}^{m} P_{ij}(a) = 1, \qquad i = 0, \dots, m. \tag{26}$$

Thus, $P_{ij}(a)$ form a stochastic matrix. Together with the initial distribution

$$P_i(0) = \Pr\{\xi(0) = Z_i\}, \qquad i = 0, \dots, m, \tag{27}$$

the transition probabilities $P_{ij}(a)$ completely determine the distribution law of the Markov process. In particular, for $t > 0$, the state probabilities

$$P_j(t) = \Pr\{\xi(t) = Z_j\}, \qquad j = 0, \dots, m \tag{28}$$

can be obtained by means of

$$P_j(t) = \sum_{i=0}^{m} P_i(0) P_{ij}(t). \tag{29}$$

Assuming for the transition probabilities $P_{ij}(t)$ to be continuous at $t = 0$, it can be proven that $P_{ij}(t)$ are also differentiable at $t = 0$. The limiting values

$$\lim_{\delta t \downarrow 0} \frac{P_{ij}(\delta t)}{\delta t} = \rho_{ij} \quad \text{for } i \neq j \quad \text{and} \quad \lim_{\delta t \downarrow 0} \frac{1 - P_{ii}(\delta t)}{\delta t} = \rho_i, \tag{30}$$

exist and satisfy

$$\rho_i = \sum_{j=0}^{m} \rho_{ij}, \qquad i = 0, \dots, m, \qquad \rho_{ii} = 0. \tag{31}$$

Equation (30) can be written in the form

$$P_{ij}(\delta t) = \rho_{ij}\, \delta t + o(\delta t) \qquad \text{and} \qquad 1 - P_{ii}(\delta t) = \rho_i\, \delta t + o(\delta t).$$

Considering that $P_{ij}(\delta t) = \Pr\{\xi(t + \delta t) = Z_j \mid \xi(t) = Z_i\}$ is true for all $t \geq 0$, the following useful interpretation for the quantities ρ_{ij} and ρ_i can be obtained (for $\delta t \downarrow 0$ and arbitrary t, in particular $t = 0$)

$$\rho_{ij}\, \delta t \approx \Pr\{\text{to jump from } Z_i \text{ to } Z_j \text{ in } (t, t + \delta t] \mid \xi(t) = Z_i\}, \tag{32}$$

$$\rho_i\, \delta t \approx \Pr\{\text{to leave } Z_i \text{ in } (t, t + \delta t] \mid \xi(t) = Z_i\}. \tag{33}$$

ρ_{ij} and ρ_i are called *transition rates*. They play a similar role in the analysis of Markov processes as the transition probabilities p_{ij} for Markov chains.

The following description of a (time-homogeneous) Markov process with initial distribution $P_i(0)$, $i = 0, \dots, m$, and transition rates ρ_{ij} and ρ_i, $i, j \in \{0, \dots, m\}$, provides better insight into the structure of a Markov process as a process with piece-wise constant sample paths (it is also the basis for investigations of Markov processes using integral equations and motivates the introduction of semi-Markov processes). Let $\xi_0, \xi_1, \dots$ be a sequence of random variables taking values in $\{Z_0, \dots, Z_m\}$ (states successively occupied) and $\eta_0, \eta_1, \dots$ a sequence of positive random variables (sojourn times between two consecutive state transitions). Define

$$p_{ij} = \frac{\rho_{ij}}{\rho_i}, \qquad i \neq j \qquad \text{and} \qquad p_{ii} = 0, \qquad i = 0, \dots, m,$$

and assume furthermore that

$$\Pr\{\xi_0 = Z_i\} = P_i(0), \qquad i = 0, \ldots, m,$$

and, for any $n = 1, 2, \ldots,$

$$\Pr\{(\xi_{n+1} = Z_j \cap \eta_n \le x) \mid (\xi_n = Z_i \cap \eta_{n-1} = t_{n-1} \cap \ldots \cap \eta_0 = t_0 \cap \xi_0 = Z_{i_0})\}$$

$$= \Pr\{(\xi_{n+1} = Z_j \cap \eta_n \le x) \mid \xi_n = Z_i\} = Q_{ij}(x) = p_{ij}\, F_{ij}(x) = p_{ij}\,(1 - e^{-\rho_i x}). \tag{34}$$

Thus, $\xi_0, \xi_1, \ldots$ is a *Markov chain* with initial distribution $P_i(0) = \Pr\{\xi_0 = Z_i\}$ and transition probabilities $p_{ij} = \Pr\{\xi_{n+1} = Z_j \mid \xi_n = Z_i\}$. From Eq. (34), $F_{ij}(x)$ follows as

$$F_{ij}(x) = \Pr\{\eta_n \le x \mid (\xi_n = Z_i \cap \xi_{n+1} = Z_j)\} = 1 - e^{-\rho_i x}. \tag{35}$$

Define now $S_0 = 0$, $S_n = \eta_0 + \ldots + \eta_{n-1}$, $n = 1, 2, \ldots$ and $\xi(t) = \xi_n$, if $S_n \le t < S_{n+1}$. From this and the *memoryless property* of the exponential distribution it follows that $\xi(t)$, $t \ge 0$ is a Markov process with initial distribution $P_i(0)$ and transition rates ρ_{ij} and ρ_i. The evolution of a (time-homogeneous) Markov process with transiton rates ρ_{ij} and ρ_i can be then described in the following way: *If at a time $t = 0$ the process enters the state Z_i, i.e. $\xi_0 = Z_i$, then the next state to be entered, say Z_j $(j \ne i)$ is selected according to the probabilities p_{ij} and the sojourn time in Z_i is a random variable (η_0) with distribution function $Pr\{\eta_0 \le x \mid (\xi_0 = Z_i \cap \xi_1 = Z_j)\} = 1 - e^{-\rho_i x}$; as the process enters Z_j, the next state to be entered, say Z_k $(k \ne j)$, will be selected with probability p_{jk} and the sojourn time (η_1) in Z_j will be distributed according to $Pr\{\eta_1 \le x \mid (\xi_1 = Z_j \cap \xi_2 = Z_k)\} = 1 - e^{-\rho_j x}$ etc.*

In practical applications, following arguments can be used to calculate the quantities $Q_{ij}(x)$, p_{ij}, and $F_{ij}(x)$ in Eq. (34): *If the process enters the state Z_i at an arbitrary time, say at $t = 0$, then a collection of independent random times τ_{ij}, $j \ne i$, are started (τ_{ij} is the sojourn time in Z_i with next jump in Z_j); the process will jump in Z_j at the time x only if $\tau_{ij} = x$ and $\tau_{ik} > \tau_{ij}$ for $k \ne j$.* In this interpretation, the quantities $Q_{ij}(x)$, p_{ij}, and $F_{ij}(x)$ are given by

$$Q_{ij}(x) = \Pr\{\tau_{ij} \le x \cap \tau_{ik} > \tau_{ij}, \quad k \ne j\}, \tag{36}$$

$$p_{ij} = \Pr\{\tau_{ik} > \tau_{ij}, \quad k \ne j\}, \tag{37}$$

$$F_{ij}(x) = \Pr\{\tau_{ij} \le x \mid \tau_{ik} > \tau_{ij}, \quad k \ne j\}. \tag{38}$$

Considering for the Markov process

$$\Pr\{\tau_{ij} \le x\} = 1 - e^{-\rho_{ij} x}$$

one obtains as before

$$Q_{ij}(x) = \int_0^x \rho_{ij}\, e^{-\rho_{ij} y} \prod_{\substack{k=0 \\ k \ne j}}^{m} e^{-\rho_{ik} y}\, dy = \frac{\rho_{ij}}{\rho_i}(1 - e^{-\rho_i x}),$$

$$p_{ij} = \frac{\rho_{ij}}{\rho_i}, \qquad \text{and} \qquad F_{ij}(x) = 1 - e^{-\rho_i x}.$$

It should be emphasized that due to the memoryless property of the Markov process, there is no difference wheter the process enters Z_i at $t = 0$ or is already there (this will not be true for the semi-Markov processes).

A useful tool when investigating a Markov process is the so-called *diagram of transition probabilities in $(t, t + \delta t]$* where δt is small and t is an arbitrary time point, e.g. $t = 0$; it is a modification of the state transition diagram, more appropriate for applications. This diagram is a directed graph with nodes labeled by states Z_i, $i = 0, \ldots, m$, and arcs labeled by transition probabilities $P_{ij}(\delta t)$, where terms of order $o(\delta t)$ are omitted. Accounting for the properties of random variables τ_{ij} introduced with Eq. (36) yields $P_{ij}(\delta t) = \rho_{ij}\, \delta t + o(\delta t)$ and $P_{ii}(\delta t) = 1 - \rho_i\, \delta t + o(\delta t)$ as with Eq. (30). Figure 1 gives the diagram of transition probabilities in $(t + \delta t]$ of two series/parallel redundant structures. The states in which the system is down are hatched on the diagram. In the state Z_0 all elements are up (operating or in reserve state).

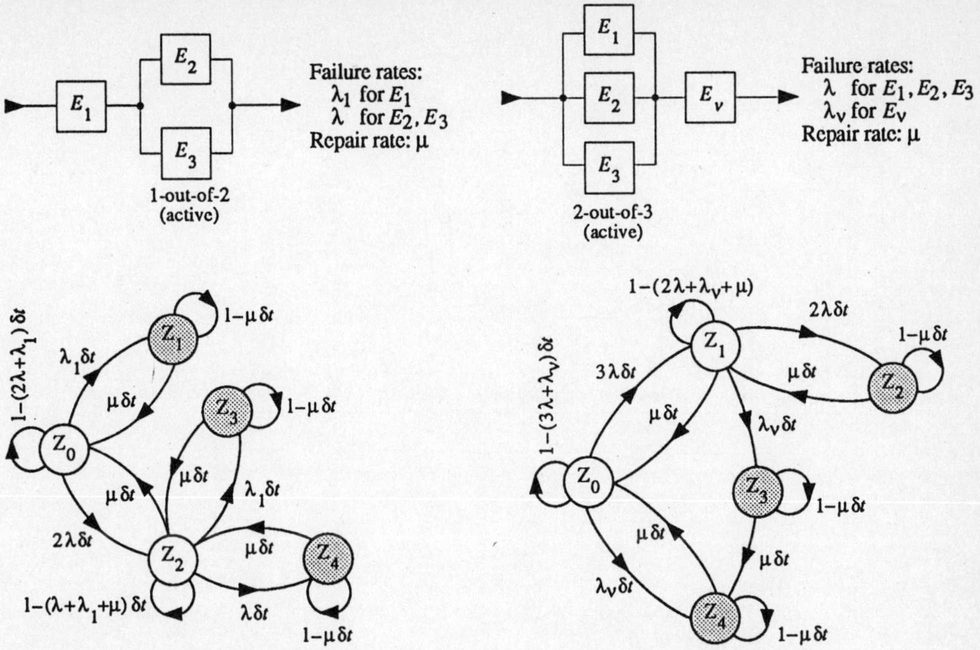

$\rho_{01} = \rho_{23} = \lambda_1;\ \rho_{02} = 2\lambda$

$\rho_{10} = \rho_{20} = \rho_{32} = \rho_{42} = \mu;\ \rho_{24} = \lambda$

$\rho_{01} = 3\lambda;\ \rho_{04} = \rho_{13} = \lambda_v;$

$\rho_{10} = \rho_{21} = \rho_{34} = \rho_{40} = \mu;\ \rho_{12} = 2\lambda$

Figure 1 Reliability block diagram, diagram of the transition probabilities in $(t, t+\delta t]$, and transition rates ρ_{ij} for a 1-out-of-2 active redundancy with switch in series and a majority redundancy 2-out-of-3 with voter (constant failure and repair rates, only one repair crew, no further failure at system down, t arbitrary (e.g. $t = 0$), $\delta t \to 0$)

In reliability theory, two important quantities are the state probabilities and the distribution function of the sojourn times in the set of the system up states. The state probabilities allow the calculation of the *point-availability*. The *reliability function* can be obtained from the distribution function of the sojourn time in the set of the system up states. Furthermore, a combination of these quantities allows (for time-homogeneous Markov processes) a simple calculation of the *interval-reliability*. It is useful in such analyses to divide the system state space into the following complementary sets U and $\overline{U}$

U = set of the system up states (up states at system level)

$\overline{U}$ = set of the system down states.

Calculation of state probabilities and of the sojourn time in U can be carried out using the method of *differential equations* or of *integral equations*. Table 1 summarizes in the 2nd and 3th row the equations necessary for the investigation of reliability models based on Markov processes, in particular those needed for the calculation of the reliability function $R_{Si}(t)$, the system's mean time to failure $MTTF_{Si}$, the point-availability $PA_{Si}(t)$, and the interval-reliability $IR_{Si}(t, t+\theta)$. The condition "Z_i is entered at $t = 0$", necessary for semi-Markov models (1st row of Tab. 1) is used (for Markov models, the condition "system is in Z_i at $t = 0$" would be sufficient). Also given in Tab. 1 are the steady-state (stationary) values of the point-availability PA_S and of the interval-reliability $IR_S(\theta)$. These values are identical to the ones obtained by investigating the *asymptotic behavior* ($t \to \infty$), whose existence can be shown [5, 8, 11] by assuming that the underlying Markov process (or the corresponding embedded Markov chain) is irreducible (each state can be reached from every other state).

	Reliability function	Point-availability	Interval-reliability
Semi-Markov processes	$R_{Si}(t) = 1 - Q_i(t) + \sum_{Z_j \in U} \int_0^t q_{ij}(x) R_{Sj}(t-x)\,dx$ $MTTF_{Si} = T_i + \sum_{Z_j \in U} p_{ij}\, MTTF_{Sj}, \qquad Z_i \in U$ $Q_{ij}(x) = \Pr\{\tau_{ij} \le x \cap \tau_{ik} > \tau_{ij},\ \ k \ne j\}$ $\quad = p_{ij} F_{ij}(x)$ $p_{ij} = \Pr\{\tau_{ik} > \tau_{ij},\ \ k \ne j\}, \qquad p_{ii} = 0$ $F_{ij}(x) = \Pr\{\tau_{ij} \le x \mid \tau_{ik} > \tau_{ij},\ k \ne j\}$	$PA_{Si}(t) = \sum_{Z_j \in U} P_{ij}(t), \qquad i = 0, \ldots, m$ $PA_S = \sum_{Z_j \in U} p_j T_j \Big/ \sum_{k=0}^{m} p_k T_k$ $P_{ij}(t) = \delta_{ij}(1 - Q_i(t)) + \sum_{k=0}^{m} \int_0^t q_{ik}(x) P_{kj}(t-x)\,dx$ $i, j = 0, \ldots, m, \quad \delta_{ii} = 1, \quad \delta_{ij} = 0$ for $j \ne i$ $Q_i(x) = \sum_{j=0}^{m} Q_{ij}(x), \quad T_i = \int_0^\infty (1 - Q_i(x))\,dx$	computation has to be problem oriented; for constant failure rates and in the stationary state one has $IR_S(\theta) = \sum_{Z_j \in U} \frac{T_j}{T_{jj}} R_{Sj}(\theta)$ $T_{jj} = \frac{1}{p_j} \sum_{k=0}^{m} p_k T_k$ p_j from $p_j = \sum_{i=0}^{m} p_i\, p_{ij}$, with $p_{ii} = 0, \quad p_j > 0,$ and $p_0 + \ldots + p_m = 1$
Time homogeneous Markov processes (method of integral equations)	$R_{Si}(t) = e^{-\rho_i t} + \sum_{Z_j \in U} \int_0^t \rho_{ij} e^{-\rho_i x} R_{Sj}(t-x)\,dx$ $MTTF_{Si} = \frac{1}{\rho_i} + \sum_{Z_j \in U} \frac{\rho_{ij}}{\rho_i} MTTF_{Sj}, \qquad Z_i \in U$ ρ_{ij} = transition rate (see definitions) $\rho_{ii} = 0, \qquad \rho_i = \sum_{j=0}^{m} \rho_{ij}$	$PA_{Si}(t) = \sum_{Z_j \in U} P_{ij}(t), \qquad i = 0, \ldots, m$ $PA_S = \sum_{Z_j \in U} p_j$ $P_{ij}(t) = \delta_{ij} e^{-\rho_i t} + \sum_{k=0}^{m} \int_0^t \rho_{ik} e^{-\rho_i x} P_{kj}(t-x)\,dx$ $i, j = 0, \ldots, m,$ $\rho_{ii} = 0, \quad \delta_{ii} = 1, \quad \delta_{ij} = 0$ for $j \ne i$	$IR_{Si}(t, t+\theta) = \sum_{Z_j \in U} P_{ij}(t) R_{Sj}(\theta),$ $i = 0, \ldots, m$ $IR_S(\theta) = \sum_{Z_j \in U} p_j R_{Sj}(\theta)$ p_j from $\rho_j p_j = \sum_{i=0}^{m} p_i \rho_{ij}$, with $p_{ii} = 0, \quad p_j > 0,$ and $p_0 + \ldots + p_m = 1$

Table 1 Basic equations for reliability and availability computation of Markov and semi-Markov models

Table 1 cont.

Time homogeneous Markov processes (method of differential equations)	$R_{Si}(t) = \sum_{Z_j \in U} P'_{ij}(t), \quad Z_i \in U$ $MTTF_{Si} = \frac{1}{\rho_i} + \sum_{Z_j \in U} \frac{\rho_{ij}}{\rho_i} MTTF_{Sj}, \quad Z_i \in U$ $P'_{ij}(t) \equiv P'_j(t)$, with $P'_j(t)$ from $\dot{P}'_j(t) = -\rho_j P'_j(t) + \sum_{Z_i \in U} P'_i(t)\rho_{ij}$, $j = 0, \ldots, m, \quad \rho_{ii} = 0$ with $P'_i(0) = 1$ and $P'_j(0) = 0$ for $j \neq i$	$PA_{Si}(t) = \sum_{Z_j \in U} P_{ij}(t), \quad i = 0, \ldots, m$ $PA_S = \sum_{Z_j \in U} p_j$ $P_{ij}(t) \equiv P_j(t)$, with $P_j(t)$ from $\dot{P}_j(t) = -\rho_j P_j(t) + \sum_{i=0}^{m} P_i(t)\rho_{ij}$ $j = 0, \ldots, m, \quad \rho_{ii} = 0$ with $P_i(0) = 1$ and $P_j(0) = 0$ for $j \neq i$	$IR_{Si}(t, t+\theta) = \sum_{Z_j \in U} P_{ij}(t) R_{Sj}(\theta),$ $i = 0, \ldots, m$ $IR_S(\theta) = \sum_{Z_j \in U} p_j R_{Sj}(\theta)$ p_j from $p_j \rho_j = \sum_{i=0}^{m} p_i \rho_{ij}$, with $\rho_{ii} = 0, \; p_j > 0$, and $p_0 + \ldots + p_m = 1$

$R_{Si}(t)$ = Pr{system up in $(0, t]$ \| Z_i is entered at $t = 0$}, $Z_i \in U$

$MTTF_{Si}$ = E[system failure free time \| Z_i is entered at $t = 0$] = $\int_0^\infty R_{Si}(t)\,dt = \tilde{R}_{Si}(0)$, $Z_i \in U$

$PA_{Si}(t)$ = Pr{system up at t \| Z_i is entered at $t = 0$}, $i = 0, \ldots, m$

PA_S = Pr{system up at t in the stationary state or for $t \to \infty$} = $\lim_{s \to 0} s\tilde{PA}_{Si}(s)$, $i = 0, \ldots, m$

AA_S = $\lim_{t \to \infty} \frac{1}{t}$ E[system up time in $(0, t]$] = PA_S

$IR_{Si}(t, t+\theta)$ = Pr{system up in $(t, t+\theta]$ \| Z_i is entered at $t = 0$}, $i = 0, \ldots, m$

$IR_S(\theta)$ = Pr{system up in $(t, t+\theta]$ in the stationary state or for $t \to \infty$}

U = set of the up states

$\overline{U}$ = set of the down states

$P_{ij}(t)$ = Pr{system in state Z_j at t \| Z_i is entered at $t = 0$}

ρ_{ij} = $\lim_{\delta t \downarrow 0} \frac{1}{\delta t}$ Pr{transition from Z_i to Z_j in $(t, t+\delta t]$ \| system in Z_i at t}

S stays for system; $\tilde{u}(s) = \int_0^\infty u(t) e^{-st}\,dt$ = Laplace transform of $u(t)$

2.4 Semi-Markov-Processes

A semi-Markov process is a pure jump process. The sequence of consecutively occurring states forms a time-homogeneous Markov chain, just as with Markov processes. The sojourn time in a state is only dependent on this state and on the following one. However, contrary to Markov processes, this sojourn time is not exponentially distributed.

To define semi-Markov processes, let $\xi_0, \xi_1, \ldots$ be the sequence of consecutively occuring states, i.e. a sequence of random variables taking values in $\{Z_0, \ldots, Z_m\}$ and $\eta_0, \eta_1, \ldots$ the sojourn times between consecutive state changes, i.e. a sequence of positive and continuous random variables. A stochastic process $\xi(t)$ with state space $\{Z_0, \ldots, Z_m\}$ is a *semi-Markov process* if for $n = 1, 2, \ldots$, arbitrary $i, j, i_0, \ldots, i_{n-1} \in \{0, \ldots, m\}$, and arbitrary positive numbers $x_0, \ldots, x_{n-1}$,

$$\begin{aligned}&\Pr\{(\xi_{n+1} = Z_j \cap \eta_n \leq x) \mid \\ &\quad (\xi_n = Z_i \cap \eta_{n-1} = x_{n-1} \cap \ldots \xi_1 = Z_{i_1} \cap \eta_0 = x_0 \cap \xi_0 = Z_{i_0})\} \\ &\qquad = \Pr\{(\xi_{n+1} = Z_j \cap \eta_n \leq x) \mid \xi_n = Z_i\} = Q_{ij}(x). \qquad (39)\end{aligned}$$

The functions $Q_{ij}(x)$ in Eq. (39) are called *semi-Markov transition probabilities*. Setting $Q_{ij}(\infty) = p_{ij}$ and $F_{ij}(x) = Q_{ij}(x) / p_{ij}$ leads to

$$Q_{ij}(x) = p_{ij} F_{ij}(x), \qquad (40)$$

with

$$p_{ij} = \Pr\{\xi_{n+1} = Z_j \mid \xi_n = Z_i\}, \qquad p_{ii} = 0 \qquad (41)$$

and

$$F_{ij}(x) = \Pr\{\eta_n \leq x \mid (\xi_n = Z_i \cap \xi_{n+1} = Z_j)\}. \qquad (42)$$

From Eq. (39), the consecutive jump points at which the process enters Z_i are regeneration points. This holds for any $i = 0, \ldots, m$. Thus, all states of a semi-Markov process are *regeneration states*.

The initial distribution, i.e. the distribution of the vector $(\xi(0), \xi_1, \eta_0)$ is given for the general case by

$$A_{ij}(x) = \Pr\{\xi(0) = Z_i \cap \xi_1 = Z_j \cap \text{residual sojourn time in } Z_i \leq x\} = P_i(0)\, p_{ij}\, F_{ij}^{\circ}(x), \qquad (43)$$

with $P_i(0) = \Pr\{\xi(0) = Z_i\}$, p_{ij} according to Eq. (41), and $F_{ij}^{\circ}(x) = \Pr\{$residual sojourn time in $Z_j \leq x \mid (\xi(0) = Z_i \cap \xi_1 = Z_j)\}$. As pointed out above, the semi-Markov process is markovian, i.e. memoryless, in general only at the transition points from one state to the other. To have the time $t = 0$ as a regeneration point, the initial condition "$\xi(0) = Z_i$", sufficient for Markov processes, must be reinforced for semi-Markov processes to "Z_i is entered at $t = 0$".

The sequence $\xi_0, \xi_1, \ldots$ forms a Markov chain (*embedded Markov chain*) with transition probabilities p_{ij} according to Eq. (41) and initial probabilities $P_i(0)$, $i = 0, \ldots, m$. $F_{ij}(x)$ is the conditonal distribution function of the sojourn time in Z_i with consequent jump in Z_j. A semi-Markov process is a Markov process if and only if $F_{ij}(x) = 1 - e^{-\rho_i x}$, for $i, j \in \{0, \ldots, m\}$. As an example for a two states semi-Markov process consider the alternating renewal process introduced in Section 2.2 ($Z_0 = up$, $Z_1 = down$, $p_{01} = p_{10} = 1$, $F_{01}(x) = F(x)$, $F_{10}(x) = G(x)$, $F_0^{\circ}(x) = F_A(x)$, $F_1^{\circ}(x) = G_A(x)$, $P_0(0) = p$, $P_1(0) = 1 - p$).

In most applications, the quantities $Q_{ij}(x)$, or p_{ij} and $F_{ij}(x)$, can be calculated according to Eqs. (36) to (38). This will be illustrated in Section 3. For the unconditional sojourn time in Z_i, the distribution function is given by

$$Q_i(x) = \Pr\{\eta_n \le x \mid \xi_n = Z_i\} = \sum_{j=1}^{m} p_{ij} F_{ij}(x) = \sum_{j=1}^{m} Q_{ij}(x), \tag{44}$$

and the mean by

$$T_i = \int_0^\infty (1 - Q_i(x))\, dx. \tag{45}$$

It is assumed in the following that the derivative of $Q_{ij}(x)$ exists for $i, j \in \{0,..., m\}$

$$q_{ij}(x) = \frac{d\,Q_{ij}(x)}{dx}. \tag{46}$$

Table 1 summarizes in the first row the equations necessary for the investigation of reliability models based on semi-Markov processes. Given in Tab. 1 are also the steady-state (stationary) values of the point-availability PA_S and of the interval-reliability $IR_S(\theta)$. These values are identical to the ones obtained by investigating the asymptotic behaviour ($t \to \infty$), whose existence can be shown by assuming continuous sojourn times with finite means and irreducible embedded Markov chain.

2.5 Semi-regenerative Processes

As pointed out in Section 2.3, the time evolution of a system can be described by a Markov process only if the failure-free operating times and repair times of all elements are exponentially distributed (the constant failure rate can depend upon the actual state).Non-exponentially distributed repair times and/or failure-free operating times lead in some cases to semi-Markov processes, but more generally to processes with only few regeneration states or even to non-regenerative processes. *To ensure that the time behaviour of a system can be described by a semi-Markov process, there must be no "running" failure-free operating time or repair time at any state transition (state change) which is not exponentially distributed* (otherwise, the sojourn time to the next state transition would depend on how long these non-exponentially distributed times have already run). Section 3.5 considers the case of a process with states Z_0 to Z_4 in which only states Z_0, Z_1, and Z_4 are regenerative. Z_0, Z_1, and Z_4 form a semi-Markov process embedded in the original process, on which the investigation can be based.

The time behavior of many systems can be described by so-called *semi-regenerative processes* or processes with an *embedded semi-Markov process*. A pure jump process $\xi(t)$ with state space $Z_0, \ldots, Z_m$ is semi-regenerative, with regeneration states $Z_0, \ldots, Z_k$, $k < m$, if the following is true: Let $\zeta_0, \zeta_1, \ldots$ be the sequence of consecutively occuring regenerative states, i.e. a sequence of random variables taking values in $\{Z_0, \ldots, Z_k\}$ and $\varphi_0, \varphi_1, \ldots$ the random time intervals between consecutive occurence of regenerative states, i.e. a sequence of positive and continuous random variables, then Eq. (39) must be fulfiled for $n = 1, 2, \ldots$, arbitrary $i, j, i_0, \ldots, i_{n-1} \in \{0, \ldots, k\}$, and arbitrary positive values $x_0, \ldots, x_{n-1}$ (where ξ_n, η_n are changed by ζ_n, φ_n). In other words, $\zeta(t)$ as given by $\zeta(t) = \zeta_n$ for $\varphi_0 + \ldots + \varphi_{n-1} \le t < \varphi_0 + \ldots + \varphi_n$ is a semi-Markov process with state space $Z_0, \ldots, Z_k$ and transition probabilities $Q_{ij}(x)$, $i, j \in \{0,..., k\}$ embedded in the original process $\xi(t)$.

The pieces $\xi(t)$, $\varphi_0 + \ldots + \varphi_{n-1} \le t < \varphi_0 + \ldots + \varphi_n$, $n = 1, 2\ldots$, of the original process are called *cycles*. Their probabilistic structure can be very complicated. The epochs at which a fixed state Z_i, $0 \le i \le k$ occurs are *regeneration points* and consecutive occurrence points of Z_i constitute a renewal process. Often $U \subseteq \{Z_0, \ldots, Z_k\}$, with U being the set of up states of the system. Integral equations can then be used to calculate the reliability function $R_{Si}(t)$ and the point-availability $PA_{Si}(t)$, see Section 3.5.

If the embedded semi-Markov process has an irreducible embedded Markov chain and continuos conditional distribution functions $F_{ij}(x) = \Pr\{\varphi_n \le x \mid (\zeta_n = Z_i \cap \zeta_{n+1} = Z_j)\}$, $i, j, \in \{0, \ldots, k\}$, then the limiting state probabilities

$$\lim_{t\to\infty} \Pr\{\xi(t) = Z_j\}, \qquad j = 0, \ldots, m \tag{47}$$

exist and do not depend on the initial distribution at $t = 0$. The proof is again based on the key renewal theorem (Eq. (13)). Denoting by $p_j > 0, j = 0, \ldots, k$ the stationary probability of the embedded Markov chain $\zeta_0, \zeta_1, \ldots$, by T_i the mean of the cycle length, provided the cycle is started with Z_i, and by T_i' the mean sojourn time of the original process $\xi(t)$ in the state Z_i, $i = 0, \ldots, k$, the limiting state probabilities according to Eq. (47) are

$$\lim_{t\to\infty} \Pr\{\xi(t) = Z_j\} = \frac{p_j T_j'}{\sum_{i=0}^{k} p_i T_i}, \qquad j = 0, \ldots, k.$$

For the steady-state value of the point-availability follows then

$$PA_S = \sum_{Z_j \in U} \frac{p_j T_j'}{\sum_{i=0}^{k} p_i T_i}. \tag{48}$$

2.6 Non-regenerative Stochastic Processes

The assumption of arbitrarily (i.e. non-exponentially) distributed failure-free operating times and repair times for the elements of a system already leads to non-regenerative stochastic processes for simple series and parallel structures. Besides problem-oriented solutions, a general method of analysis exists which consists in the transformation of the given stochastic process in a semi-Markov or a Markov process by means of a suitable state space extension. The following possibilities are known:

1. Approximation of distribution functions: Approximation of the distribution functions of failure-free operating times and/or repair times by *Erlang distributions* allows transformation of the original process in a semi-Markov process, or a time-homogeneous Markov process, by introducing additional (artificial) states.
2. Introduction of supplementary variables: By introducing *supplementary variables* (for every element with time-dependent failure and/or repair rates, the failure-free operating time since the last repair and/or the repair time since the last failure), the original process can be transformed in a Markov process with state space consisting of discrete and continuous parameters. The investigations usually lead to a system of partial differential equations which can be solved with corresponding boundary conditions.

The first method is best used when the involved failure or repair rates are monotonically increasing from zero to a final value, its application is simple and easy to understand. The second method is very general, however, difficulties in solving the partial differential equations limit its use in practice; this method was introduced in [4] and has been applied to some particular problems.

Assuming that each element of the system considered would have its own (independent) repair crew, the method of Boolean functions could be applied directly to calculate the system's point-availability on the basis of the element's point-availabilities and of the system's structure. In the investigation of the time-dependent behaviour, conditional failure rates $\lambda_i = \lambda_i(t$, history of element E_i in $(0, t])$ and repair rate $\mu_i = \mu_i(t$, history of Element E_i in $(0, t])$ can be used [3].

3. Applications to Reliability Problems

3.1 One-item Repairable Structures

One-item repairable structures are generally investigated by assuming that their stochastic behaviour can be described by an alternating renewal process. This assumption has to be verified for complex items, because it implies that after each repair the *item is like new*. Some important results for such an item in steady-state are given in Tab. 2.

3.2 Systems Without Redundancy

From a reliability point of view, systems without redundancy are *series structures*. Assuming that the system consists of n independent elements with failure rates λ_{0i} ($i = 1, 2, \ldots, n$), arbitrarily distributed repair times with mean $MTTR_i$, and that no further failure can occur at system down, one obtains for the reliability function

$$R_{S0}(t) = e^{-\lambda_{S0} t} \quad \text{with} \quad \lambda_{S0} = \sum_{i=1}^{n} \lambda_{0i}\,. \tag{49}$$

	Failure and Repair Rate: arbitrary	Failure and Repair Rate: constant	Remarks, assumptions
1. Pr{item up at $t=0$} (p)	$\frac{MTTF}{MTTF+MTTR}$	$\frac{\mu}{\lambda+\mu}$	$MTTF = E[\tau_i],\ i \ge 1$ $MTTF = E[\tau_i'],\ i \ge 1$
2. Distribution function of the up time starting at $t=0$ ($F_A(t) = \Pr\{\tau_0 \le t\}$)	$\frac{1}{MTTF}\int_{\theta}^{\infty}(1-F(x))\,dx$	$1-e^{-\lambda t}$	It holds also $F_A(x) = \Pr\{\tau_{Ru}(t) \le x\}$
3. Distribution function of the down time starting at $t=0$ ($G_A(t) = \Pr\{\tau_0' \le t\}$)	$\frac{1}{MTTR}\int_{\theta}^{\infty}(1-G(x))\,dx$	$1-e^{-\mu t}$	It holds also $G_A(x) = \Pr\{\tau_{Rd}(t) \le x\}$
4. Renewal density $h_{du}(t)$, $h_{ud}(t)$	$\frac{1}{MTTF+MTTR}$	$\frac{\lambda\mu}{\lambda+\mu}$	$h_{du}(t) = p h_{duu}(t) + (1-p) h_{dud}(t)$ $h_{ud}(t) = p h_{udu}(t) + (1-p) h_{udd}(t)$ p according to point 1
5. Point-availability	$\frac{MTTF}{MTTF+MTTR}$	$\frac{\mu}{\lambda+\mu}$	$PA = \Pr\{\text{item up at } t\},\ t \ge 0$
6. Average- availability	$\frac{MTTF}{MTTF+MTTR}$	$\frac{\mu}{\lambda+\mu}$	$AA = (1/t)\,E[\text{up time in } (0, t]],\ t > 0$
7. Interval-reliability	$\frac{1}{MTTF+MTTR} \cdot \int_{\theta}^{\infty}(1-F(x))\,dx$	$\frac{\mu}{\lambda+\mu} e^{-\lambda\theta}$	$IR(\theta) = \Pr\{\text{item up in } (t, t+\theta]\},\ t \ge 0$
8. Joint-availabilaty	$\frac{MTTF \cdot PA_{us}(\theta)}{MTTF+MTTR}$	$(\frac{\mu}{\lambda+\mu})^2 + \frac{\lambda\mu e^{-(\lambda+\mu)\theta}}{(\lambda+\mu)^2}$	$JA(\theta) = \Pr\{\text{item up at } t \cap \text{item up at } t+\theta\}$, $PA_{us}(\theta) = 1 - F_A(\theta) + \int_0^{\theta} h_{duu}(x)(1-F(\theta-x))\,dx)$

Table 2 Main results for a one-item repairable structure in steady-state (λ = failure rate; μ = repair rate)

For the stationary value of the point-availability it follows

$$PA_S = AA_S = \frac{1}{1+\sum_{i=1}^{n} \lambda_{0i}\, MTTR_i}, \tag{50}$$

and for that of the interval reliability

$$IR_S(\theta) = PA_S\, e^{-\lambda_{S0}\theta}. \tag{51}$$

Investigations of the point-availability and of the interval reliability for the case of arbitrary failure rates are still in progress (for the reliability function, $R_S(t) = R_1(t) \dots R_n(t)$ still holds).

3.3 One-out-of-two Redundancies

One-out-of-two redundancies have been largely investigated in the literature. Table 3 gives the results for some cases for which a regenerative process still exist (see [3] for greater details and for a further generalization in the case of warm redundancy).

			1-out-of-2 standby redundancy	1-out-of-2 warm redundancy		1-out-of-2 active redundancy
Elements E_1, E_2	Distribution of the failure-free time	OS	$F(t)$	$1-e^{-\lambda t}$	$F(t)$	$1-e^{-\lambda t}$
		RS	–	$1-e^{-\lambda_r t}$	$1-e^{-\lambda_r t}$	$1-e^{-\lambda t}$
	Destribution of the repair time	OS	$G(t)$	$G(t)$	$G(t)$	$G(t)$
		RS	–	$G(t)$	$W(t)$	$G(t)$
	Mean time to failure		$MTTF = \int_0^\infty (1-F(t))dt$	$\frac{1}{\lambda}$ bzw. $\frac{1}{\lambda_r}$	$MTTF$ bzw. $\frac{1}{\lambda_r}$	$\frac{1}{\lambda}$
	Mean time to repair		$MTTR = \int_0^\infty (1-G(t))dt$	$MTTR$	$MTTR$ bzw. $MTTR_W$	$MTTR$
1-out-of-2 redundancy	Mean time to system failure ($MTTF_{S0}$, Z_0 is entered at $t=0$)		$MTTF + \frac{MTTF}{1-\int_0^\infty f(t)G(t)dt}$	$\frac{1}{\lambda}+\frac{1}{(\lambda+\lambda_r)(1-\tilde{g}(\lambda))} \approx \frac{1}{\lambda}(1+\frac{1}{(\lambda+\lambda_r)MTTR})$	$MTTF + \frac{MTTF\int_0^\infty u_3(t)dt}{1-\int_0^\infty u_1(t)dt}$	$\frac{1}{\lambda}+\frac{1}{2\lambda(1-\tilde{g}(\lambda))} \approx \frac{1}{\lambda}(1+\frac{1}{2\lambda MTTR})$
	Point- and average-availability ($PA_S = AA_S$)*		$\frac{MTTF}{1-\int_0^\infty t\,d(F(t)G(t))}$	$\frac{\lambda+\lambda_r(1-\tilde{g}(\lambda))}{\lambda(\lambda+\lambda_r)MTTR+\lambda\tilde{g}(\lambda)}$	$\frac{MTTF}{\int_0^\infty t(u_1(t)+u_2(t))dt}$	$\frac{2-\tilde{g}(\lambda)}{2\lambda MTTR+\tilde{g}(\lambda)}$
	Interval-reliability ($IR_S(\theta)$)*		$\approx R_{S0}(\theta)$	$\approx R_{S0}(\theta)$	$\approx R_{S0}(\theta)$	$\approx R_{S0}(\theta)$

OS = operating state; RS = reserve state; *asymptotical and stationary value

Table 3 Main results for repairable one-out-of-two redundancies (only one repair crew)

The case of two different elements in active redundancy with constant failure rates λ_1 and λ_2, and constant repair rates μ_1 and μ_2 leads to the mean time to system failure [3]

$$MTTF_{S0} = \frac{(\lambda_1+\mu_2)(\lambda_2+\mu_1)+\lambda_1(\lambda_1+\mu_2)+\lambda_2(\lambda_2+\mu_1)}{\lambda_1\lambda_2(\lambda_1+\lambda_2+\mu_1+\mu_2)} \tag{52}$$

and to the stationary value of the point availability

$$PA_S = \frac{\mu_1\mu_2(\mu_1\mu_2+(\lambda_1+\lambda_2)(\lambda_1+\lambda_2+\mu_1+\mu_2))}{\mu_1^2\mu_2^2+\mu_1\mu_2(\lambda_1+\lambda_2)(\lambda_1+\lambda_2+\mu_1+\mu_2)+\lambda_1\lambda_2(\mu_1^2+\mu_2^2+(\lambda_1+\lambda_2)(\mu_1+\mu_2))}. \tag{53}$$

3.4 *k*-out-of-*n* Redundancies

In most practical applications it is assumed that all elements of a *k-out-of-n redundancy* are identical with constant failure rate λ in operating state and λ_r in reserve state, and constant repair rate μ. The stochastic process governing the system behaviour is then a *birth and death process*. Assuming only one repair crew and no further failure at system down, the results depend on λ, λ_r, μ and $n-k$ only. Table 4 gives these results explicitly for $n-k=1$ and $n-k=2$. The generalization of the repair rate leads to regenerative stochastic processes with only two regeneration states. The generalization of the failure and repair rates leads to non-regenerative stochastic processes.

3.5 Series/Parallel Structures

As an example of a *series/parallel structure* let us consider a majority redundancy with $n = 1$, i.e. a 2-out-of-3 redundancy in series with a voter (Fig. 1). Let λ be the failure rate of the three parallel elements and λ_v that of the voter. The repair times are distributed according to G(x) with density g(x) and mean *MTTR*. The system has only one repair crew, no repair priority on the voter and no further failure at system down. Because of the assumptions made the process has 5 states (Z_0 to Z_4) of which only states Z_0, Z_1 and Z_4 are regenerative (use Fig. 1 only to visualize the states). The corresponding semi-Markov transition probabilities are, according to Tab. 1,

$$Q_{01}(x) = \int_0^x 3\lambda e^{-3\lambda y} e^{-\lambda_v y} dy \qquad Q_{12}(x) = \int_0^x 2\lambda e^{-2\lambda y} e^{-\lambda_v y}(1-G(y))dy$$

$$Q_{10}(x) = \int_0^x g(y) e^{-(2\lambda+\lambda_v)y} dy \qquad Q_{121}(x) = \int_0^x g(y)\frac{2\lambda}{2\lambda+\lambda_v}(1-e^{-(2\lambda+\lambda_v)y})dy$$

$$Q_{13}(x) = \frac{\lambda_v}{2\lambda} Q_{12}(x) \qquad Q_{134}(x) = \frac{\lambda_v}{2\lambda} Q_{121}(x)$$

$$Q_{04}(x) = \frac{\lambda_v}{3\lambda} Q_{01}(x) \qquad Q_{40}(x) = G(x). \tag{54}$$

$Q_{12}(x)$ and $Q_{13}(x)$ will be used to compute the reliability function. $Q_{121}(x)$ and $Q_{134}(x)$ take care of a transition throughout the non-regenerative states Z_2 and Z_3 respectively. They will be used to compute the point-availability.

		Mean time to system failure ($MTTF_{S0}$, Z_0 is entered at $t=0$)	asymptotical and stationary value of the point-availability and average-availability ($PA_S = AA_S$)	interval-reliability ($IR_S(\theta)$)
$n-k=1$	gen. case	$\frac{\nu_0+\nu_1+\mu}{\nu_0\,\nu_1}$	$\frac{\nu_0\mu+\mu^2}{\nu_0\,\nu_1+\nu_0\mu+\mu^2}$	$\approx R_{S0}(\theta)^*$
$n-k=1$	$n=2$, $k=1$	$\frac{2\lambda+\lambda_r+\mu}{\lambda(\lambda+\lambda_r)}$	$\frac{\mu(\lambda+\lambda_r+\mu)}{(\lambda+\lambda_r)(\lambda+\mu)+\mu^2}$	$\approx R_{S0}(\theta)^*$
$n-k=1$	$n=3$, $k=2$	$\frac{4\lambda+\lambda_r+\mu}{2\lambda(2\lambda+\lambda_r)}$	$\frac{\mu(2\lambda+\lambda_r+\mu)}{(2\lambda+\lambda_r)(2\lambda+\mu)+\mu^2}$	$\approx R_{S0}(\theta)^*$
$n-k=2$	gen. case	$\frac{\nu_2(\nu_0+\nu_1+\mu)}{\nu_0\,\nu_1\,\nu_2}+\frac{\mu(\nu_0+\mu)+\nu_0\,\nu_1}{\nu_0\,\nu_1\,\nu_2}$	$\frac{\nu_0\,\nu_1\mu+\nu_0\mu^2+\mu^3}{\nu_0\,\nu_1\,\nu_2+\nu_0\,\nu_1\mu+\nu_0\mu^2+\mu^3}$	$\approx R_{S0}(\theta)^*$
$n-k=2$	$n=3$, $k=1$	$\frac{1}{\lambda}+\frac{\lambda(2\lambda+3\lambda_r+\mu)}{\lambda(\lambda+\lambda_r)(\lambda+2\lambda_r)}+\frac{\mu(\lambda+2\lambda_r+\mu)}{\lambda(\lambda+\lambda_r)(\lambda+2\lambda_r)}$	$\frac{\mu((\lambda+2\lambda_r)(\lambda+\lambda_r)+\mu(\lambda+2\lambda_r)+\mu^2)}{(\lambda+2\lambda_r)(\lambda(\lambda+\lambda_r)+\mu(\lambda+\lambda_r)+\mu^2)+\mu^3}$	$\approx R_{S0}(\theta)^*$
$n-k=2$	$n=5$, $k=3$	$\frac{1}{3\lambda}+\frac{3\lambda(6\lambda+3\lambda_r+\mu)}{(3\lambda+2\lambda_r)(3\lambda+\lambda_r)3\lambda}+\frac{\mu(3\lambda+2\lambda_r+\mu)}{(3\lambda+2\lambda_r)(3\lambda+\lambda_r)3\lambda}$	$\frac{\mu(3\lambda+2\lambda_r)(3\lambda+\lambda_r+\mu)+\mu^2}{(3\lambda+2\lambda_r)(3\lambda+\lambda_r)(3\lambda+\mu)+\mu^2(3\lambda+2\lambda_r)+\mu^3}$	$\approx R_{S0}(\theta)^*$
$n-k$ arbitrary		$MTTF_{Sj}=\int_0^\infty R_{Sj}(t)dt$	$PA_S = AA_S = \sum_{j=0}^{n-k} p_j$; $p_j=\pi_j / \sum_{i=0}^{n-k+1}\pi_i$, $\pi_0=1$ and $\pi_i=\frac{\nu_0\,\nu_1 \dots \nu_{i-1}}{\mu^i}$	$IR_S(\theta)=\sum_{j=0}^{n-k} p_j R_{Sj}(\theta)$

$\nu_i = k\lambda + (n-k-i)\,\lambda_r$, $i=0, \dots, n-k$; λ, λ_r = failure rates ($\lambda_r=\lambda \rightarrow$ active, $\lambda_r \equiv 0 \rightarrow$ standby); μ = repair rate; *see [3, 1985] for exact equations

Table 4 Main results for repairable *k*-out-of-*n* redundancies (constant failure and repair rates, one repair crew, no further failure at system down)

For the reliability functions $R_{S0}(t)$ and $R_{S1}(t)$ one obtains, according to Tab. 1,

$$R_{S0}(t)=e^{-(3\lambda+\lambda_v)t}+\int_0^t 3\lambda e^{-(3\lambda+\lambda_v)x}R_{S1}(t-x)\,dx$$

$$R_{S1}(t)=e^{-(2\lambda+\lambda_v)t}(1-G(t))+\int_0^t g(x)e^{-(2\lambda+\lambda_v)x}R_{S0}(t-x)\,dx. \tag{55}$$

The Laplace transform of $R_{S0}(t)$ is then given by (Tab.1)

$$\tilde{R}_{S0}(s) = \frac{s + 5\lambda + \lambda_v - 3\lambda\,\tilde{g}(s + 2\lambda + \lambda_v)}{(s + 2\lambda + \lambda_v)(s + \lambda_v + 3\lambda(1 - \tilde{g}(s + 2\lambda + \lambda_v)))} \tag{56}$$

and the mean time to system failure $MTTF_{S0}$ by (Tab. 1)

$$MTTF_{S0} = \frac{5\lambda + \lambda_v - 3\lambda\,\tilde{g}(2\lambda + \lambda_v)}{(2\lambda + \lambda_v)(\lambda_v + 3\lambda(1 - \tilde{g}(2\lambda + \lambda_v)))}. \tag{57}$$

The point-availability $PA_{S0}(t) = P_{00}(t) + P_{01}(t)$ can be computed from Tab. 1 with $U = \{Z_0, Z_1\}$ and $P_{ij}(t)$ from the following systems of integral equations

$$P_{00}(t) = e^{-(3\lambda+\lambda_v)t} + \int_0^t 3\lambda e^{-(3\lambda+\lambda_v)x} P_{10}(t-x)\,dx + \int_0^t \lambda_v e^{-(3\lambda+\lambda_v)x} P_{40}(t-x)\,dx$$

$$P_{10}(t) = \int_0^t g(x) e^{-(2\lambda+\lambda_v)x} P_{00}(t-x)\,dx + \int_0^t \frac{2\lambda}{2\lambda+\lambda_v}(1 - e^{-(2\lambda+\lambda_v)x}) g(x) P_{10}(t-x)\,dx$$

$$+ \int_0^t \frac{\lambda_v}{2\lambda+\lambda_v}(1 - e^{-(2\lambda+\lambda_v)x}) g(x) P_{40}(t-x)\,dx$$

$$P_{40}(t) = \int_0^t g(x) P_{00}(t-x)\,dx$$

and

$$P_{01}(t) = \int_0^t 3\lambda e^{-(3\lambda+\lambda_v)x} P_{11}(t-x)\,dx + \int_0^t \lambda_v e^{-(3\lambda+\lambda_v)x} P_{41}(t-x)\,dx$$

$$P_{11}(t) = e^{-(2\lambda+\lambda_v)t}(1 - G(t)) + \int_0^t g(x) e^{-(2\lambda+\lambda_v)x} P_{01}(t-x)\,dx$$

$$+ \int_0^t \frac{1}{2\lambda+\lambda_v}(1 - e^{-(2\lambda+\lambda_v)x}) g(x)(2\lambda P_{11}(t-x) + \lambda_v P_{41}(t-x))\,dx$$

$$P_{41}(t) = \int_0^t g(x) P_{01}(t-x)\,dx. \tag{58}$$

The interval-reliability $IR_{S0}(t, t+\theta)$ can be computed by

$$IR_{S0}(t, t+\theta) \approx P_{00}(t) R_{S0}(\theta) + P_{01}(t) R_{S1}(\theta), \tag{59}$$

with $P_{00}(t)$ and $P_{01}(t)$ from Eq. (58) and $R_{S0}(\theta)$ and $R_{S1}(\theta)$ from Eq. (55). The approximation, which assumes that state Z_1 is regenerative at each time point, is good if $MTTR \ll 1/\lambda$ holds.

Asymptotic behavior exists, independent of initial conditions at $t = 0$ and leads to

$$PA_S = \frac{2\lambda + \lambda_v + \lambda(1 - \tilde{g}(2\lambda + \lambda_v))}{(2\lambda + \lambda_v)(1 + (3\lambda + \lambda_v)MTTR) + \lambda(\lambda_v\, MTTR - 2)(1 - \tilde{g}(2\lambda + \lambda_v))}, \tag{60}$$

$$\mathrm{IR}_S(\theta) \approx p_0\, \mathrm{R}_{S0}(\theta)$$
$$= \frac{((2\lambda+\lambda_v) - 2\lambda(1-\tilde{g}(2\lambda+\lambda_v)))\mathrm{R}_{S0}(\theta)}{(2\lambda+\lambda_v)(1+(3\lambda+\lambda_v)MTTR) + \lambda(\lambda_v\, MTTR - 2)(1-\tilde{g}(2\lambda+\lambda_v))}. \tag{61}$$

Equation (61) only considers the first term of (59). This approximation holds for $MTTR \ll 1/\lambda$. Generalization of repair and failure rates leads to non-regenerative stochastic processes.

4. Complex Structures

Investigation of complex structures, also only of large series/parallel structures, may become very time consuming, even if constant failure and repair rates are assumed. In these cases, an approximation of the system reliability and availability can be obtained by assuming that each element in the reliability block diagram (at system level) works, fails and is repaired independently of any other element. The approximation is good for practical applications in which for each element $MTTR\,/\,MTTF \ll 1$ holds (ratios less than 0.01 are frequent). Such an assumption does not anyway change the reliability function for the cases of a series structure and of a one-out-of-two redundancy. For the steady-state point-availability it holds in this cases

$$PA_S = PA_1 \cdot PA_2 \cdot \ldots \cdot PA_n \tag{62}$$

for the serie connection of Elements E_1 to E_n and

$$PA_S = PA_1 + PA_2 - PA_1 \cdot PA_2 \tag{63}$$

for the one-out-of-two active redundancy of elements E_1 and E_2 (Eqs. (62) and (63) remain true also for $PA_i = PA_i(t)$). It is not difficult to see that equations (50) and (62), as well as (53) and (63), will give the same approximation for $MTTR_i \ll 1\,/\,\lambda_i$. Table 5 summarizes the basic models used for such investigations and give also the corresponding exact and approximating equations.

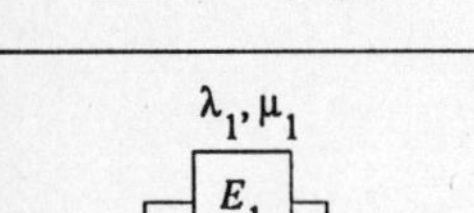	$\lambda_S = \lambda,\quad \mu_S = \mu,\quad PA_S = \frac{1}{1+\lambda_S/\mu_S} \approx 1 - \frac{\lambda_S}{\mu_S}$ $\Rightarrow\ \mu_S = \frac{\lambda_S\, PA_S}{1-PA_S} \approx \frac{\lambda_S}{1-PA_S}$
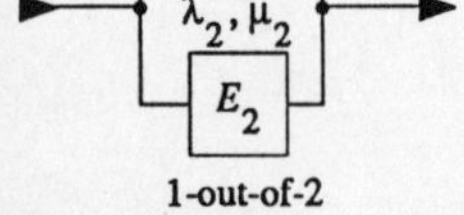	$PA_S = PA_1\, PA_2 = \frac{1}{1+\lambda_1/\mu_1} \cdot \frac{1}{1+\lambda_2/\mu_2} \approx 1 - (\frac{\lambda_1}{\mu_1} + \frac{\lambda_2}{\mu_2})$ $\lambda_S = \lambda_1 + \lambda_2\ \Rightarrow\ \mu_S \approx \frac{\lambda_S}{1-PA_S} \approx \frac{\lambda_1+\lambda_2}{\lambda_1/\mu_1 + \lambda_2/\mu_2}$
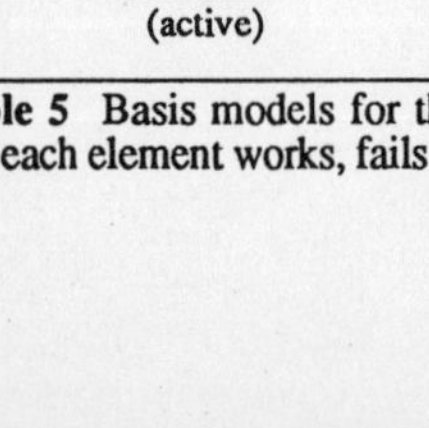 1-out-of-2 (active)	$PA_S = PA_1 + PA_2 - PA_1\, PA_2$ $\approx 1 - \frac{\lambda_1 \lambda_2}{\mu_1\mu_2 + \mu_1\lambda_2 + \mu_2\lambda_1} \approx 1 - \frac{\lambda_1\lambda_2}{\mu_1\mu_2}$ $MTTF_{S0} = \frac{(\lambda_1+\mu_2)(\lambda_2+\mu_1) + \lambda_1(\lambda_1+\mu_2) + \lambda_2(\lambda_2+\mu_1)}{\lambda_1\lambda_2(\lambda_1+\lambda_2+\mu_1+\mu_2)} = \frac{1}{\lambda_S}$ $\rightarrow\ \lambda_S \approx \frac{\lambda_1\lambda_2(\mu_1+\mu_2)}{\mu_1\mu_2}\quad \Rightarrow\ \mu_S \approx \frac{\lambda_S}{1-PA_S} \approx \mu_1 + \mu_2$

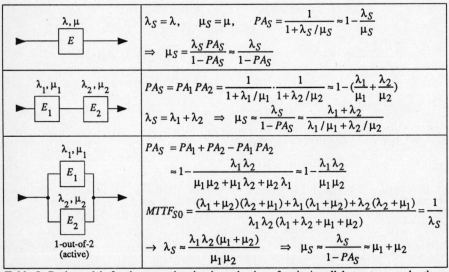

Table 5 Basis models for the approximating investigation of series/parallel structures under the assumption that each element works, fails and is repaired independently of any other element

As an example let us consider the system given by the reliability block diagram of Fig. 2.

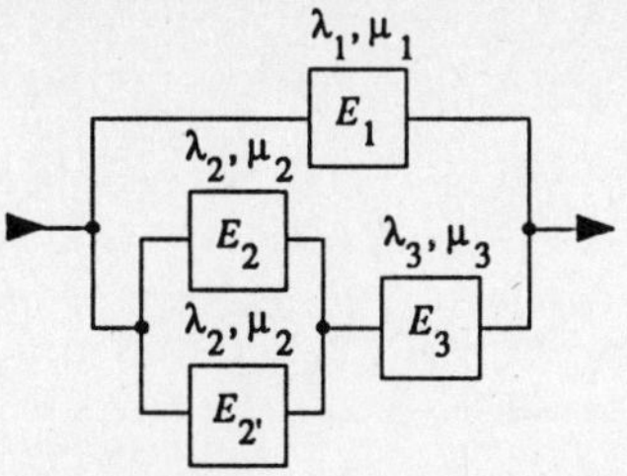

Figure 2 Reliability block diagram of a moderately complex system (active redundancy)

With λ_1 to λ_3 as constant failure rates and μ_1 to μ_3 as constant repair rates of element E_1 to E_3 it follows for the mean time to system failure and for the steady-state value of the point-availability (see Fig. 3 for details)

$$MTTF_{S0} = 1/\lambda_s \qquad \text{with} \qquad \lambda_S \approx \lambda_1 ((\frac{\lambda_2}{\mu_2})^2 + \frac{\lambda_3}{\mu_3} + \frac{2\lambda_2^2}{\mu_1 \mu_2} + \frac{\lambda_3}{\mu_1}) \tag{64}$$

and

$$PA_S \approx 1 - \frac{\lambda_S}{\mu_S} = 1 - \frac{\lambda_1}{\mu_1}(\frac{\lambda_3}{\mu_3} + (\frac{\lambda_2}{\mu_2})^2). \tag{65}$$

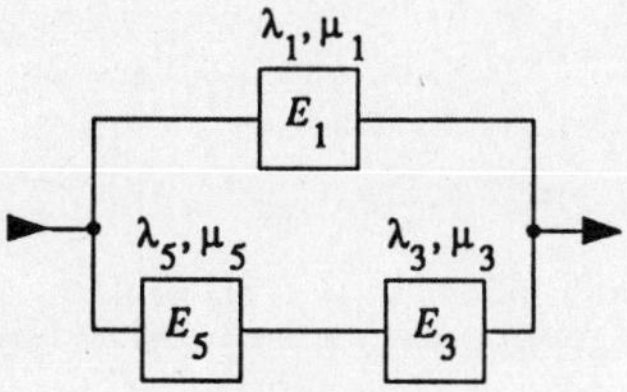

$$\lambda_5 \approx \frac{2\lambda_2^2}{\mu_2}, \qquad \mu_5 \approx 2\mu_2$$

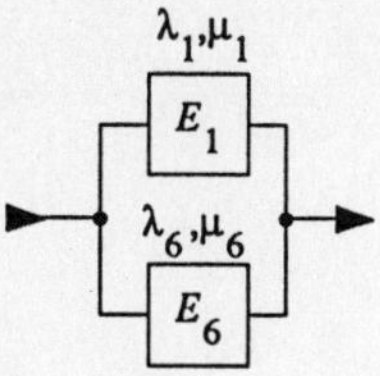

$$\lambda_6 \approx \lambda_3 + \lambda_5, \quad \mu_6 \approx \frac{\lambda_5 + \lambda_3}{\lambda_5 / \mu_5 + \lambda_3 / \mu_3}$$

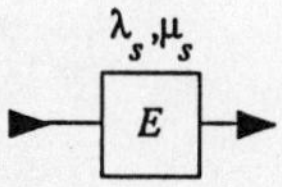

$$\lambda_S \approx \frac{\lambda_1 \lambda_6 (\mu_1 + \mu_6)}{\mu_1 \mu_6}, \qquad \mu_S \approx \mu_1 + \mu_6$$

Figure 3 Step by step calculation of the reliability structure according to Fig. 2.

The investigation of the structure given by Fig. 2 by assuming no further failure at system down and repair priority according to the sequence E_1, E_3, E_2 leads to the diagram of transition probabilities in $(t, t+\delta t]$ given in Fig. 4.

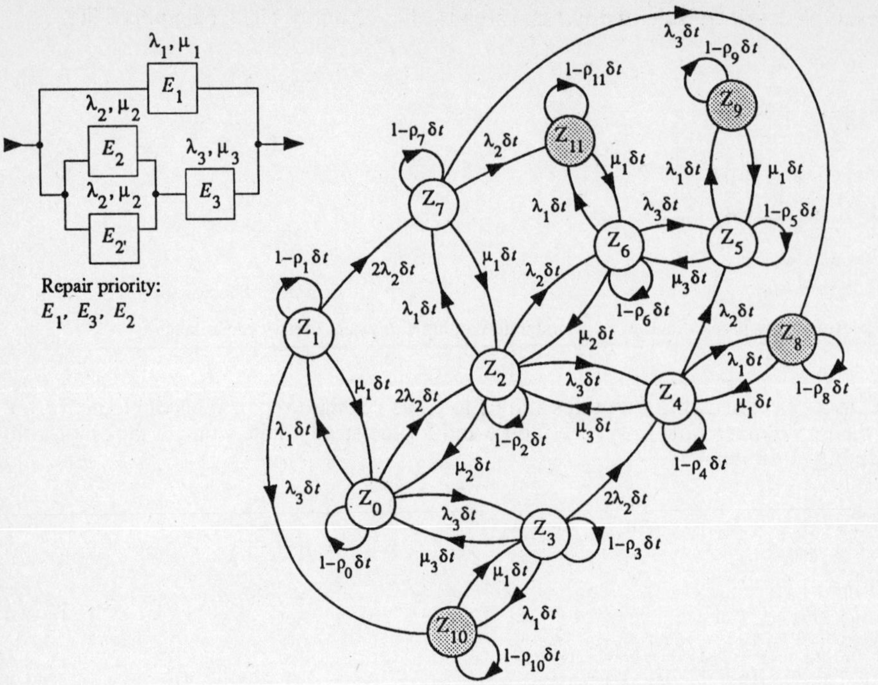

$\rho_{01}=\lambda_1$; $\rho_{02}=2\lambda_2$; $\rho_{03}=\lambda_3$; $\rho_{10}=\mu_1$; $\rho_{17}=2\lambda_2$; $\rho_{1\,10}=\lambda_3$; $\rho_{20}=\mu_2$;
$\rho_{24}=\lambda_3$; $\rho_{26}=\lambda_2$; $\rho_{27}=\lambda_1$; $\rho_{30}=\mu_3$; $\rho_{34}=2\lambda_2$; $\rho_{3\,10}=\lambda_1$; $\rho_{42}=\mu_3$;
$\rho_{45}=\lambda_2$; $\rho_{48}=\lambda_1$; $\rho_{56}=\mu_3$; $\rho_{59}=\lambda_1$; $\rho_{62}=\mu_2$; $\rho_{65}=\lambda_3$; $\rho_{6\,11}=\lambda_1$;
$\rho_{72}=\mu_1$; $\rho_{78}=\lambda_3$; $\rho_{7\,11}=\lambda_2$; $\rho_{84}=\mu_1$; $\rho_{95}=\mu_1$; $\rho_{10\,3}=\mu_1$; $\rho_{11\,6}=\mu_1$

Figure 4 Reliability block diagram and diagram of the transition probabilities in $(t, t+\delta t]$ of the system given by Fig. 2 (constant failure and repair rates, only one repair crew, repair priority E_1, E_3, E_2, no further failure at system down, t arbitrary (e.g. $t = 0$), $\delta t \to 0$)

From Fig. 4 it follows (using Tab. 1 and with ρ_i as in Fig. 4)

$$MTTF_{S0} = \frac{a_5 + a_6(a_8 + a_9 a_{10}) + a_7 a_{10}}{1 - a_6 a_{12} - a_{11}(a_7 + a_6 a_9)}, \tag{66}$$

with

$$a_1 = \frac{1}{\rho_4} + \frac{\lambda_2}{\rho_4 \rho_5}\left(1 + \mu_3 \frac{\lambda_3 + \rho_5}{\rho_5 \rho_6 - \lambda_3 \mu_3}\right), \quad a_2 = \frac{\lambda_2 \mu_2 \mu_3}{\rho_4(\rho_5 \rho_6 - \lambda_3 \mu_3)} + \frac{\mu_3}{\rho_4}, \quad a_3 = \frac{1}{\rho_3}(1 + 2\lambda_2 a_1),$$

$$a_4 = \frac{2\lambda_2}{\rho_3} a_2, \quad a_5 = \frac{1 + \lambda_3 a_3}{\rho_0 - \lambda_3 \mu_3 / \rho_3}, \quad a_6 = \frac{\lambda_1}{\rho_0 - \lambda_3 \mu_3 / \rho_3}, \quad a_7 = \frac{2\lambda_2 + \lambda_3 a_4}{\rho_0 - \lambda_3 \mu_3 / \rho_3},$$

$$a_8 = \frac{1 + 2\lambda_2 / \rho_7}{\rho_1}, \quad a_9 = \frac{2\lambda_2 \mu_1}{\rho_1 \rho_7}, \quad a_{10} = \frac{1 + \lambda_3 a_1 + \dfrac{\lambda_2 \lambda_3 + \lambda_2 \rho_5}{\rho_5 \rho_6 - \lambda_3 \mu_3} + \dfrac{\lambda_1}{\rho_7}}{\rho_2 - \lambda_3 a_2 - \dfrac{\lambda_2 \mu_2 \rho_5}{\rho_5 \rho_6 - \lambda_3 \mu_3} - \dfrac{\lambda_1 \mu_1}{\rho_7}},$$

$$a_{11} = \frac{\mu_2}{\rho_2 - \lambda_3 a_2 - \dfrac{\lambda_2 \mu_2 \rho_5}{\rho_5 \rho_6 - \lambda_3 \mu_3} - \dfrac{\lambda_1 \mu_1}{\rho_7}}, \quad a_{12} = \frac{\mu_1}{\rho_1}.$$

and

$$PA_S = p_0(1+b_1+b_2+b_3+b_4+b_5+b_6+b_7), \tag{67}$$

with $p_0 = 1/(1+b_1+\ldots+b_{11})$ and

$$b_1 = \frac{\lambda_1}{\rho_1}, \qquad b_2 = \frac{\rho_0 - \lambda_1\mu_1/\rho_1}{\mu_2} - \frac{\mu_3\lambda_3(1+\lambda_1/\rho_1)}{(\mu_3+2\lambda_2)\mu_2}, \qquad b_3 = \frac{\lambda_3(1+\lambda_1/\rho_1)}{\mu_3+2\lambda_2},$$

$$b_4 = \frac{\lambda_3 b_2(1+\lambda_1/\rho_7)+2\lambda_2 b_3 + \dfrac{2\lambda_1\lambda_2\lambda_3}{\rho_7\rho_1}}{\rho_4-\lambda_1},$$

$$b_5 = \frac{\lambda_2 b_4 + \dfrac{\lambda_2\lambda_3}{\rho_7(\mu_2+\lambda_3)}(b_2(\rho_7+\lambda_1)+2\lambda_1\lambda_2/\rho_1)}{\rho_5-\lambda_1-\dfrac{\mu_3\lambda_3}{\mu_2+\lambda_3}},$$

$$b_6 = \frac{\lambda_2}{\mu_2+\lambda_3}\left(b_2+\frac{2\lambda_1\lambda_2}{\rho_1\rho_7}+\frac{\lambda_1}{\rho_7}b_2+\frac{\mu_3}{\lambda_2}b_5\right), \qquad b_7 = \frac{2\lambda_1\lambda_2}{\rho_1\rho_7}+\frac{\lambda_1}{\rho_7}b_2, \qquad b_8 = \frac{\lambda_1}{\mu_1}b_4+\frac{\lambda_3}{\mu_1}b_7,$$

$$b_9 = \frac{\lambda_1}{\mu_1}b_5, \qquad b_{10} = \frac{\lambda_3\lambda_1}{\mu_1\rho_1}+\frac{\lambda_1}{\mu_1}b_3, \qquad b_{11} = \frac{\lambda_1}{\mu_1}b_6+\frac{\lambda_2}{\mu_1}b_7.$$

Comparing the exact results of Eqs. (66) and (67) with the approximating values given by Eqs. (64) and (65) one obtains for some values of λ_1, λ_2, λ_3, μ_1, μ_2, and μ_3 (λ, μ in h^{-1} and $MTTF$ in h)

λ_1	1/100	1/100	1/1'000	1/1'000
λ_2	1/1'000	1/1'000	1/10'000	1/10'000
λ_3	1/10'000	1/10'000	1/100'000	1/100'000
μ_1	1	1/5	1	1/5
μ_2	1/5	1/5	1/5	1/5
μ_3	1/5	1/5	1/5	1/5
$MTTF_{S0}$ (Eq. (64))	$1.575\cdot10^{+5}$	$9.302\cdot10^{+4}$	$1.657\cdot10^{+7}$	$9.926\cdot10^{+6}$
$MTTF_{S0}$ (Eq. (66))	$1.589\cdot10^{+5}$	$9.332\cdot10^{+4}$	$1.658\cdot10^{+7}$	$9.927\cdot10^{+6}$
$1-PA_S$ (Eq. (65))	$5.250\cdot10^{-6}$	$2.625\cdot10^{-5}$	$5.025\cdot10^{-8}$	$2.513\cdot10^{-7}$
$1-PA_S$ (Eq. (67))	$6.574\cdot10^{-6}$	$5.598\cdot10^{-5}$	$6.060\cdot10^{-8}$	$5.062\cdot10^{-7}$

From this it follows that for $\lambda_i/\mu_i \ll 1$, approximating expressions obtained by assuming independent elements (Tab. 5) can be successfully used in many practical applications (research in this field, also using macro structures, is still in progress [3, 3rd Ed.]).

Further aspects which can be considered in the context of complex systems are the influence of preventive maintenance, of hidden failures, and of imperfect switching. Extensive literature exists on these subjects, some basic models can be found in [3].

5. Conclusions

Stochastic processes are powerful tools for the investigation of reliability problems, in praticular of the time behaviour of repairable equipment and systems. If all elements have constant failure and repair rates, time-homogeneous Markov processes with a finite state space can be used. Markov processes also arise if the repair and/or failure-free times have an Erlang distribution (supplementary states). If repair rates are general, but the system has only one repair crew, the involved process contains an embedded semi-Markov process. A further generalization leads to processes with only one (of a few) regeneration state(s), or even to non-regenerative stochastic processes. For complex structures, analytical solutions become more and more difficult, even if constant failure and repair rates are assumed. Useful approximating expressions for practical applications can be obtained by assuming that each element of the reliability block diagram at system level has its own repair crew or by using macro structures.

References

[1] Ascher, H. and Feingold, H., *Repairable Systems Reliability*. New York: Marcel Dekker, 1984.

[2] Barlow, R.E. and Proschan, F., *Mathematical Theory of Reliability*. New York: Wiley, 1965; *Statistical Theory of Reliability and Life Testing*, New York: Holt, Rinehart & Winston, 1975.

[3] Birolini, A., *On the Use of Stochastic Processes in Modeling Reliability Problems*. Springer-Verlag, Lect. Notes in Ec. and Math. Syst. Nr. 252, 1985; *Qualität und Zuverlässigkeit technischer Systeme*. Berlin: Springer-Verlag, 2nd Ed. 1988 (3rd Ed. with English translation in press).

[4] Cox, D.R., The analysis of non Markovian stochastic processes by the inclusion of supplementary variables. *Proc. Cambridge Phil. Soc.*, 51 (1955), pp. 433-441.

[5] Feller, W., *An Introduction to Probability Theory and its Applications*. New York: Wiley, vol. I 1957, vol. II 1966.

[6] Gnedenko, B.V., Beljajev J.K. and Soloviev, A.D., *Mathematical Methods of Reliability Theory*. New York: Academic, 1969.

[7] Osaki, S. and Hatoyama Y. (eds.), *Stochastic Models in Reliability Theory*. Berlin: Springer Verlag, Lect. Notes in Ec. and Math. Syst. Nr. 235, 1984.

[8] Parzen, E., *Stochastic Processes*. San Francisco: Holden-Day, 1962.

[9] Smith, W.L., Asymptotic renewal theorems. *Proc. Roy. Soc. Edinbourgh*, 64 (1954), pp. 9-48; Regenerative stochastic processes. *Proc. Int. Congress Math. Amsterdam*, 3 (1954), pp. 304-305; Regenerative stochastic processes. *Proc. Royal Soc. London,* Ser. A, 256 (1960), pp. 496-501.; Renewal theory and its ramifications. *J. Roy. Stat Soc.*, Ser. B, 20 (1958), pp. 243-302

[10] Soloviev, A.D., Standby with rapid renewal. *Eng. Cybernetics*, (1970) 1, pp. 49-64; Asymptotic behavior of the time of first occurence of a rare event. *Eng. Cybernetics*, (1971) 6, pp. 1038-1048.

[11] Srinivasan, S.K. and Mehata K.M., *Stochastic Processes*. New Delhi: McGraw-Hill, 2nd Ed. 1988.

[12] Störmer, H., *Semi-Markoff-Prozesse mit endlich vielen Zuständen*. Berlin: Springer Verlag, Lect. Notes in Op. Res. and Math. Syst. Nr. 34, 1970.

STOCHASTIC PROCESSES AND OPTIMIZATION PROBLEMS IN ASSEMBLAGE SYSTEMS

Franz Ferschl
Seminar für Angewandte Stochastik, Universität München
Akademiestrasse 1/IV, 8000 München 19

ABSTRACT

Assemblage systems e.g. arise from production systems, where k different pieces are delivered by k parallel production lines. At some place this pieces are assembled in order to form the desired good, taking exactly one piece from every of the k single lines. According to random fluctuations of the production process parallel queues are formed by single pieces waiting to be processed at the assemblage station. If one neglects the time needed to transform a complete group of k different parts into the final assembled good, a pure assemblage system emerges. It is always possible, to separate an assemblage system into two stages: firstly a pure assemblage system, followed by a queueing system with a single waiting line of complete sets of k pieces. In this paper pure assemblage systems are considered. If such systems have unlimited waiting–room for each line, an equilibrium distribution of queue lenghts never exists. Therefore control measures such as limitation of the waiting–room or partial reduction of production speed are taken and give raise to various optimization problems. With few exceptions, such problems seem to be rather difficult for $k > 2$ production lines.

In this article the specific reasons of the difficulties of multiline systems should be pointed out and the following results should be given: Firstly, an algorithm in matrix form for the calculation of the equilibrium distribution in the case of three production lines; secondly, approximations for the distribution of the number of single parts waiting in front of the assemblage system; this approximations are needed for explicit handling of optimization problems.

1. INTRODUCTION

Assemblage systems arise mostly from production systems, where ℓ different *parts* or *components* are delivered by ℓ parallel production lines. At some place these parts are assembled in order to form a desired good, called *item*, taking exactly one part from each of the ℓ single lines. According to random fluctuations of the production process parallel queues are formed by single parts waiting to be processed at a *service station.* The server processes only complete sets of ℓ parts, otherwise it is idle.

There are several possibilities to look at the system of parallel queues in front of the service station. If one neglects the time needed to transform a complete group of ℓ different parts into the final item, a so called *pure assemblage system* emerges. We then use the expression of an *assemblage point* rather than of a service station. One characteristic feature of a pure assemblage system is that at least one of the parallel queues must be empty, indicating the production line (resp.lines), which supplied the smallest number of parts up to the observed moment. It is always possible to separate an assemblage system into two stages: firstly a pure assemblage system terminated by an assemblage point, followed by a queueing system with a single waiting line of complete sets of ℓ pieces, waiting for treatment at the service station.

As HARRISON [6] showed in much detail, if no measures are taken to control the length of some or of all the parallel queues, an equilibrium of the system never exists. Different types of control are considered in the literature. For instance, LATOUCHE [9] puts no limits on the waiting-room for the second stage; he does not separate the pure assemblage system but rather looks at the *excess* of

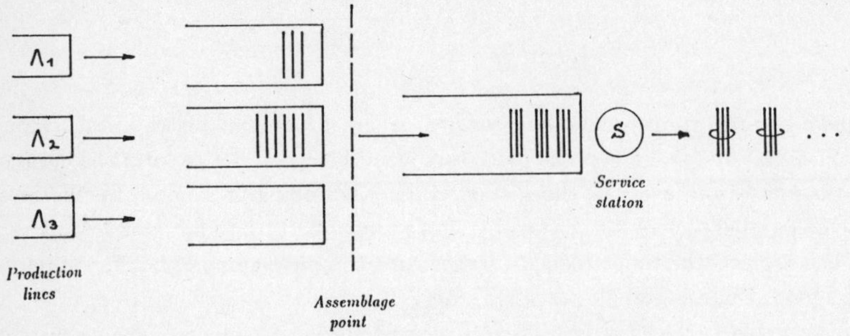

Figure 1.1: The two stages of an assemblage system.

a waiting line of single parts over another line. This is equivalent to control only the pure assemblage system, whereas the queue of complete sets has unlimited waiting-room. In such a case one also speaks of an ***assembly-like queue.***

In two more recent articles LIPPER and SENGUPTA [10] and BHAT [1] investigate ***finite capacity assembly-like queues.*** In this type of assemblage system each of the ℓ lines has its own, separated waiting-room of possibly different size.

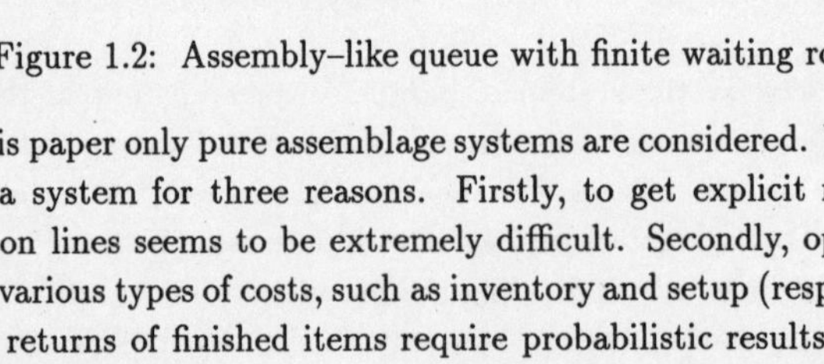

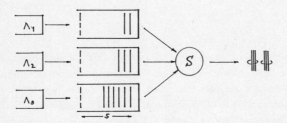

Figure 1.2: Assembly-like queue with finite waiting room

In this paper only pure assemblage systems are considered. We concentrate upon the investigation of such a system for three reasons. Firstly, to get explicit results in the case of more than two production lines seems to be extremely difficult. Secondly, optimization problems which take into account various types of costs, such as inventory and setup (resp. switching) costs, and which consider also the returns of finished items require probabilistic results as precise as possible. Finally, most of the peculiar features of an assemblage system appear – at least in principle – already in pure assemblage systems.

Various optimization problems in assemblage systems are considered in FERSCHL [2], [3], [4], [5] and in HTOO [7]. They all take into account an essential modification of the concept of a limited waiting–room, which is better adapted to production systems. In such systems it does not make much sense to "reject" incoming single parts, which find their waiting–room occupied. It is more appropriate, to introduce critical levels of the queue length to reduce and restore the production speed, following so rather the procedure in inventory models. But satisfactory solutions for this modification could be found only for assemblage systems with $\ell = 2$ parallel lines.

With the exception of asymptotic results, as in HARRISON [6] and an early article of KASHYAP [8], the models considered were fully governed by the exponential distribution. The production of the single parts is represented by independent Poisson streams, and the service station handles complete sets of ℓ parts with exponential distributed service time.

These Markovian assumptions are made also in the present paper, in which we will look more closely at pure assemblage systems with $\ell = 3$ lines. In the following sections we consider the problem of computing the state probabilities for a system with limited waiting–room of size s for all three lines. The computation of these probabilities is the first step towards handling optimization problems in connection with the production process. In section 3 an algorithm for straightforward numerical computation is developed, but parametrical solutions depending on the decision variable s seem to be out of reach. Therefore we consider in section 5 an approximation, which allows closed algebraic expressions of the target function in the variable s. This function is derived in section 4.

2. DESCRIPTION OF THE SYSTEM

As a fundamental building block of a realistic production system of assembly–type we consider a pure assemblage system with limited waiting–room. For this we make the following assumptions.

Three lines $\Lambda_i, i = 1,2,3,$ working with equal speed send single parts for assemblage to an assemblage point. Their output consists of three independent Poisson streams of equal intensity $\lambda = 1$ (without loss of generality). Let the waiting–room in front of the assemblage point be of equal size s for each of the three lines.

Let

$$R_i(t) := \text{ number of parts of type } i \text{ in the system.}$$

Imagine, that the line Λ_i is switched off, as soon as $R_i(t)$ takes the value s; production is set up again, if the queue length drops to $s-1$.

The state of the system can be described by

$$R(t) = (R_1(t),\ R_2(t),\ R_3(t)) \tag{2.1}$$

whereby at least one of the components is zero. This is a typical feature of pure assemblage systems. The states of the system are therefore triples

$$(n_1, n_2, n_3)\ ,\ \text{with } 0 \le n_i \le s,\ \ i = 1,2,3 \quad \text{and } \min_i n_i = 0. \tag{2.2}$$

It is easily seen that $(R(t)|t \ge 0)$ is a Markov process. The corresponding flow diagram (Markov–graph) is shown in Fig. 2.1 for the case $s = 2$.

For most applications, knowledge of the steady state probabilities

$$P_{n_1,n_2,n_3} := \lim_{t\to\infty} P\{R_1(t) = n_1, R_2(t) = n_2, R_3(t) = n_3\} \tag{2.3}$$

will be sufficient. They can be derived using a directed graph of the form shown in Fig. 2.1, but some simplifications beforehand are appropriate.

To begin with, the symmetries of the model allow sets of equations of the form

$$P_{a,b,0} = P_{b,a,0} = \ldots = P_{0,a,b} = P_{0,b,a} =: P_{a,b}. \tag{2.4}$$

These equations especially show, that only two indices are needed to characterize states of different type. With the simplified notation we arrive at a slightly modified diagram. Fig. 2.2 stresses the threefold symmetry of the problem, which in turn allows to concentrate upon a triangular domain in the graph, as indicated by the broken line in Fig. 2.2.

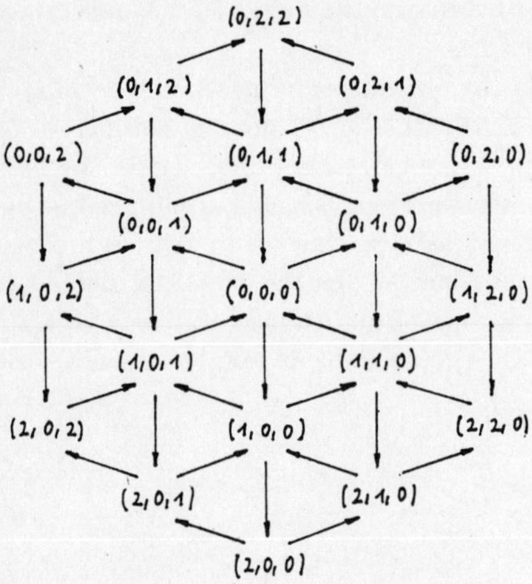

Fig. 2.1: Markov–graph of a three–line system with limited waiting room

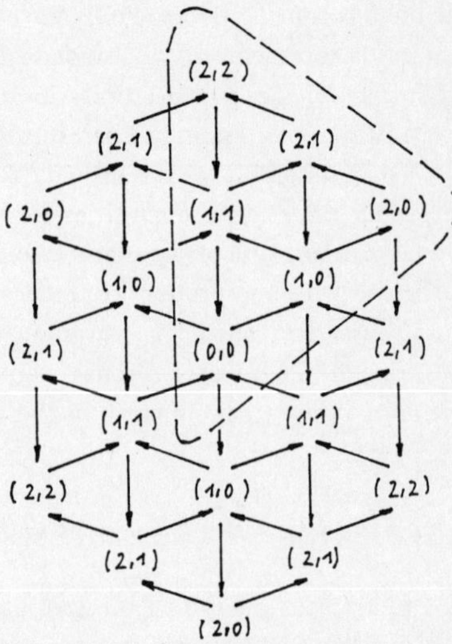

Fig. 2.2: The modified graph of the three–line system

The balance equations for the steady–state probabilities can be derived by taking inflows– and outflows of every state equal. If one wishes to utilize fully the reduction to the system of the $P_{a,b}'s$,

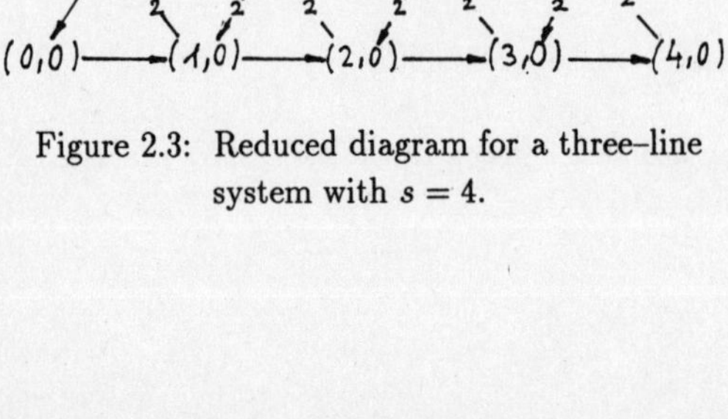

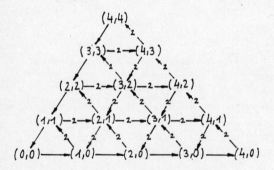

Figure 2.3: Reduced diagram for a three–line system with $s = 4$.

one may use a triangular flow graph, which corresponds to the encircled domain in Fig. 2.2. But now one has to take into account the flows across the boundary of this domain. It is possible, however to compensate for these flows, and we arrive at a final reduced flow graph, which is shown – for greater clearness with $s = 4$ – in Fig. 2.3. Notice the doubled intensities in the reduced picture. It can be easily checked that the technique of equating inflows– and outflows of a node leads to the same equations for the probabilities $P_{a,b}$ as the picture of Fig. 2.2.

3. AN ALGORITHM FOR THE COMPUTATION OF THE PROBABILITIES $P_{a,b}$

From Fig. 2.3 a system of equations for the steady–state probabilities is derived easily:

$$P_{0,0} = P_{1,1} \tag{3.1a}$$

$$\begin{aligned} 3P_{1,0} &= 2P_{2,1} + P_{0,0} \\ 3P_{1,1} &= P_{2,2} + 2P_{1,0} \end{aligned} \tag{3.1b}$$

$$\begin{aligned} 3P_{k,0} &= 2P_{k+1,1} + P_{k-1,0} \\ &\cdots \\ 3P_{k,j} &= P_{k+1,j+1} + P_{k,j-1} + P_{k-1,j} \qquad j = 1, \cdots, k-1 \\ &\cdots \\ 3P_{k,k} &= P_{k+1,k+1} + 2P_{k,k-1} \qquad k = 2, \cdots, s-1 \end{aligned} \tag{3.1c}$$

$$\begin{aligned} 2P_{s,0} &= P_{s-1,0} \\ &\cdots \\ 2P_{s,j} &= P_{s-1,j} + P_{s,j-1} \qquad j = 1, \cdots, s-1 \\ &\cdots \\ P_{s,s} &= 2P_{s,s-1} \end{aligned} \tag{3.1d}$$

Now the computation of the steady–state probabilities from the System (3.1) starts from the "outside", that means the side opposite to State (0,0) in the triangular shaped flow diagram (see Fig.2.3) and works toward states with smaller queue lengths. Therefore we take the probabilities

$$P_{s,0}\ ,\ P_{s,1}, \ldots,\ P_{s,s} \tag{3.2}$$

as preliminary parameters and express by those in a first step the other state probabilities. To this end probability vectors of varying dimension are introduced

$$P_k{}^T := \ (P_{k,0}\ ,\ P_{k,1}, \ldots,\ P_{k,k})\ ,\ k = 1, \ldots, s \quad . \tag{3.3}$$

To write the System (3.1) in matrix form, we define furthermore a sequence of rectangular matrices

$$A_s = \begin{pmatrix} 2 & & & & 0 \\ -1 & 2 & & O & \\ & \ddots & \ddots & & \vdots \\ O & & 2 & & \\ & & -1 & 2 & 0 \end{pmatrix} \quad \text{type } (s, s+1) \tag{3.4a}$$

$$B_k = \begin{pmatrix} 3 & & & & 0 \\ -1 & 3 & & O & \\ & \ddots & \ddots & & \vdots \\ O & & 3 & & \\ & & -1 & 3 & 0 \end{pmatrix} \quad \text{type } (k, k+1) \tag{3.4b}$$

$$C_k = \begin{pmatrix} 0 & -2 & & & & 0 \\ 0 & 0 & -1 & & O & 0 \\ \vdots & & \ddots & \ddots & & \vdots \\ \vdots & O & & -1 & & \vdots \\ 0 & & & 0 & -1 & 0 \end{pmatrix} \quad \text{type } (k, k+2) \ . \tag{3.4c}$$

$$k = s-1, \ldots, 2 \ .$$

Now the backward recursion for the vectors $P_{s-1}, \ldots, P_1$ is seen more clearly:

$$P_{s-1} = A_s P_s \tag{3.5a}$$

$$P_{k-1} = B_k P_k + C_k P_{k+1} \ , \quad k = s-1, \ldots, 2 \ . \tag{3.5b}$$

The solution of (3.5) has the form

$$P_{k-1} = M_k P_s \ , \quad k = s, \ldots, 2 \tag{3.6}$$

where the matrices M_k depend on A_s, B_j and C_j with $j \geq k$.

It remains to find the values of $P_{s,0}$, $P_{s,1}, \ldots, P_{s,s}$ and of $P_{0,0}$. Notice, that the matrices A_s, B_k and C_k are constructed in a way, that the equations from the original System (3.1)

$$P_{s,s} = \qquad 2P_{s,s-1} \tag{3.7a}$$

$$3P_{k,k} = P_{k+1,k+1} + 2P_{k,k-1} \qquad k = s-1, \ldots, 1 \tag{3.7b}$$

are cut off by their zero colums. Inserting the corresponding parts of the solution (3.6) into (3.7) gives s equations for $s+1$ variables. In the case of $s = 4$ for example this second-stage system has the form:

$$
\begin{aligned}
-104P_{4,0} + 88P_{4,1} - \ 2P_{4,2} + \ P_{4,3} &= 0 \\
13P_{4,0} - 27P_{4,1} + 21P_{4,2} - 5P_{4,3} &= 0 \\
2P_{4,1} - \ 7P_{4,2} + 6P_{4,3} &= P_{4,4} \\
2P_{4,3} &= P_{4,4}
\end{aligned}
\tag{3.8}
$$

All the probabilities then can be expressed in terms of $P_{s,s}$, $s \geq 1$. $P_{0,0}$ does appear neither in (3.5) nor (3.8), but we have $P_{0,0} = P_{1,1}$.

In order to determine $P_{s,s}$ we go back to the original flow diagram, as shown in Fig. 2.2. Remembering the symmetries in the system, we sum up all state probabilities to get the norming equation, which eventually yields the value of $P_{s,s}$

$$
P_{0,0} + 3\sum_{k\geq 1}(P_{k,k} + P_{k,0}) + 6\sum_{k>}\sum_{j>0} P_{k,j} = 1 \ . \tag{3.9}
$$

For illustration we show *relative* values of the state probabilities $P_{k,j}$ for $s = 4$ in the following table

	values of k				
j	0	1	2	3	4
0	1,000	0,960	0,844	0,652	0,326
1		1,000	0,940	0,786	0,556
2			1,080	0,976	0,766
3				1,359	1,063
4					2,125

To give a clearer picture in (3.10) the *ratios* $P_{k,j}/P_{0,0}$ were listed rather than the probabilities themselves. Especially size and location of deviations from the uniform distribution, which will be used later as an approximation, can be seen much better.

One has to recognize a jump of magnitude of complexity if one passes from the case of two parallel production lines to that of three lines. As shown in FERSCHL [2], the steady–states are uniformly distributed for $\ell = 2$, whereas one sees in the case $\ell = 3$ marked deviations from uniformity, especially for the extreme cases $P_{s,0}$ and $P_{s,s}$.

4. THE OPTIMIZATION PROBLEM

Let us consider the stochastic model called "assemblage system" as a facility which should be run with maximal profit. We take into consideration two sources of costs, namely inventory costs which arise from parts waiting in front of the assemblage point and switching costs connected with every stop–and–go–decision at each production line. The ultimate goal are assembled *items*; this is reflected by the introduction of gains for outgoing items. Let

$c :=$ inventory cost per part and time unit

$u :=$ switching cost per manipulation at any line

$g :=$ returns for an assembled item which leaves the assemblage point.

Furthermore, let indicating the dependence on s:

$J(s) :=$ Mean inventory of single parts
$U(s) :=$ Mean number of switchings per time unit
$G(s) :=$ Mean number of departing items per time unit.

Then the economic goal is to maximize

$$f(s) := \quad g \cdot G(s) - c \cdot J(s) - u \cdot U(s) \tag{4.1}$$

If the steady–state probabilities are known for a given s, the computation of the functions $G(s)$, $J(s)$ and $U(s)$ will be an easy task.

In order to get comfortable expressions, we introduce the following quantities, suggested by the symmetries indicated in (2.4).

$$\begin{aligned} p_{0,0}: &= P_{0,0} \\ p_{k,k}: &= 3P_{k,k} \quad 1 \le k \le s \\ p_{k,0}: &= 3P_{k,0} \quad 1 \le k \le s \\ p_{k,j}: &= 6P_{k,j} \quad 1 \le j < k \le s \end{aligned} \tag{4.2}$$

Then we get the following formulas

$$U(s) = 2(\sum_{j=1}^{s} p_{s,j} + p_{s,s}) \tag{4.3}$$

$$J(s) = \sum_{k=o}^{s} \sum_{j=o}^{k} (k+j) p_{k,j} \tag{4.4}$$

$$G(s) = 1 - \sum_{k=o}^{s} p_{k,0} \tag{4.5}$$

Now we give some hints concerning the derivation of (4.3) – (4.5).

First take into account, that in the steady state the mean number of stop– and–go–decisions must be equal. In formula (4.3) the setup decisions therefore are counted twice. But a setup of one of the production lines occurs, if its queue has length s and therefore has been stopped, one of the other queues is not empty, and a new part arrives at the empty line. If two waiting–rooms are fully occupied, an arrival at the third line causes *two* switches.

The mean inventory is simply the expectation of the number of parts in the system, as denoted in (4.4).

To compute $G(s)$ we consider the probability, that a part delivered from any line causes an output. In the states of type $p_{k,0}$ no item is assembled. If any other state is met, the probability is one third, that the empty line is visited. Per unit of time three single parts in the mean arrive at all, therefore the mean number of the output equals the probability, to meet a state of type $p_{k,j}$, $k \ge j > 0$. Therefore (4.5) is the desired quantity.

Closed expressions depending on the parameter s are not available, unlike the simpler case of two production lines. Only numerical evaluations of the formulas (4.3) – (4.5) are possible. We give a short table of such results in the next section, where they are compared with approximate solutions.

5. AN APPROXIMATING APPROACH TO THE OPTIMIZATION PROBLEM

A look at Table (3.10) may justify a simple attempt towards approximate solutions: we assume, that the steady–state distribution of the system is not too far from the uniform distribution, which is exact in the case of $\ell = 2$ lines. In the following, we compute quantities $J_a(s), G_a(s)$ and $U_a(s)$ which serve as approximation for $J(s), G(s)$ and $U(s)$ for an arbitrary number ℓ of production lines.

The total number of distinct states can be easily determined. It is $(s+1)^\ell - s^\ell$. Each line has $s+1$ possibilities for the length of its queue. The nature of a pure assemblage system does not allow configurations, where all lines are occupied, therefore, the corresponding number s^ℓ of configurations, which are contained in $(s+1)^\ell$, must be subtracted.

The mean inventory in the case of the uniform distribution can be determined at similar lines. We calculate the total weight of all $(s+1)^\ell - s^\ell$ configurations, each of them weighted with the number of its parts waiting. The total weight then will be calculated with an averaging argument. If all $(s+1)^\ell$ configurations are considered, the mean queue length at a line is $s/2$. So in the average the total number for all (including the not allowed) configurations is

$$\ell \cdot \frac{s}{2} \cdot (s+1)^\ell \tag{5.1}$$

The total weight of the configurations not allowed configurations is

$$\ell \cdot \frac{s+1}{2} \cdot s^\ell \tag{5.2}$$

because the mean length at a line of such a configuration would be $(s+1)/2$.
The difference between (5.1) and (5.2), divided by the total number of states, gives eventually the desired approximate mean inventory

$$J_a(s) = \ell \cdot \frac{s(s+1)}{2} \frac{[(s+1)^{\ell-1} - s^{\ell-1}]}{(s+1)^\ell - s^\ell} \tag{5.3}$$

With $\ell = 3$ we get

$$J_a(s) = \frac{3}{2} \frac{s(s+1)(2s+1)}{(s+1)^3 - s^3} \tag{5.4}$$

For the approximation of the mean output of assembled items, we consider a specific production line, say the first. We consider the proportion of those parts from this line, which leave the system as a part of an assembled item. The stream of these successful parts is clearly identical with the output stream of items from the assemblage point. Therefore the complement of the rejection probability at a fixed line equals the desired proportion. In the case of the uniform distribution the rejection probability at a line is equal to the proportion of the number of states, for which the waiting room at this line is filled up with s parts. But this proportion equals $((s+1)^{\ell-1} - s^{\ell-1})/((s+1)^\ell - s^\ell)$ and therefore we obtain the approximation

$$G_a(s) = 1 - \frac{(s+1)^{\ell-1} - s^{\ell-1}}{(s+1)^\ell - s^\ell} \tag{5.5}$$

for the mean output. In the case of $\ell = 3$ we have

$$G_a(s) = \frac{3s^2 + s}{3s^2 + 3s + 1} \tag{5.6}$$

Note, that in the case of a Poisson stream of parts, rejections and stop–decisions lead to the same result. The mean number of switches must be the same for each of the production lines. Furthermore, at a specific line, stop–and–go–decision occur alternately, therefore with the same intensity. A production line is shut down, if it sends a part to a queue of length $s-1$. The probability of the state $s-1$ at a specific line is the same as given above for the state s under the assumption of uniform distribution. Hence the total number of switches per time unit is then approximately given by

$$U_a(s) = 2\ell \cdot \frac{(s+1)^{\ell-1} - s^{\ell-1}}{(s+1)^{\ell} - s^{\ell}} \tag{5.7}$$

The special case $\ell = 3$ gives

$$U_a(s) = \frac{12s+6}{3s^2+3s+1} \tag{5.8}$$

It should be remarked, that the reasoning in the foregoing section, which led us to exact probabilities, runs partly in a different manner. With the exact values of the probabilities, the different approaches would give the same results. But interestingly this is not the case for an approximation under the assumption of uniform distribution. A closer inspection shows, that the method chosen in this section gives a better approximation than a method analogous to that of section 4 and is also easier to handle for arbitrary ℓ.

In the following table we compare the exact values with the approximate values in the case of uniformly distributed states for small values of s in the three–line–model.

s	$J(s)$	$U(s)$	$G(s)$	$J_a(s)$	$U_a(s)$	$G_a(s)$
1	1,364	2,182	0,546	1,286	2,571	0,571
2	2,513	1,575	0,713	2,368	1,579	0,737
3	3,617	1,195	0,790	3,405	1,135	0,811
4	4,705	0,957	0,835	4,426	0,885	0,853
5	5,787	0,797	0,864	5,440	0,725	0,879
6	6,863	0,682	0,884	6,449	0,614	0,898
7	7,938	0,595	0,899	7,456	0,533	0,911
8	9,011	0,528	0,911	8,461	0,470	0,922
9	10,083	0,475	0,920	9,465	0,421	0,930
10	11,154	0,431	0,927	10,468	0,381	0,937

At a first glance the approximations seem to work quite well, but unfortunately they show no uniform behaviour.

The deviations from the uniform distribution seemingly have different effects on the quality of the approximation. Further calculations covering a wider range of s are therefore needed.

6. CONCLUDING REMARKS

The central formula for the investigation of the three–line pure assemblage system is the matrix recursion (3.5). It has the form of a difference equation of second degree with quasiconstant matrix coefficients B_k and C_k and shows some resemblance to the matrix–geometric approach, especially the case of quasi–birth–and–death processes (see for instance NEUTS [11], p.81f.). The real problem is, that the dimension of the matrices grow with the value of s. The simple structure of the matrices in (3.5) will allow rather straightforward calculations for not too great s, but the growing dimension seems to exclude results which explicitly depend on the parameter s. Approximations which converge in a satisfactory manner to the exact values are therefore of great interest. A first simple approach using the idea of the uniform distribution demands further improvement.

REFERENCES

[1] U.N. Bhat, Finite capacity assembly–like queues, Queueing Systems 1 (1986), 85–101.

[2] F. Ferschl, Über eine Optimierungsaufgabe in Assemblage–Systemen, Statistische Hefte 8 (1967), 2–27.

[3] F. Ferschl, Assemblage– (Zusammensetz–) Systeme, Research Report, Bonn (1968).

[4] F. Ferschl, Über Zusammensetz– (Assemblage–) Systeme, eine Klasse von stochastischen Produktionsmodellen, Economicko–Matematicky Obzor 5 (1969), 304–324.

[5] F. Ferschl, An optimization problem in assemblage systems, in: *Operations Research and Economic Theory* , Springer–Verlag Berlin (1984).

[6] J.M. Harrison, Assembly–like queues, J.Appl.Prob.10 (1973), 354–367.

[7] T. Htoo, Optimierungsprobleme in Assemblage–Systemen, Dissertation München (1990).

[8] B.R.K. Kashyap, Further results for the double–ended queue, Metrika 11 (1967), 168–186.

[9] G. Latouche, Queues with paired customers, J.Appl.Prob.18 (1981), 684–696.

[10] E.H. Lipper and B. Sengupta, Assembly–like queues with finite capacity: Bounds, asymptotics and approximations, Queueing Systems 1 (1986), 67–83.

[11] M.F. Neuts, Matrix–geometric solutions in stochastic models. John Hopkins, Baltimore (1981).

A SOFTWARE PACKAGE TOOL FOR MARKOVIAN COMPUTING MODELS WITH MANY STATES: PRINCIPLES AND ITS APPLICATIONS

Satoshi Fukumoto and Shunji Osaki
Department of Industrial and Systems Engineering,
Faculty of Engineering, Hiroshima University,
Higashi-Hiroshima-shi, 724 Japan

ABSTRACT

A computing system can be formulated by modeling a continuous-time Markov chain with many states, and be evaluated by using the reliability/performance measures. The randomization technique is discussed to derive the transient solution for a Markov chain. A software package tool is implemented by using the randomization technique and introducing a new idea of identifying when the transient solution converges to the steady-state solution in advance. Numerical examples are illustrated by using our software package tool to evaluate the optimal maintenance policies for computing systems. Some interesting maintenance policies are suggested from the numerical examples.

1. INTRODUCTION

It is of great interest and importance to operate a computing system with high reliability and performance [1, 2]. To evaluate such a system, we should derive analytically and/or numerically reliability/performance measures by formulating a stochastic model of the system [3, 4, 5]. A continuous-time Markov chain is one of the most powerful stochastic processes to analyze the system. In particular, we are very much interested in a continuous-time Markov chain with many states since modeling a Markov chain yields many states in practice [6]. We develop a software package tool for calculating the transient state probabilities for a continuous-time Markov chain with many states. Several performance/reliability measures can be calculated by using the state probabilities.

We adopt the randomization technique [7] for calculating the transient state probabilities as well as the steady-state probabilities. It is assumed that the transient state probabilities converge the steady-state probabilities as time tends to infinity under certain assumputions. In principle, it is possible to calculate the transient and steady state transition probabilities. However, it is quite difficult to do so if there are many states such as some hundreds or thousands of states.

For our software package tool, we specify the initial state probability vector $\pi(0)$. Once the initial state probability vector $\pi(0)$ is specified, we can calculate the transient state probability vector $\pi(t)$ at time t. However, it is quite difficult in advance to identify when the transient state probability vector converges to the steady-state probability vector with enough precision. We propose a new idea of calculating the convergence time t_s of the steady-state probability in advance from the knowledge of the randomization technique.

In this paper, we discuss our software package tool and its applications. In §2 we discuss the randomization technique for calculating the transient solutions as well as the steady-state solutions for a continuous-time Markov chain with many states. We propose a new idea of the convergence time which will be implemented in our software package tool. We further present two examples of maintenance policies for a computing system in §3, and show how our software package tool is useful. The first example is maintenance policies based on retries for a computing system. Calculating the transient and steady-state availabilities, we can obtain the effective maintenance policies. The second example discusses maintenance policies for a hardware and software system. We propose two software maintenance policies for a two-unit hardware system and compare them.

2. MATHEMATICAL PRELIMINARIES

2.1 Randomization Technique

Let us briefly sketch the randomization technique (see Ross [6], pp 141-183). There are several techniques of calculating the exponential of matrix [8], since they are quite famous as the eigen value problems of the matrices.

Consider a continuous-time Markov chain with N states. Let

$$\pi(t) = \{\pi_1(t), \pi_2(t), \dots, \pi_N(t)\} \tag{1}$$

be the state probability vector at time t, where the initial state vector

$$\pi(0) = \{\pi_1(0), \pi_2(0), \dots, \pi_N(0)\} \tag{2}$$

is prespecified. Let Q be the infinitestimal generator for the continuous-time Markov chain. Then, the matrix differential equation is given by

$$\frac{d\pi(t)}{dt} = \pi(t)\mathbf{Q}, \tag{3}$$

where the initial condition $\pi(0)$ is given. Note that each element of the infinitestimal generator is given by

$$q_{ij} = \lim_{\Delta t \to 0} \frac{P\{X(t+\Delta t) = j | X(t) = i\}}{\Delta t}, \quad (i \neq j) \tag{4}$$

$$q_{ii} = -\sum_{i \neq j}^{N} q_{ij}. \quad (i = j) \tag{5}$$

It is easy to solve the matrix differential equation (3). We have

$$\begin{aligned} \pi(t) &= \pi(0)e^{\mathbf{Q}t} \\ &= \pi(0)[\mathbf{I} + \sum_{n=1}^{\infty} \frac{(\mathbf{Q}t)^n}{n!}], \end{aligned} \tag{6}$$

where $\mathbf{I}$ is an identity matrix.

Specifing Λ such that $\Lambda = \max_i |q_i|$, we transform the matrix $\mathbf{Q}$ into the matrix $\mathbf{P}$:

$$\mathbf{P} = \mathbf{Q}/\Lambda + \mathbf{I}. \tag{7}$$

We notice that the matrix $\mathbf{P}$ is a transition probability matrix and the properties of all states are preserved under the transformation (7). Introduce the n-step transition probability vector $\phi(n)$ for a discrete-time Markov chain with transition probability matrix $\mathbf{P}$. That is,

$$\phi(0) = \pi(0), \tag{8}$$

$$\phi(n+1) = \phi(n)\mathbf{P}. \quad (n \geq 0) \tag{9}$$

Substituting (7) into (6), we have

$$\pi(t) = \sum_{n=0}^{\infty} \{ \frac{(\Lambda t)^n}{n!} e^{-\Lambda t} \cdot \phi(n) \}. \tag{10}$$

The right-hand side of (10) is the infinite series of the product of the probability mass function of the Poisson distribution with parameter Λt and the n-step transition probability $\phi(n)$.

In practice, instead of infinite series in (10), we adopt the finite series

$$\pi^{\varepsilon}(t) \equiv \sum_{n=0}^{T(\varepsilon,t)} \{ \frac{(\Lambda t)^n}{n!} e^{-\Lambda t} \cdot \phi(n) \} \tag{11}$$

where

$$T(\varepsilon, t) = \min[k : \sum_{n=0}^{k} \frac{(\Lambda t)^n}{n!} e^{-\Lambda t} > 1 - \varepsilon]. \tag{12}$$

and ε is an acceptable error which is enough small. Applying (11) with the prespecified acceptable error, we can calculate $\pi^{\varepsilon}(t)$, which is the transition probabibility vector at time t with enough precision. Fig. 1 shows an illustration of how to calculate $\pi^{\varepsilon}(t)$.

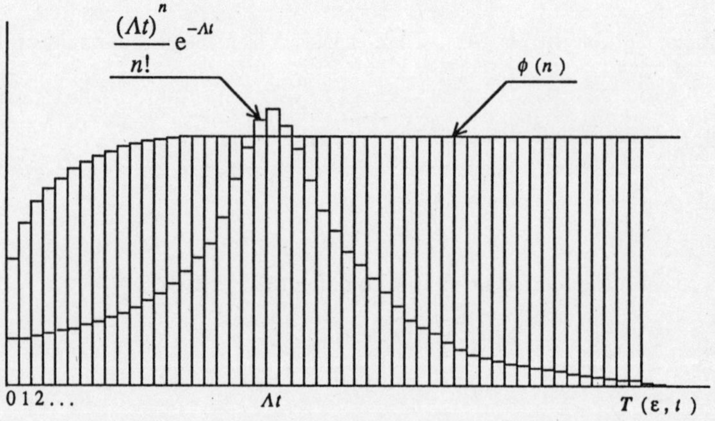

Fig. 1 An illustration of how to calculate $\pi^{\varepsilon}(t)$.

2.2 Steady-State Solution

If the continuous-time Markov chain under consideration is regular, there exists the steady-state probability vector π whose solution is given by

$$\pi \mathbf{Q} = 0, \tag{13}$$

$$\sum_{i=1}^{N} \pi_i = 1, \tag{14}$$

where

$$\pi = \{\pi_1, \pi_2, \cdots, \pi_N\}. \tag{15}$$

In principle, it is analytically easy to solve the linear simultaneous equations in (13) and (14). However, if we consider many states such as several hundreds or thousands of states, we have to consider the efficient method of solving equations (13) and (14) numerically.

As shown in (10), we have to calculate $\phi(n)$. It is easier to obtain the steady-state vector $\phi = \lim_{n\to\infty} \phi(n)$ since the properties of all states are preserved both for the discrete-time Markov chain and the continuous-time Markov chain. That is

$$\lim_{t\to\infty} \pi(t) = \lim_{n\to\infty} \phi(n). \tag{16}$$

It is evident that $\phi = \phi \mathbf{P}$. Multiplying ϕ for both sides of (7), we have

$$\phi \mathbf{Q} = 0, \tag{17}$$

and $\sum_i \phi_i = 1$, which is a unique steady-state vector π. Let n_s be the minimum step number in which $\phi(n_s)$ approximates ϕ with enough precision. That is,

$$\phi(n_s) = \pi(0)\mathbf{P}^{n_s} \simeq \phi = \pi, \tag{18}$$

where n_s is the minimum number of the step n such that

$$[\max_j |\phi_j(n) - \phi_j(n-1)|] < \varepsilon_2, \tag{19}$$

$$\phi = \{\phi_1, \phi_2, \cdots, \phi_N\}. \tag{20}$$

Note that ε_2 is an acceptable error which is prespecified and enough small. From these facts, we can calculate $\phi(n_s)$ instead of π. For the randomization technique, we have to calculate $\phi(n)$ with enough steps which is approximately regarded as the steady-state. We should apply $\phi(n_s)$ in practice.

2.3 Convergence Time of the Steady-State Solutions

The randomization technique is available for calculating the transient solutions for a continuous-time Markov chain. However, it involves an important and difficult problem of identifying when the transient solution converges to the steady-state solution with enough precision. Otherwise, we have to calculate the transient solution by applying the randomization technique in (10) which is enormous calaulations as time tends to infinity.

Let t_s denote the convergence time such that

$$\pi(t_s) \simeq \pi. \tag{21}$$

Then we should calculate $\pi(t)$ for $0 \leq t \leq t_s$. If we have to calculate $\pi(t)$ for $t > t_s$, we should use π instead of $\pi(t)$ in order to avoid the unnecessary calculation.

From (18), we assume

$$\phi(n) = \pi \tag{22}$$

where $n \geq n_s$. If t_s satisfies $\pi(t_s) \simeq \pi$, almost all the probability mass functions of the Poisson distribution with parameter Λt_s is distributed on n such that $n \geq n_s$. That is,

$$\sum_{n=n_s}^{\infty} \frac{(\Lambda t_s)^n}{n!} e^{-\Lambda t_s} \simeq 1. \tag{23}$$

Hence we have the following approximate equation

$$\begin{aligned} \pi(t_s) &\simeq \sum_{n=n_s}^{\infty} \frac{(\Lambda t_s)^n}{n!} e^{-\Lambda t_s} \cdot \phi(n), \\ &= \sum_{n=n_s}^{\infty} \frac{(\Lambda t_s)^n}{n!} e^{-\Lambda t_s} \cdot \pi, \\ &= \pi \sum_{n=n_s}^{\infty} \frac{(\Lambda t_s)^n}{n!} e^{-\Lambda t_s}, \\ &\simeq \pi. \end{aligned} \tag{24}$$

Let t_s satisfy

$$\Lambda t_s = n_s + k\sqrt{\Lambda t_s}, \tag{25}$$

(see Fig. 2), where Λt_s and $\sqrt{\Lambda t_s}$ are the mean and standard deviation of the Poisson distribution with parameter Λt_s, and k is a positive real constant. Solving t_s in (25), we have

$$t_s = \frac{2n_s + k^2 + k\sqrt{k^2 + 4n_s}}{2\Lambda}, \tag{26}$$

That is, t_s is expressed in terms of n_s. In other words, once n_s and k are specified, we can obtain t_s in (26). Let us consider how to specify a constant k in (25) or (26). It is well-known that the Poisson distribution is approximated by the normal distribution when the parameter Λt_s is enough large. Noting this fact, we have

$$\int_{-4}^{\infty} \frac{1}{\sqrt{2\pi}} e^{-\frac{t^2}{2}} dt = 0.9999864\,. \tag{27}$$

In practice, if we assume that $k = 4$, equation (23) is approximated by

$$\sum_{n=n_s}^{\infty} \frac{(\Lambda t_s)^n}{n!} e^{-\Lambda t_s} = 0.9999864\,. \tag{28}$$

which is a good approximation from the viewpoint of round error.

We summarize the convergence time t_s. Once n_s is specified by calculating $\phi(n_s)$ which is approximately the steady-state probability vector, we can calculate t_s from (26). That is, if $t > t_s$, we should use $\pi = \phi$ instead of calculating $\pi(t)$ for each t. We emphasize that t_s can be calculated in advance when we implement the randomization technique in our software package tool.

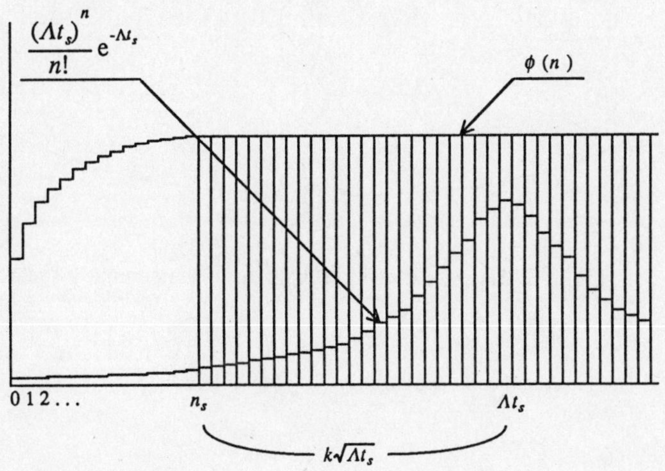

Fig. 2 The Poisson distribution and the discrete-time Markov chain at time t_s.

3. APPLICATIONS

3.1 Maintenance Policies for a Computing System with Retries

In this subsection, we discuss maintenance policies for a computing system with retries, by applying our software package tool.

It is generally considered that a computing system has two kinds of faults from the viewpoint of maintenance [9]. One is an *intermittent* fault that is only occasionally present. The other is a *permanent* fault that is continuous and stable. In order to avoid the system downs caused by the intermittent faults, the computing system execute retries several times. We present a maintenance policy based on retries for a computing system.

We assume that for a fault occurrence in the system, if retries are executed and do not succeed, we identify the fault as a permanent fault and remove the fault, and if retries are executed and succeed, we identify the fault as an intermittent fault, and remove the fault after the same actions are observed N times. Let the constant failure rate, repair rate for a permanent fault be λ and μ_0, respectively, and let the repair rate for an intermittent fault be μ_1, and the failure rate is assumed to be proportional to the number of successful retries. Let P^{k+1} be the probability with which the retry is successful after k times successful retries.

The state transition diagram is shown in Fig. 3, where each state is defined in the following:

State 0 : The system is operating.

State k : Success of retries are observed k times.

State D_k : Repair for a permanent fault starts (system down).

State D : Success of retries are observed N times and the repair for an intermittent fault starts (system down).

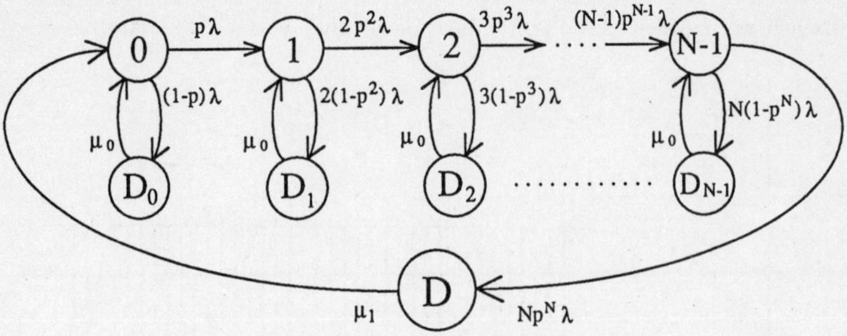

Fig. 3 The state transition diagram of the computer system.

We can obtain the instantaneous availability $A_v(t)$ as follows:

$$A_v(t) = \sum_{k=0}^{N-1} P_k(t),$$

where $P_k(t)$ is the state probability at time t.

Specifying all the parameters, we can numerically calculate the instantaneous availability $A_v(t)$. Let $\lambda = 0.01$, $\mu_0 = 1$, $\mu = 0.2$, and $P = 0.9$. We further specify parameter $N = 1 \sim 20$. Using our new idea in our software package tool, it is shown that the steady-state availability A_v attains the maximum 0.976 at $N = 5$ among all possible N. The convergence time t_s at $N = 5$ is 596. Therefore, we should calculate the transient state availabilities for $0 \leq t \leq 600$.

Fig. 4 shows the transient state availability $A_v(t)$. It is interesting that the more N increases, the more the availability $A_v(t)$ increases, for $0 \leq t \leq 230$, in contrast with the steady-state availability.

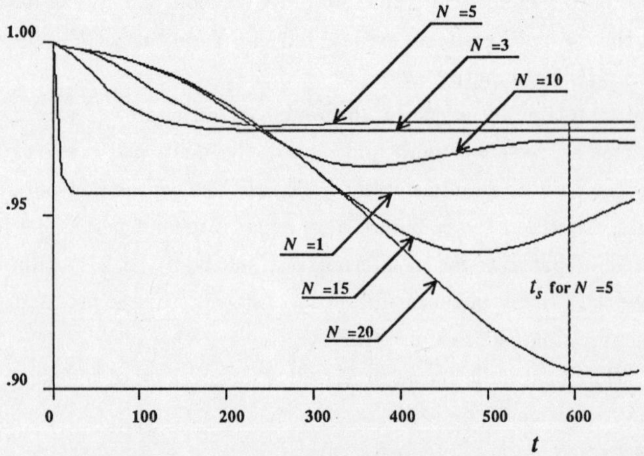

Fig. 4 The behavior of $A_v(t)$ for each N.

Referring to the above results, we can conclude that we should repair and remove the intermittent fault after successes of retries are observed 5 times, in operation of the system for a long term. However, in operation of the system for a short term like for $0 \leq t \leq 200$, we should not repair the intermittent fault.

3.2 Maintenance Policies for a Hardware-Software System

In this subsection we discuss a two-unit hardware system, propose two software maintenance policies [4], and compare them. Considering maintenance for a computing system from the viewpoint of software error detection, we present the following model.

A hardware system is composed of two units. The system can function if only one of the two units functions. Each unit is assumed to be repairable, and the constant failure rates and maintenance rates for each unit are assumed to be λ_0 and μ_0, respectively.

We assume that any software error causes the system down, and that there are N software errors latent at the installation of the software system. The detection rate and the maintenance rate of each software error are assumed to be constant and proportional to the remaining number of errors (see Jelinski and Moranda [10]). Let λ_s and μ_s be the detection and maintenance rate for each software error. It is assumed that there is a single repair or maintenance facility. Here a repair or maintenance is assumed perfect. All the states of the system are defined as follows.

State 0_n : A system starts operating.

State 1_n : A hardware failure of a unit takes place and its repair starts.

State 2_n : Through state 1_n, a hardware failure of the remaining unit takes place (system down).

State 3_n : Through state 0_n, a software error takes place (system down).

State 4_n : Through state 1_n, a software error takes place (system down).

Here n denotes the number of remaining error at that time, where $n = 0, 1, \cdots, N$.

We are now ready to introduce the following software maintenance policies through state 4_n.

Model 1 : After the repair completion of hardware, the software maintenance starts.

Model 2 : The software maintenance starts even if the repair of hardware is interrupted; the interrupted hardware repair restarts after the software error maintenance completion.

From the above definitions, for Model 1 the process can move to state 3_n from state 4_n, and for Model 2 the process can move to state 1_{n-1} from state 4_n. The state transition diagram is shown in Fig. 5.

We can obtain the instantaneous availability $A_v(t)$ as follows:

$$A_v(t) = \sum_{k=0}^{N} [P_{0_k}(t) + P_{1_k}(t)]$$

where $P_{0_k}(t)$ or $P_{1_k}(t)$ is the state probability at time t.

Specifying all the parameters, we can numerically calculate the instantaneous availability $A_v(t)$. Let $N = 10$, $\lambda_s = 0.02$, $\mu_s = 0.05$, $\lambda_0 = 0.05$, and $\mu_0 = 0.025$. Since t_s for Model 1 is 1003 and t_s for Model 2 is 871 by our software package tool, we should calculate the transient probabilities for $0 \leq t \leq 1000$.

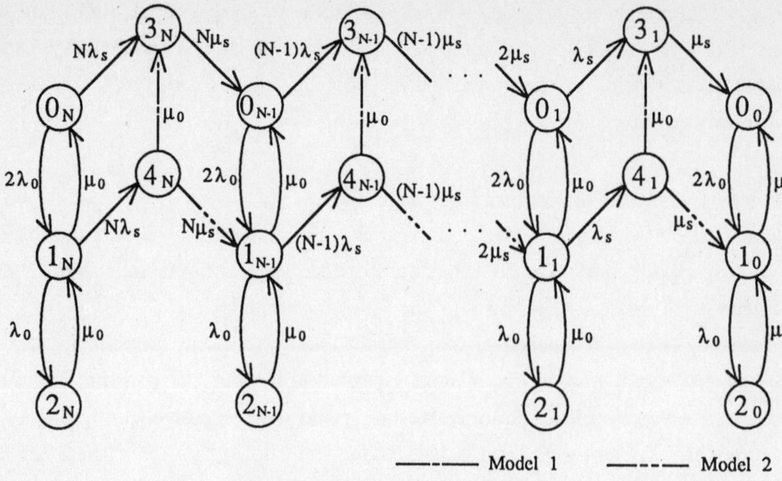

Fig. 5 The state transition diagram for each model.

Fig. 6 shows the availability $A_v(t)$. After the start of the operation, the availability decreases and attains the minima 0.54 and 0.67 for Models 1 and 2, respectively. It is obvious that the availability of Model 2 is better and approaches stability faster than that of Model 1.

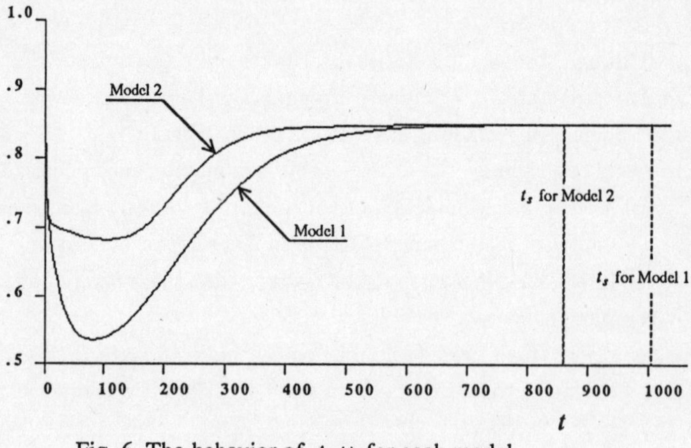

Fig. 6 The behavior of $A_v(t)$ for each model.

4. CONCLUSION

In this paper, we have discussed the randomization technique for calculating the transient solutions as well as the steady-state probability for a continuous-time Markov chain. Two models of maintenance policies for a computing system are presented by applying our software packege tool. Our software package tool is useful in a case where the Markovian model for a computing system has many states such as two examples above. In particular, the convergence time t_s is of great importance to calculate the transient probability solutions in practice. Our analytical results for convergence time is of great use for analyzing such a computing system, since it is difficult to identify when the transient solution converges to the steady-state solution in advance. Our software package tool implemented the analytical results on convergence time is of great use to calculate reliability/performance measures for a computing system in practice.

REFERENCES

[1] A. Avizienis, "Fault-tolerant system", *IEEE Trans. Computers*, vol. C-25, pp. 1304-1312, Dec. 1976.

[2] D. P. Siewiorek and R. S. Swarz, *The Theory and Practice of Reliable System Design*, Digital Press, Bedford, Massachusetts, 1982.

[3] S. V. Makam and A. Avizienis, "ARIES: A reliability and life-cycle evaluation tool for foult-tolerant system", in *Proc. 12th Int. Symp. on Fault-Tolerant Computing*, Santa Monica, California, pp. 267-274, 1982.

[4] A. Avizienis and J. C. Laprie, "Dependable Computing: From Concepts to Design Diversity", *Proc. IEEE*, vol. 74, pp. 629-638, May 1986.

[5] J. Arlat, K. Kanoun and J. C. Laprie, "Dependability evaluation of software fault-tolerance", in *Proc. 18th Int. Symp. on Fault-Tolerant Computing*, Tokyo, pp. 142-147, June 1988.

[6] S. M. Ross, *Stochastic Processes*, John Wiley and Sons, New York, 1983.

[7] D. Gross and D. R. Miller, "The randomization technique as a modeling tool and solution procedure for transient Markov processes", *Operations Research*, vol. 32, pp. 343-361, Mar.-Apr. 1984.

[8] C. Moler and C. Van Loan, "Nineteen dubious ways to compute the exponential of matrix", *SIAM Review*, vol. 20, pp. 801-836, Oct. 1978.

[9] S. Y. H. Su, I. Koren, Y. K. Malaiya, "A continuous-parameter Markov model and detection procedures for intermittent faults", *IEEE Trans. Computers*, vol. C-27, pp. 567-570, June 1987.

[10] Z. Jelinski and P. B. Moranda, "Software reliabilty research", in *Statistical Computer Performance Evaluation*, W. Freiberger (ed.), Accademic Press, New York, pp. 465-484, 1972.

RELIABILITY ASSESSMENT MEASUREMENT FOR REDUNDANT SOFTWARE SYSTEMS

Jun Hishitani, Shigeru Yamada, and Shunji Osaki
Department of Industrial and Systems Engineering,
Faculty of Engineering, Hiroshima University,
Higashi-Hiroshima-shi, 724 Japan

ABSTRACT

It is important to make a software system more reliable. We discuss reliability assessment for redundant software systems developed by multiversion programming, i.e., N-version programming and recovery blocks systems, during the operation phase. In this paper, the reliability assessment measures such as software reliability function, hazard rate function, and MTTF (mean time to failure) are obtained by using two software reliability growth models, i.e., Jelinski-Moranda and NHPP models. Also the results of reliability assessment based on the two software reliability growth models are numerically compared for each redundant software system.

1. INTRODUCTION

It is important to develop a reliable computer system. Many technologies have been developed and improved from both hardware and software management viewpoints. In particular, the break-downs of a computer system caused by software faults are becoming increasingly visible and important. In general, there are two approaches which make a software system be reliable: fault avoidance and fault tolerance approaches.

Fault avoidance is the primary and effective approach, which avoids the introduction of faults into the software system in the development process and assists in the removal of faults during the testing phase. The software development process is composed of the following four phases: specification, design, coding and testing. For software systems, the system reliability is increasing as the number of faults latent in the system is decreasing. In fact, however, it is impossible to remove perfectly latent faults from the software system in the limited development period and resources. Then, fault tolerance approach becomes an important tequnique to realize a more highly reliable software system. In a fault tolerant design, the influence of failures upon the whole system is minimized as much as possible by means of redundancy, even if failures will be occurred in the system.

In this paper, we discuss reliability assessment for the redundant software systems during the operation phase. We adopt N-version programming and recovery blocks systems as the fault tolerant approaches for a software system. The reliability measures to assess and predict system reliability quantitatively are obtained by using two software reliability growth models.

2. REDUNDANT SOFTWARE SYSTEMS

2.1 Multiversion software

In the hardware reliability technologies, redundancy is used as one of the fault tolerant techniques to prevent the system from being broken down by a component failure. The concept of redundancy is that a system consists of redundancy schemes for fault tolerant. When one of these operational components fails, the whole system can function properly by the redundancy, as if it seems that the failure has not occurred from outside. Using this technique, a highly reliable system can be achieved. For example, we consider the computer system with N components which are arranged in a parallel configuration. In this system, at least one of N identical and independent components is required to work without failure. Let R denote the reliability of a component. Then, the system reliability R_{system} is given by

$$R_{system} = 1 - (1-R)^{N} > R \qquad (0 < R < 1). \tag{1}$$

Thus, we can confirm that the reliability of the parallel redundant system is greater than that of single component.

Multiversion is an application of this hardware reliability technique to a software system. With respect to multiversion software, more than one version which satisfy the same requirement specification are independently developed. However, there is a wide difference between software and hardware redundancy techniques.

In the case of hardware redundancy, all components are developed through the same process (i.e., the same design and production phases). In other word, a hardware redundant system is composed of some of duplicate components. On the other hand, in the case of software redundancy, the system reliability with N duplicate versions corresponds to that of single version. That is, in the parallel redundant system, for example, the system reliability R_{system} in (1) is coincident with R. This is the main difference nature of hardware and software reliability.

In general, since a hardware failure is caused by wearout phenomenon, the probability that a failure occurs in all of N duplicate components at the same time is negligible. On the other hand, a software failure is caused by software faults latent in the software system, which are introduced in the development process. Therefore, duplicate versions contains the same faults. When the N versions arranged in a parallel configuration is put into an incorrect state logically, a software failure occurs certainly. That is, as discussed above, the probability that a failure occurs in all of N duplicate version at the same time is coincident with that of single version.

To establish multiversion software, independent N versions which contains no common faults each other are required. Under the same requirement specification, the independent versions can be developed through different design, different coding and different testing phases by the different development teams. Then, it is possible to provide redundancy to a software system. N-version programming and recovery blocks are typical redundant software systems composed of such independently developed versions [1].

2.2 N-version programming

N-version programming is a realization of the parallel configuration for software. N-version programming system is composed of independent N versions of a program and a driver module. In this system, first, the driver activates all versions providing the same input data and then collects the individual outputs from the versions. In the simplest case, the driver performs a majority vote in order to determine the overall output from the N version programs and to mask software faults. Therefore, a software failure which occurs in any one version is avoided.

Consider 3-version programming system (N=3) with version 1, 2 and 3. In this system, 2 out of 3 versions are required to function properly. When more than 2 versions produce software failures, it causes a system failure. Define the following:

$F_i(t)$ = the probability that version i produces a software failure in the time interval (0, t]

$R_i(t)$ = the probability that version i functions properly in the time interval (0, t]

$= 1 - F_i(t) \qquad (i = 1, 2, 3)$.

Assuming that the driver module could not fail, the reliability function $R(t)$ for the whole system is given by

$$R(t) = R_1(t)\,R_2(t)\,F_3(t) + R_1(t)\,F_2(t)\,R_3(t) + F_1(t)\,R_2(t)\,R_3(t) + R_1(t)\,R_2(t)\,R_3(t)$$
$$= R_1(t)\,R_2(t) + R_2(t)\,R_3(t) + R_3(t)\,R_1(t) - 2\,R_1(t)\,R_2(t)\,R_3(t) \,. \tag{2}$$

2.3 *Recovery blocks*

Recovery blocks system is composed of independent N version programs (i.e., a primary module and other alternate modules) and acceptance test programs. In this system, first, the primary module is executed and the acceptance test checks the state of the program for successful operation. If the acceptance test fails, the program is restored to the recovery point taken on entry to the recovery block, the first alternate is executed and the acceptance test is applied again. This process continues until an acceptable output is obtained. When all versions fails through N times acceptance test, it comes to system failure. In the recovery blocks system, therefore, only one version are required to function properly for the system to function if the acceptance test program would never fail.

Consider the recovery blocks system with version 1 (the primary module), 2 and 3 (the alternate modules), and a perfect acceptance test program. Let $F_i(t)$, $R_i(t)$ denote the definitions in section 2.2. Then, the reliability function for the whole system is given by

$$R(t) = R_1(t) + F_1(t) * R_2(t) + F_1(t) * F_2(t) * R_3(t)$$
$$= 1 - F_1(t) * F_2(t) * F_3(t) \,, \tag{3}$$

where $*$ means a Stieltjes convolution defined for arbitrary functions $A(t)$ and $B(t)$ as

$$A(t) * B(t) = \int_0^t A(t-x)\,dB(x) \,. \tag{4}$$

3. SOFTWARE RELIABILITY GROWTH MODEL

Software reliability measurement and assessment during the testing phase which is the last stage of the software development process have been often discussed. The purpose of software testing is to detect and remove software faults latent in the software system. During the testing phase, many software failures are observed and recorded. A software failure is defined as an unacceptable departure of program operation caused by a fault in the software system. We assume that all detected faults are removed and no new faults are introduced into the program. Under this assumption, the cumulative number of detected faults is increasing as these are corrected, and the time-interval between software failures becomes longer. This means that the probability of software failure-occurrence is decreasing, i.e., the software reliability is increasing, as the testing goes on.

A mathematical tool which treats such a software failure-occurrence phenomenon during the testing phase is called a *software reliability growth model* (Ramamoorthy and Bastani [2] and Yamada[3]). Many software reliability growth models have been developed for describing the failure-occurrence and fault-detection phenomena and assessing the software reliability (e.g., Goel and Okumoto [4], Jelinski and Moranda [5], Littlewood [6], Musa [7], Musa and Okumoto [8], and Yamada and Osaki [9]). As quantitative measures for software reliability assessment, we can derive the expected number of faults remaining in the system, the mean time-interval between software failures, the software reliability function, and so on, from the software reliability growth models. In this paper, we adopt *Jelinski-Moranda* and *NHPP models*.

3.1 Jelinski-Moranda model

Let X_k denote a random variable representing the time-interval between (k-1)-st and k-th failures (k=1, 2, ···). In Jelinski-Moranda model, the probability distribution function of X_k is given by

$$F(x_k) = 1 - \exp[-\phi\,(N-k+1)\,x_k] \quad (k=1,2,\cdots,N;\ N>0,\ \phi>0)\,, \tag{5}$$

where N is the initial fault content in the system (or the total expected number of faults to be eventually detected) and ϕ is the fault-detection rate per fault remaining in the system. Equation (5) means that X_k obeys an exponential distribution with failure rate $\phi\,(N-k+1)$.

Then, the software reliability function of X_k which represents the probability that a software failure does not occur is given by

$$\begin{aligned} R(x_k) &\equiv 1 - F(x_k) \\ &= \exp[-\phi\,(N-k+1)\,x_k]\,, \end{aligned} \tag{6}$$

and the mean time to software failure (MTTF) is given by

$$\begin{aligned} E[X_k] &\equiv \int_0^\infty R(x_k)\,dx_k \\ &= \frac{1}{\phi\,(N-k+1)}\,. \end{aligned} \tag{7}$$

3.2 NHPP model

Let $\{N(t),\ t\geq 0\}$ denote a counting process representing the cumulative number of faults detected up to testing time t. Then, a software reliability growth model for the fault-detection (or the failure occurrence) phenomenon can be generally described by an NHPP as (see Goel and Okumoto [4] and Yamada and Osaki [9]):

$$\Pr\{N(t)=n\} = \frac{\{H(t)\}^n}{n!}\exp[-H(t)] \qquad (n=0,1,2,\cdots)\,, \tag{8}$$

$$H(t) = \int_0^t h(x)\,dx\,, \tag{9}$$

where $H(t)$ is a mean value function which means the expected cumulative number of faults detected during time-interval $[0, t)$, and $h(t)$ is an intensity function which means the fault-detection rate at testing time t.

From this software reliability growth model, we can derive the residual fault content in the software system and the software reliability function which are useful as software reliability assessment measures. The expected number of faults remaining at testing time t is represented by

$$n(t) = a - H(t)\,, \tag{10}$$

where a is the expected initial fault content or the expected cumulative number of faults to be eventually detected. Let X_k represent the time-interval between (k-1)-st and k-th software failures ($k = 1, 2, \cdots$). When the (k-1)-st failure has occurred at testing time t, the conditional survival probability of X_k, i.e. the conditional reliability function, is given by

$$\begin{aligned} R(x|t) &\equiv \Pr\{X_k > x \mid \sum_{i=1}^{k-1} X_i = t\} \\ &= \exp[-\{H(t+x) - H(t)\}] \quad (t \geq 0, x \geq 0)\,, \end{aligned} \tag{11}$$

which is called software reliability.

Assuming that the expected number of detected faults at an arbitrary testing-time is proportional to the current fault content, we have

$$\frac{dH(t)}{dt} = b\,[a - H(t)] \quad (a > 0, b > 0)\,, \tag{12}$$

where b is the fault-detection rate per fault at testing time t. Solving (12) in terms of mean value function yields

$$H(t) = a(1 - e^{-bt}) \qquad (a > 0, b > 0)\,. \tag{13}$$

From (11) and (13), the software reliability function is given by

$$\begin{aligned} R(x|t) &= \exp[-a(e^{-bt} - e^{-b(t+x)})] \\ &= \exp[-e^{-bt} H(x)]\,. \end{aligned} \tag{14}$$

The mean time to software failure (MTTF) is given as the mean of conditional probability that a failure occurs in $(t, t+x]$, i.e.

$$\begin{aligned} F(x|t) &= 1 - R(x|t) \\ &= 1 - \exp[-e^{-bt} H(x)]\,. \end{aligned} \tag{15}$$

However it is shown that

$$\left.\begin{aligned} &F(0|t) = 0 \\ &F(\infty|t) = 1 - \exp[-a e^{-bt}] < 1 \end{aligned}\right\}, \tag{16}$$

which implies that the conditional distribution $F(x|t)$ is improper and there does not exist the mean. Consequently, there does not exit MTTF for NHPP model.

4. ASSESSMENT MEASURES FOR REDUNDANT SOFTWARE SYSTEMS

Using the software reliability growth model, software reliability assessment measures such as the expected number of remaining faults in each version, the software reliability function of each version, and so on, can be estimated. Based on the test data of each version, we can derive several reliability assessment measures of the redundant software systems. In this paper, we adopt Jelinski-Moranda and NHPP models as the software reliability growth models which describe a software fault-occurrence phenomenon during the operation phase.

4.1 Software Reliability Function

We consider the behavior of software reliability based on Jelinski-Moranda model. For the version i of a redundant software system (or single version system), the software reliability function at the operation time s is given by, from (6)

$$R_i(s) = \exp[-\phi N_i s] = \exp[-\lambda_i s], \tag{17}$$

$$\lambda_i = \phi_i N_i \qquad (i = 1, 2, \cdots), \tag{18}$$

where N_i is the initial fault content in the version i of a redundant software system and ϕ_i is the fault-detection rate per fault remaining in the version i of the system. In the 3-version programming system with version i=1, 2 and 3, the software reliability function for the system is given by, from (2), (17) and (18)

$$R(s) = \exp[-(\lambda_1+\lambda_2)\,s] + \exp[-(\lambda_2+\lambda_3)\,s] + \exp[-(\lambda_3+\lambda_1)\,s] - 2\exp[-(\lambda_1+\lambda_2+\lambda_3)\,s]\,. \tag{19}$$

In the recovery blocks system, the software reliability function for the system is given by, from (3), (17) and (18)

$$R(s) = -\frac{\lambda_2\lambda_3\exp[-\lambda_1 s]}{(\lambda_1-\lambda_2)(\lambda_3-\lambda_1)} - \frac{\lambda_3\lambda_1\exp[-\lambda_2 s]}{(\lambda_2-\lambda_3)(\lambda_1-\lambda_2)} - \frac{\lambda_1\lambda_2\exp[-\lambda_3 s]}{(\lambda_3-\lambda_1)(\lambda_2-\lambda_3)}\,. \tag{20}$$

On the other hand, based on NHPP model, the software reliability function are given as the conditional reliability function. For the version i of a redundant software system (or single version system), the software reliability function at the operation time s is given by, from (14)

$$\left.\begin{aligned} R_i(s|t_i) &= \exp[-e^{-b_i t_i} H_i(s)] \\ H_i(s) &= a_i(1-e^{-b_i s}) \qquad (i = 1, 2, \cdots) \end{aligned}\right\}, \tag{21}$$

where a_i is the initial fault content in the version i of a redundant software system, b_i is the fault-detection rate per fault remaining in the version i of the system, and t_i is the time when to stop testing for version i of the software system. In the 3-version programming system with version i=1, 2 and 3, the software reliability function for the system is given by, from (2) and (21)

$$\left.\begin{aligned} R(s) &= R_1(s|t_1)\,R_2(s|t_2) + R_2(s|t_2)\,R_3(s|t_3) + R_3(s|t_3)\,R_1(s|t_1) \\ &\quad - 2R_1(s|t_1)\,R_2(s|t_2)\,R_3(s|t_3) \\ R_i(s|t_i) &= \exp[-a_i\,e^{-b_i t_i}(1-e^{-b_i s})] \qquad (i = 1, 2, 3) \end{aligned}\right\}. \tag{22}$$

In the recovery blocks system, the software reliability function for the system is given by, from (3) and (21)

$$\left.\begin{aligned} R(s) &= 1 - F_1(s|t_1) * F_2(s|t_2) * F_3(s|t_3) \\ F_i(s|t_i) &= 1 - \exp[\,-a_i\, e^{-b_i t_i}(1 - e^{-b_i s})\,] \qquad (i = 1, 2, 3) \end{aligned}\right\}. \tag{23}$$

4.2 *MTTF*

Based on Jelinski-Moranda model, the MTTF for the 3-version programming system is given by, from (7) and (19)

$$\begin{aligned} \text{MTTF} &= \int_0^\infty \{ \exp[-(\lambda_1 + \lambda_2)\, s] + \exp[-(\lambda_2 + \lambda_3)\, s] + \exp[-(\lambda_3 + \lambda_1)\, s] \\ &\qquad - 2\exp[-(\lambda_1+\lambda_2+\lambda_3)s] \}\, ds \\ &= \frac{1}{\lambda_1 + \lambda_2} + \frac{1}{\lambda_2 + \lambda_3} + \frac{1}{\lambda_3 + \lambda_1} - \frac{2}{\lambda_1 + \lambda_2 + \lambda_3}. \end{aligned} \tag{24}$$

In the recovery blocks system, the MTTF for the system is given by, from (7) and (20)

$$\begin{aligned} \text{MTTF} &= \int_0^\infty \{ -\frac{\lambda_2 \lambda_3 \exp[-\lambda_1 s]}{(\lambda_1 - \lambda_2)(\lambda_3 - \lambda_1)} - \frac{\lambda_3 \lambda_1 \exp[-\lambda_2 s]}{(\lambda_2 - \lambda_3)(\lambda_1 - \lambda_2)} - \frac{\lambda_1 \lambda_2 \exp[-\lambda_3 s]}{(\lambda_3 - \lambda_1)(\lambda_2 - \lambda_3)} \}\, ds \\ &= \frac{1}{\lambda_1} + \frac{1}{\lambda_2} + \frac{1}{\lambda_3}. \end{aligned} \tag{25}$$

4.3 *Hazard Rate Function*

In general, hazard rate function at time t is given by

$$z(t) \equiv \frac{-\frac{d}{ds} R(t)}{R(t)}, \tag{26}$$

where $R(t)$ is the software reliability function at time t. Based on Jelinski-Moranda model, the hazard rate function for the 3-version programming system at the operation time s is given by, from (19) and (26)

$$\begin{aligned} z(s) = \{ & (\lambda_1+\lambda_2) \exp[-(\lambda_1+\lambda_2)\, s] + (\lambda_2+\lambda_3) \exp[-(\lambda_2+\lambda_3)\, s] \\ & + (\lambda_3+\lambda_1) \exp[-(\lambda_3+\lambda_1)\, s] - 2\,(\lambda_1+\lambda_2+\lambda_3) \exp[-(\lambda_1+\lambda_2+\lambda_3)\, s] \} \\ & / \{ \exp[-(\lambda_1+\lambda_2)\, s] + \exp[-(\lambda_2+\lambda_3)\, s] + \exp[-(\lambda_3+\lambda_1)\, s] \\ & - 2\exp[-(\lambda_1+\lambda_2+\lambda_3)\, s] \}. \end{aligned} \tag{27}$$

In the recovery blocks system, the hazard rate function for the system is given by, from (20) and (26)

$$\begin{aligned} z(s) = \lambda_1\lambda_2\lambda_3 \{ & (\lambda_2-\lambda_3) \exp[-\lambda_1 s] + (\lambda_3-\lambda_1) \exp[-\lambda_2 s] + (\lambda_1-\lambda_2) \exp[-\lambda_3 s] \} \\ & / \{ \lambda_2\lambda_3 (\lambda_2-\lambda_3) \exp[-\lambda_1 s] + \lambda_3\lambda_1 (\lambda_3-\lambda_1) \exp[-\lambda_2 s] \\ & + \lambda_1\lambda_2 (\lambda_1-\lambda_2) \exp[-\lambda_3 s] \}. \end{aligned} \tag{28}$$

5. NUMERICAL ILLUSTRATIONS

To examine the characteristics of software system reliability of the N-version programming and recovery blocks systems, we present numerical illustrations of the software reliability assessment measures based on Jelinski-Moranda model discussed in this paper. Let $\lambda_1 = 0.11$, $\lambda_2 = 0.10$ and $\lambda_3 = 0.09$ in (18). Fig. 1 shows the software reliability functions given by (19) and (20) of the N-version programming and recovery blocks systems. To make clear the effect of software redundancy, we show the software reliability function of the single version system in Fig. 1. From (24) and (25), we have MTTF = 8.36 for N-version programming system and MTTF = 30.20 for recovery blocks system, respectively. Fig. 2 shows the hazard rate function given by (27) and (28) of the N-version programming and recovery blocks systems. Under the assumptions in this paper, it is obvious that the reliability of the recovery blocks system is higher than that of the N-version programming system. In practical, however, we should consider the failure probability of the driver module of N-version programming system or the acceptance test module of recovery blocks system to compare the two systems with respect to software system reliability.

6. CONCLUSIONS

In this paper, we have discussed reliability assessment for the redundant software systems during the operation phase. N-version programming and recovery blocks systems have been adopted as typical fault tolerant techniques for a software system. The software redundant system has been discussed from the view point of different characteristics between hardware and software reliability. Describing the fault-occurrence phenomenon of each version by using the software reliability growth models i.e., Jelinski-Moranda and NHPP models, we have discussed the system reliability of the redundant software systems during the operation phase. As the software reliability assessment measures of the systems, the software reliability function, MTTF and hazard rate function have been derived. We have also shown numerical illustrations of the software reliability assessment measures for the N-version programming and recovery blocks systems.

In this paper, each measure has been derived for the redundant system composed of 3 versions (functional module) and driver module or acceptance test module as an example. Of course, it is possible to derive measures for the redundant systems composed of more versions. In N-version programming system, the software reliability function at the operation time s is given by

$$R(s) = 1 - \{ \prod_{i=1}^{N} F_i(s) + \sum_{i=1}^{N} (\frac{R_i(s)}{F_i(s)} \prod_{j=1}^{N} F_j(s)) \} . \tag{29}$$

In recovery blocks system with N-version, the software reliability function at the operation time s is given by

$$R(s) = \sum_{i=1}^{N} (R_i(s) * \prod_{j<i}^{*} F_j(s)) , \tag{30}$$

where $\prod^*$ means a Stieltjes convolution product defined as

$$\prod_{j<i}^{*} F_j(s) = F_1(s) * F_2(s) * \cdots * F_{i-1}(s) . \tag{31}$$

By using software reliability growth models, i.e., Jelinski-Moranda, NHPP and the other models, to describe a software fault-occurrence phenomenon during the operation phase, we will be able to discuss reliability assessment measurement for redundant software system similarly.

In future, we are going to discuss the behavior of system reliability considering the failure probability of the driver module or the acceptance test module. For example, as a simplest case, denote p be the probability that driver module performs properly. Then, the system reliability for N-version programming and recovery blocks systems are given by

$$R(s) = p\,[1 - \{ \prod_{i=1}^{N} F_i(s) + \sum_{i=1}^{N} (\frac{R_i(s)}{F_i(s)} \prod_{j=1}^{N} F_j(s)) \}], \tag{32}$$

$$R(s) = \sum_{i=1}^{N} [\, p\, R_i(s) * \prod_{j<i}^{*} \{ (1-p)\, R_j(s) + p\, F_j(s) \}], \tag{33}$$

respectively.

REFERENCES

[1] U. Voges (ed.), *Software diversity in computerized control systems*, Springer-Verlag, Wien (1988).
[2] C. V. Ramamoorthy and F. B. Bastani, "Software reliability – Status and perspectives," *IEEE Trans. Software Engineering*, Vol. SE-8, pp. 354-371 (1982).
[3] S. Yamada, *Software Reliability Assessment Technology* (in Japanese), HBJ Japan, Tokyo (1989).
[4] A. L. Goel and K. Okumoto, "Time-dependent error-detection rate model for software reliability and other performance measures," *IEEE Trans. Reliability*, Vol. R-28, pp. 206-211 (1979).
[5] Z. Jelinski and P. B. Moranda, "Software reliability research", *in Statistical Computer Performance Evaluation*, ed. W. Freiberger, Academic Press, New York, pp. 465-484 (1972).
[6] B. Littlewood, "Theories of software reliability: How good are they and how can they be improved?" *IEEE Trans. Software Engineering*, Vol. SE-6, pp. 489-500 (1980).
[7] J. D. Musa, "The measurement and management of software reliability," *Proc. IEEE*, Vol. 68, pp. 1131-1143 (1980).
[8] J. D. Musa and K. Okumoto, "A logarithmic Poisson execution time model for software reliability measurement," *Proc. 7th Int. Conf. Software Engineering*, pp. 230-238 (1984).
[9] S. Yamada and S. Osaki, "Software reliability growth modeling: Models and applications," *IEEE Trans. Software Engineering*, Vol. SE-11, pp. 1431-1437 (1985).

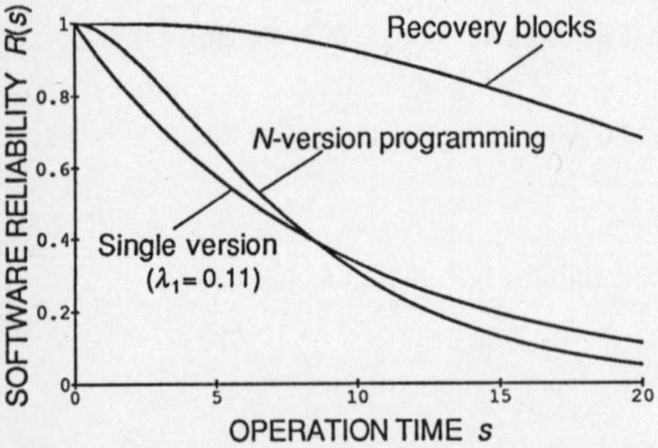

Fig. 1 The software reliability functions for N-version programming and recovery blocks ($\lambda_1 = 0.11$, $\lambda_2 = 0.10$, $\lambda_3 = 0.09$).

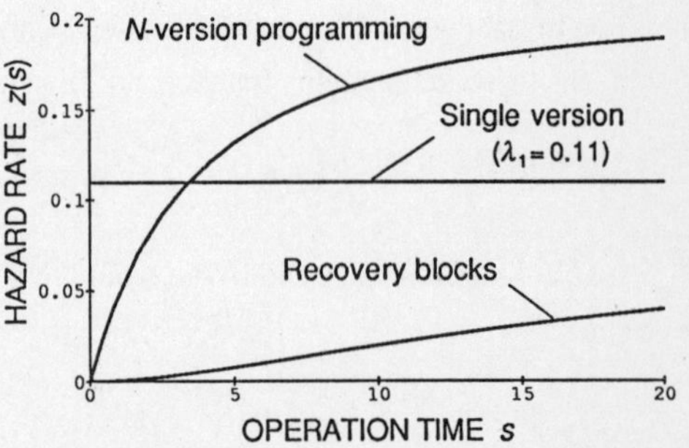

Fig. 2 The hazard rate functions for N-version programming and recovery blocks ($\lambda_1 = 0.11$, $\lambda_2 = 0.10$, $\lambda_3 = 0.09$).

A LOST SALES INVENTORY SYSTEM WITH MULTIPLE REORDER LEVELS

S. KALPAKAM
Department of Mathematics
Indian Institute of Technology
Madras 600 036, INDIA

G. ARIVARIGNAN
School of Mathematics
Madurai Kamaraj University
Madurai 625 021, INDIA

ABSTRACT

In the analysis of continuous review inventory systems with lost sales, all the models studied so far impose the restriction $s < S-s$ or $s = S-1$ to make the analysis tractable. This paper deals with a general lost sales inventory system with renewal demands and exponential lead times without any condition on the values of s and S. The inventory level distribution and the mean reorder and shortage rates are obtained. A computational procedure to calculate the various limiting values along with numerical illustration is provided.

INTRODUCTION

The continuous review (s,S) inventory system with lost sales have been analyzed in the literature either for the case of atmost one outstanding order i.e., $S-s > s$ [1,4,10] or for one-to-one ordering policy i.e., $S-s=1$ which is more suitable for a slow moving, high cost product [3,4,5,8]. In the general situation which allows more than one pending order, it is difficult to derive the exact equation for the lost sales case, as unlike the backorder situation, it is not possible to treat the changes in the amount on hand plus on order independent of the amount on hand [4, p 197]. This article deals with the general case without any restriction on the values of s and S other than $S > s \geq 0$.

PROBLEM FORMULATION

Consider a single item inventory system with a maximum stock of S units from which items are drawn to meet unit demands generated by a renewal process. The probability density function of the time interval between successive demands is denoted by f(t) with mean $m = \int_0^\infty tf(t)\, dt$. The operating policy when described in terms of on hand inventory will be to place orders for $Q = S-s$ items as and when Q demands are met. In other words orders are placed for Q $(0 < Q \leq S)$ items whenever the inventory level drops from above to any one of the reorder levels, $S-Q, S-2Q, \ldots, S-\ell Q$ where ℓ is the largest integer less than or equal to S/Q. Since more than one order can be pending in the case of $S-s < s$, it is assumed that orders can cross (i.e., lead times are independent). The lead times are assumed to be exponentially distributed with parameter μ (>0). Demands occurring during the stock out periods are lost.

We use the following notation:

$\bar{F}(t)$: $1- \int_0^t f(u)\, du$

$[x]^+$: Largest integer less than or equal to x

D : $\left\{ 0,1,2,\ldots,S \right\}$

$\mathbf{N}^0$: $\left\{ 0,1,2,\ldots.. \right\}$

$\langle i \rangle$: $[(S-i)/Q]^+ \quad i \in \mathbf{D}$

$b(i,j,t)$: $\binom{i}{j}(1-e^{-\mu t})^{i-j}\, e^{-\mu t j}, \quad i,j \in \mathbf{N}^0$ and $j < i$

I : Identity matrix

e : Column vector with all elements equal to unity

Θ^*_α : $\int_0^t \Theta(t)\, e^{-\alpha t} dt$, Re $\alpha > 0$ for any matrix $\Theta(t)$

δ_{ij} : $\begin{cases} 1 & \text{if } i=j \\ 0 & \text{otherwise} \end{cases} \quad i,j \in \mathbf{N}^0$

PRELIMINARY RESULTS

Let $0 = T_0 < T_1 < T_2 \ldots$ denote the sequence of demand epochs and $L(t)$ be the inventory level at time t and it assumes values in the set **D**. If $L_n = L(T_n-)$, it can be shown that $(L,T) = \left\{ L_n, T_n;\ n \in \mathbf{N}^0 \right\}$ is a Markov renewal process (MRP) with state space $\mathbf{D} \times [0,\infty)$. The semi-Markov kernel $A(i,j,t)$ of this process is defined as

$$A(i,j,t) = P\left[L_{n+1}=j, T_{n+1}-T_n < t \mid L_n=i\right],\quad i,j \in \mathbf{D}, n \in \mathbf{N}^0 \text{ and } t>0$$

The derivative of $A(i,j,t)$ with respect to t, denoted by $a(i,j,t)$ is given in the following theorem in a compact form which is convenient to handle in the analysis.

Theorem:

The function $a(i,j,t)$ are given by

$$a(i,j,t) = \begin{cases} f(t)\ b(\langle i-1\rangle, \langle j\rangle, t) & i \neq 0, j-i+1 \equiv 0 \bmod(Q) \\ f(t)\ b(\ell, \langle j\rangle, t) & i = 0, \quad j \equiv 0 \bmod(Q) \\ 0 & \text{otherwise} \end{cases}$$

where $a \equiv b \bmod(c)$ implies $a=b+kc$ for some $k \in \mathbf{N}^0$.

Proof: We first note that, $\langle i\rangle$ denotes the number of orders pending at any time when the inventory level is i. Given a demand at T_n, the next demand occurs in $(T_n+t, T_n+t+\Delta)$ with probability $f(t)\Delta$.

Case (i) $i \neq 0$

Since $L(T_n-)=i$, we have $L(T_n+)=i-1$. If $k\ (>0)$ orders are delivered, in order that $L(T_n+t-)=j$, we should have $j=i-1+kQ$ which implies that $j-i+1$ is a multiple of Q. In which case k should be equal to $\langle i-1\rangle - \langle j\rangle$.

Case (ii) $i=0$

In this case $L(T_n-)=L(T_n+)=0$ and ℓ orders are pending at the instant T_n. Arguments similar to the earlier case conclude that $L(T_n+t-)=j$ only when j is a multiple of k.

It can be seen that no transition other than those listed above can occur. Also the probability of receiving any pending order in $(0,t)$ is $(1-e^{-\mu t})$ and the probability of receiving exactly $(i-j)$ out of i pending orders follows the binomial probability law $b(i,j,t)$.

As the demand and lead time processes are independent, the above arguments yield the results stated in the theorem.

ANALYSIS

Inventory Level Distribution

Define

$$\phi(i,j,t) = P\Big[L(t) = j \mid L_0 = i\Big] \qquad i,j \in \mathbf{D}$$

Since the lead times are exponential, once the level at $T_n = \text{Sup}\left\{ T_i < t \right\}$ is known, the history of L(t) prior to T_n loses its predictive value. Hence $Z=\left\{L(t), t>0\right\}$ with state space **D** is a semi-regenerative process with embedded MRP **(L,T)**. As such the function $\phi(i,j,t)$ satisfies the following Markov renewal equation (MRE)

$$\phi(i,j,t) = \psi(i,j,t) + \sum_{r\in \mathbf{D}} \int_0^t a(i,r,u)\, \psi(r,j,t-u)\, du \qquad (1)$$

where

$$\psi(i,j,t) = P\Big[L(t) = j, T_1 > t \mid L_0 = i\Big]$$

$$= \bar{F}(t)\, P\Big[\, L(t) = j \mid T_1 > t, L = i \Big], \quad i,j \in \mathbf{D}$$

and is given by

$$\psi(i,j,t) = \begin{cases} \bar{F}(t)\, b(\langle i-1\rangle, \langle j\rangle, t) & i \neq 0, \quad j \equiv (i-1) \bmod (Q) \\ \bar{F}(t)\, b(\ell, \langle j\rangle, t) & i = 0, \quad j \equiv 0 \bmod (Q) \\ 0 & \text{otherwise} \end{cases}$$

The above result is derived by employing the following arguments:

If L(0-)=i-1 (i≠0) and given that $T_1 > t$, in order that L(t) = j (>i-1), exactly r = <i-1>-<j> out of <i-1> orders pending at 0+ must be delivered before t and it occurs with probability b(<i-1>,<j>,t). Also j = i-1+rQ, which implies that j-i+1 is a multiple of Q. If i=0, we have L(0+)=0, reasons similar to the preceding case yield that exactly ℓ-<j> orders must be executed in (0,t) and that j should be a multiple of Q.

Let **A**(t), Φ(t) and Ψ(t) denote square matrices of order (S+1) with a(i,j,t), ϕ(i,j,t) and Ψ(i,j,t) as their $(i,j)^{th}$ elements respectively. It may be noted that **A**(t) and Ψ(t) are almost upper triangular (A matrix $((\, a_{ij}\,))$ is almost upper triangular, if $a_{ij} = 0$ for $i > j+1$).

Taking Laplace transform on both sides of the MRE(1), we obtain,

$$\Phi^*_\alpha = R^*_\alpha \Psi^*_\alpha$$

where $R^*_\alpha = (I-A^*_\alpha)^{-1}$ is the Laplace transform of the Markov renewal kernel of the MRP (L,T) and it exists for Re $\alpha > 0$.

Reorders and Shortages

Let events ξ_r denote the occurrence of a reorder triggered by a demand and the stock level dropping to S-rQ $(r=1,2,\dots,\ell)$ and $\{N_r(t), t > 0\}$ be the counting process associated with ξ_r events. The conditional first order product density function h(i,r,t) of ξ_r events defined by

$$h(i,r,t) = \lim_{\Delta\to 0} \frac{1}{\Delta} P\left[N_r(t+\Delta)-N_r(t)=1 \mid L_0=i \right] \qquad r=1,2,\dots,\ell; i \in D$$

satisfy the following MRE's:

$$h(i,r,t) = k(i,r,t) + \sum_{j\in D} \int_0^t a(i,j,u)\, h(j,r,t-u)\, du \tag{2}$$

where

$$k(i,r,t) = \lim_{\Delta\to 0} \frac{1}{\Delta} P\left[L(T_1-) > L(T_1+) = S-rQ\ , t<T_1<t+\Delta \mid L_0=i \right]$$

$$= \lim_{\Delta\to 0} \frac{1}{\Delta} P\left[L_1=S-rQ+1, t<T_1<t+\Delta \mid L_0=i \right]$$

$$= a(i,S-rQ+1,t) \tag{3}$$

Define $\mathbf{H}(t)$ and $\mathbf{K}(t)$, matrices of order $S \times \ell$, with entries h(i,r,t) and k(i,r,t) respectively. Then the Laplace transform of the MRE's (2) yield

$$H^*_\alpha = R^*_\alpha K^*_\alpha$$

It may be noted that the columns of K^*_α are those of A^*_α corresponding to the states S-rQ+1 $(r=1,2,\dots,\ell)$.

The conditional first order product density of the point process corresponding to the epochs of reorders is given by

$$h(i,t) = \sum_{r=1}^{\ell} h(i,r,t)$$

Similarly, if $\left\{ N_\beta(t), t >0 \right\}$ denotes the counting process of shortages, then the conditional first order product density function g(i,t) of the process is given by

$$g(i,t) = \lim_{\Delta \to 0} \frac{1}{\Delta} P\left[N_\beta(t+\Delta)-N_\beta(t)=1 \mid L_0=i\right] \qquad i \in \mathbf{D}$$

and satisfy the equation

$$g(i,t) = f(t)\, e^{-\mu t \ell} \{\delta_{i0}+\delta_{i1}\} + \sum_{j\in D} \int_0^t a(i,j,u)\, g(j,t-u)du \tag{4}$$

The above equation is obtained using the classification that the shortage at t is either for the first demand or a subsequent one. Also, the first demand to result in a shortage, the inventory level should have been zero soon after the initial demand and none of the pending orders should have materialized upto t.

From equation (4), we obtain

$$g^*_\alpha(i) = f^*_{\alpha+\ell\mu} \{\delta_{i0}+\delta_{i1}\} + \sum_{j\in D} a^*_\alpha(i,j)\, g^*_\alpha(j) \tag{5}$$

If $\mathbf{G}^*_\alpha$ denotes the (S+1) component column vector with entries $g^*_\alpha(i)$ $(i=0,1,2,..,S)$, then from (4) we obtain,

$$\mathbf{G}^*_\alpha = \mathbf{R}^*_\alpha \left(f^*_{\alpha+\ell\mu}, f^*_{\alpha+\ell\mu}, 0, \ldots, 0 \right)^T$$

STEADY STATE RESULTS

Consider the finite Markov chain (MC) $\mathbf{L} = \left\{L_n, n \in \mathbf{N}^0\right\}$ with state space $\mathbf{D}$, whose transition probability matrix $\mathbf{P}$ is given by

$$\mathbf{P} = ((p_{ij})) = \int_0^\infty \mathbf{A}(t)\, dt.$$

From the theorem given in the previous section, we have for $i,j \in \mathbf{D}$

$$p_{ij} = \begin{cases} p(\langle i-1\rangle, \langle j\rangle) & , \quad i \neq 0,\ j \equiv (i-1) \bmod (Q) \\ p(\ell, \langle j\rangle) & , \quad i = 0,\ j \equiv 0 \bmod (Q) \\ 0 & , \quad \text{otherwise} \end{cases}$$

where $p(m,n) = \int_0^\infty f(t) b(m,n,t)\, dt$.

From the structure of $\mathbf{P}$ it can be seen that the finite MC $\mathbf{L}$ is irreducible. Hence there exists a unique stationary distribution $\Pi=(\pi_0,\pi_1,\ldots,\pi_S)$ which is obtained as (for derivation refer to the appendix).

$$\pi_1 = \pi_0 \,(1-f^*_{\mu\ell}) \Big/ f^*_{\mu\ell} \tag{6}$$

$$\pi_{j+1} = \frac{1}{f^*_{\mu<j>}} \left[\pi_j - (1-\sum_{m=1}^{Q-1} \delta_{jm}) \sum_{r=1}^{d} \pi_{j+1-rQ}\, p(<j-rQ>,<j>) \right]$$

$$S>j\neq rQ, r=1,2,\ldots,\ell;\quad d=[(j-1)/Q]^+ \tag{7}$$

$$\pi_{j+1} = \frac{1}{f^*_{\mu<j>}} \left[\pi_j - (\pi_0+\pi_1)p(\ell,<j>) - (1-\delta_{r1}) \sum_{n=1}^{r-1} \pi_{nQ+1}\, p(<nQ>,<j>) \right]$$

$$j = rQ, r=1,2,\ldots,\ell \tag{8}$$

Although the above set of equations determine π_j's except for a multiplication constant, which is determined from the normalizing condition, an efficient numerical method to compute the stationary distribution is provided in the last section.

Limiting Inventory Distribution

Let $\phi_j = \lim_{t\to\infty} \phi(i,j,t) \qquad i,j \in \mathbf{D}$

The MRP **(L,T)** is irreducible and recurrent as the underlying MC **L** is irreducible and recurrent. As $a(i,j,t)$ exists, it is also aperiodic. Further the functions $\psi(i,j,t)$ are non-negative and Riemann integrable. Hence using the limit theorem for MRPs to equation (1), we obtain,

$$\phi_j = \sum_{i\in \mathbf{D}} \pi_i \int_0^\infty \psi(i,j,t)dt \Big/ \sum_{k\in \mathbf{D}} \pi_k m_k$$

where m_k is the mean sojourn time of state k in **D** and is equal to m, the mean inter demand time for all $k \in \mathbf{D}$ (refer [2]). Thus we have

$$\phi_j = \frac{1}{m} \sum_{i=0}^{j+1} \pi_i\ \psi_0^*(i,j)$$

The mean inventory level $\bar{L}$, in the steady state, is given by

$$\bar{L} = \sum_{j\in \mathbf{D}} j\phi_j.$$

Mean Reorder and Shortage Rates

Let $h_r = \lim_{t\to\infty} h(i,r,t)$

Applying the limit theorem to the MREs (2), we obtain

$$h_r = \frac{1}{m} \sum_{i\in D} \pi_i \int_0^\infty k(i,r,t)\, dt$$

$$= \frac{1}{m} \sum_{i\in D} \pi_i \; p(i, S-rQ+1) \qquad \text{(from (3))}$$

$$= \frac{1}{m} \pi_{S-rQ+1} \qquad \text{(using } \Pi P = \Pi\text{)}$$

Hence the expected reorder rate γ, in the steady state is given by

$$\gamma = \sum_{r=1}^{\ell} h_r = \frac{1}{m} \sum_{r=1}^{\ell} \pi_{S-rQ+1}$$

Similarly the expected shortage rate β, in the stationary case can be obtained from (4) as follows

$$\beta = \lim_{t\to\infty} g(i,t)$$

$$= \frac{1}{m} (\pi_0+\pi_1)\, f^*_{\ell\mu}$$

$$= \frac{1}{m} \pi_0 \qquad \text{(from (6))}$$

Remarks (Special cases)

1) The case ℓ=1 corresponds to the condition S-s > s and the various measures of system performance along with optimal cost analysis are independently derived by the authors as a special case of an inventory system with Markov renewal demand in [6].

2) Similarly the case s=S-1, has been dealt in [5], in which the existing results in the literature for Poisson demands is obtained as a special case.

NUMERICAL PROCEDURE

Since the limiting values ϕ_i, γ and β are expressed in terms of π_i, we describe a convenient procedure for the numerical computation of the latter.

The system of equations **xP** =**x** shall be solved as a first step to obtain Π. The above equation implies that **x**(I-P) = 0. Since (I-P) does not have full rank (a necessary condition for the existence of non-trivial solution) and the system of equations is consistent, the solution is given by

$$\mathbf{x} = \mathbf{b}(\ (I-P)(I-P)^{-} - I)$$

where $(I-P)^{-}$ is a generalized inverse of (I-P) and **b** is an arbitrary row vector (Rao[7]).

The matrix (I-P) can be partitioned as

$$(I-P) = \begin{bmatrix} v & a \\ Z & u \end{bmatrix} \begin{matrix} 1 \\ S \end{matrix}$$
$$\qquad\qquad\qquad S \quad 1$$

where Z is a lower triangular matrix with non-zero elements along its diagonal. If we rearrange the columns of (I-P) to have the matrix

$$\begin{bmatrix} a & v \\ u & Z \end{bmatrix}$$

then its generalized inverse is given by

$$\begin{bmatrix} a & v \\ u & Z \end{bmatrix}^{-} = \begin{bmatrix} 0 & 0 \\ 0 & Z^{-1} \end{bmatrix}$$

Hence we have, from (6)

$$\mathbf{x} = \mathbf{b}\left(\begin{bmatrix} 0 & vZ^{-1} \\ 0 & I \end{bmatrix} - I\right)$$

$$= \mathbf{b}\begin{bmatrix} -1 & vZ^{-1} \\ 0 & 0 \end{bmatrix}$$

Since **b** is arbitrary, choosing it to be a vector with first component to be a non-negative constant and using the condition **xe** = 1, we obtain $\Pi = [-1+vZ^{-1}e]^{-1}(-1, vZ^{-1})$

It may be noted that as Z is a lower triangular matrix, Z^{-1} can be easily computed.

Numerical Illustration

For purposes of numerical illustration, we consider the case when f(t) is exponential with parameter λ (>0). Numerical results are provided for fixed values of S and μ/λ. In Table-1, the stationary distribution are given for S=10 and $\mu/\lambda=0.3$ and for specific values of Q, viz., 1, 4, 7 and 9. Table-2 provides the reorder rate γ, the shortage rate β, and the mean inventory level $\bar{L}$, for various ordering quantities. It may be noted that from these tables γ and β can be computed for various combinations of μ and λ, provided their ratio is equal to the value given.

Table-1

S =10 and μ/λ = 0.3

	Stationary distribution (π_j's)			
j \ Q	1	4	7	9
0	0.0017	0.0607	0.1781	0.2217
1	0.0050	0.0364	0.0534	0.0665
2	0.0135	0.0583	0.0694	0.0864
3	0.0324	0.0933	0.0903	0.0864
4	0.0680	0.1213	0.1174	0.0864
5	0.1224	0.1213	0.1174	0.0864
6	0.1837	0.1358	0.1174	0.0864
7	0.2204	0.1415	0.1174	0.0864
8	0.1983	0.1135	0.0639	0.0864
9	0.1190	0.0771	0.0489	0.0864
10	0.0356	0.0407	0.0270	0.0199

Table-2

S =10, μ/λ =0.25

Q	γ/λ	β/λ	$\bar{L}$
1	1.0000	0.0053	6.020
2	0.4889	0.0220	5.600
3	0.3159	0.0520	5.209
4	0.2260	0.0940	4.830
5	0.1768	0.1159	4.695
6	0.1309	0.2145	4.036
7	0.1105	0.2263	3.905
8	0.0947	0.2424	3.833
9	0.0819	0.2626	3.830

APPENDIX

The stationary distribution $\Pi = (\pi_0, \pi_1, \ldots, \pi_S)$ satisfies

$$\Pi \mathbf{P} = \Pi \text{ , and } \sum_{i=0}^{S} \pi_i = 1$$

The first equation can be rewritten as

$$\pi_j = \sum_{i=0}^{j+1} \pi_i P_{ij} \qquad j=0,1,\ldots,S-1$$

case (i) $j = 0$

$$\pi_0 = \pi_0 p_{00} + \pi_1 p_{10}$$

$$= \pi_0 p(\ell,\ell) + \pi_1 p(\ell,\ell) = (\pi_0+\pi_1) f^*_{\mu\ell} \tag{A.1}$$

case (ii) $j=rQ$ $(<S), r=1,2,\ldots,\ell$

$$\pi_j = \sum_{i=1}^{j+1} \pi_i p_{ij}$$

$$= \pi_0 p_{0j} + \pi_1 p_{1j} + \sum_{i=2}^{j+1} \pi_i p_{ij}$$

$$= \pi_0 p(\ell,<j>) + \pi_1 p(\ell,<j>) + \sum_{n=1}^{r} \pi_{nQ+1} p(<nQ>,<j>)$$

Thus

$$\pi_j = (\pi_0+\pi_1) p(\ell,<j>) + \sum_{n=1}^{r-1} \pi_{nQ+1} p(<nQ>,<j>) + \pi_{j+1} f^*_{\mu<j>} \tag{A.2}$$

case (iii) $j\neq0,S$; and j is not a multiple of Q

We have

$$\pi_j = \sum_{i=2}^{j+1} \pi_i p_{ij}, \qquad S > j > 1 \text{ and } j \not\equiv 0 \bmod (Q)$$

We note that $p_{ij} > 0$ for $j-i+1 \equiv 0 \bmod (Q)$ as $i > 2$. This implies that $j-i+1 \equiv rQ$ for $r=0,1,2,\ldots,[(j-1)/Q]^+$. Hence we have

$$\pi_j = \sum_{r=0}^{d} \pi_{j+1-rQ}\, p(<j-rQ>,<j>)$$

$$= \pi_{j+1} f^*_{\mu<j>} + \left(1-\sum_{m=1}^{Q-1} \delta_{jm}\right) \sum_{r=1}^{d} \pi_{j+1-rQ}\, p(<j-rQ>,<j>) \tag{A.3}$$

where $d = [(j-1)/Q]^+$.

Equations (A.1),(A.2) and (A.3) in turn yield the results given in (6),(7) and (8).

REFERENCES

1. Archibald B.C.,Continuous Review (s,S) Policies with Lost Sales,*Mgmt.Sci.*,**27**,1171-1177,1981.
2. Cinlar E.,*Introduction to Stochastic Processes*,Prentice Hall, Englewood Cliffs,N.J.,1975.
3. Feeney G.J. and Sherbrooke C.,The (s-1,s) Inventory Policy under Compound Poisson Demand,*Mgmt.Sci.*,**12**,391-411,1966.
4. Hadley G. and Whitin T.M.,*Analysis of Inventory Systems*,Prentice Hall,Englewood Cliffs,N.J.,1963.
5. S.Kalpakam and G.Arivarignan, The (S-1,S) Inventory System with Lost Sales, *Proceedings of the ICMMST,* Madras, India, 205-212, 1988
6. S.Kalpakam and G.Arivarignan, The (s,S) Inventory System with Lost Sales and Markov Renewal Demands, *Mathl. Comput. Modelling,* **12,** 1511-1520, 1989
7. Rao C.R.,*Linear Statistical Inference and its Applications*,Second edition,John Wiley and Sons,New York,1983
8. Smith A.S.,Optimal Inventories for an (S-1,S) System With No Backorders *Mgmt.Sci.*,**23**,522-528,1977.
9. Srinivasan S.K.,*Stochastic Point Processes and their Applications*, Griffin,London,1974.
10. Srinivsan S.K.,General Analysis of s-S Inventory Systems with Random Lead Times and Unit Demands ,*J.Math.Phy.Sci.,* **13,**107-129,1979.

QUEUEING MODELS IN HIERACHICAL PLANNING SYSTEMS

Klaus-Peter Kistner

Faculty of Economics
University of Bielefeld
POB 8640 D 4800 Bielefeld 1, Germany

ABSTRACT

Frequently, modern systems of production management are characterized by a hierarchical structure: Decisions on a higher level set targets to be considered by production planning and carried out by production control on subordinate levels. Setting targets on the superior level requires information about the performance, such as net capacities available on the subordinate level. As its performance depends on inventory planning and scheduling, decisions on the subordinate level which, in turn, depend on the targets to be fulfilled, this information is not available for planning on the superior level. Capacities available and other measures of performance may, however, be derived using appropriate mathematical models of the planning process.

In this paper, several queueing theoretical approaches estimating the capacity and the throughput of subordinate manufacturing units in a hierarchical production planning surrounding are presented. Considering the assumptions necessary to find operable solutions, e.g. appropriate Markovian properties, it can be shown that models of open and closed queueing networks may be applied in the case of single item manufacturing and production of small batches. Whereas hybrid approach with a continuous flow of production and rates switching according to a semi-Markov process may be applied in the case of large batches.

1. INTRODUCTION

The structure of queueing models is characteristic for many problems arising in production planning: At the first glance, it seems to be reasonable to model lot-sizing and batch production by queueing models with arrivals in groups and batch service (cf. BAILEY [1954], FERSCHL [1961]). Restrictions of the waiting room (cf. MORSE [1959]) or impatient customers (cf. BARRER [1957]) facilitate the introduction of capacity restrictions. Furthermore, priority rules (cf. COBHAM [1954]; WHITE/CHRISTIE [1958]) as well as breakdowns (cf. White/CHRISTIE [1958], GAVER [1962], KISTNER [1974]) and maintenance of machines (cf. PALM [1947]) may be considered, too.

Queueing models with many servers should be applied to consider multiple facility systems, which are predominant in modern manufacturing. Systems with parallel channels [cf. MORSE [1959]) are available to model spare machines and different types of machines with similar functions; series of servers (cf. HUNT [1956]) may describe flowshop production and assembly lines (cf. FERSCHL [1964, p. 124]); finally, queueing networks [JACKSON [1957,1965] may be suitable models of jobshop production.

During the last decades, these basic models of queueing theory have been considerably refined and generalized; there are, however, only few applications in the practice of manufacturing planning. Apart from the reluctance of industry to apply refined mathematical planning models, a certain inconsistency between the general assumptions of queueing theory and the goals of production planning may be responsible for the gap between the plethora of appropriate queueing models and their application in manufacturing control: Most queueing models, which may either be solved explicitly or give numerical values for those quantities used in production control, have to assume that at least interarrival times are exponentially distributed. In contrast, manufacturing planning aims at a well-balanced use of capacity and a regular flow of material; as measures of operations control reduce stochastic fluctuations, the assumption of exponentially distributed characteristics of queueing models seems to be inconsistent with planning and control of production. Hence, queueing models are considered to be of little help in applied manufacturing planning.

In spite of the restrictive assumptions of queueing theory, there are, however, some recent developments in the techniques of production planning, which open possible fields of application for waiting line models. Due to the complexity of modern manufacturing systems, the large number of products to be processed on many machines, and the amount of interactions between these elements, mathematical models of production planning, integrating lot-sizing or batch-sizing with routing and sequencing, are too large and too complicated to find optimum solutions with reasonable computational effort. In order to support decision making in manufacturing, monolithic models of production have to be decomposed into smaller manageable models of special aspects of production planning, such as single product lot-sizing and sequencing for separated production units or workshops. In order to coordinate the planning process, supervisory departments have to develop targets for decentralized planning based on aggregate planning models.

As the introduction of a hierarchy in production planning is consistent with existing organizational structures, and reflects planning processes applied in manufacturing, particularly in computerized systems of production planning and control (PPS), hierarchical production planning may be helpful to increase the acceptance of mathematical models. However, in order to set targets for subordinate planning, its performance has to be considered by aggregate planning. For example, deriving production targets for a workshop, its net capacity has to be known; net capacity is, however, dependent on idle times of machines which are determined by scheduling decisions on the subordinate level. These decisions have to rely on the product mix and the amount of items to be processed, which, in turn, are dependent on targets set by aggregate planning. In order to break this circle, aggregate planning has to forecast the probable performance of subordinate planning; this may be achieved by the application of appropriate models, describing the structure and the decision process of decentralized decision units.

Recently, several approaches integrating queueing models into hierarchical planning systems have been proposed: SOLBERG [1977] suggested a queueing network for the ex ante evaluation of flexible manufacturing systems, in order to compare alternative structures with respect to their performance and their effective capacity. This model has been refined by SURI/HILDEBRANDT [1984], by BUZACOTT/SHANTIKUMAR [1985], and by TEMPELMEIER [1988]. KISTNER/ STEVEN-SWITALSKI [1990a] and KISTNER/STEVEN [1990b] applied this approach to hierarchical production planning, and calculated performance and capacity of decentralized workshops. DE KOSTER [1989] used an approach to analyse semi-continuous waiting line models of production due to WIJNGAARD [1979] for a capacity orientated analysis and design of production systems.

The common feature of these approaches is the analysis of hierarchically structured decision systems, using queueing models to predict the behaviour of subordinate productive units, in order to supply data for decisions on a superior, aggregated level. At this level, decisions concerning capacity allocation have to be made before sufficient information about the actual load of the production system are available. To derive these data from queueing models, simple assumptions on the input process and the product mix have to be introduced. As no specific information about the distributions of theses characteristics are available, it may be justified to assume independence and exponential or multinomial distributions.

In the present paper, applications of queueing models in hierarchical production planning will be analysed. Though the models will be presented briefly, the main purpose will be to check whether the assumptions facilitating the analysis of the basic queueing models have to be rejected a priori, or whether they are acceptable as a first approximation of the real processes. In particular, waiting line models will be confronted with special structures of the production process, namely single item manufacturing, production of small and large batches, and mass production.

2. QUEUEING NETWORKS IN PRODUCTION PLANNING

A workshop or a similar production unit may be modelled by a queueing network: The nodes represent machines and other units such as transportation facilities or buffers etc., the routing of a product corresponds to a path through the network; customers are orders or batches to be processed by the system. Depending on the aim of the study, two types of queueing networks may be applied:

(1) *Open networks* can be used to estimate the behaviour of the system - e.g. the number of orders waiting at a machine, their waiting time and the time spent in the system - as a function of capacity utilization, measured by the input rate.

(2) *Closed networks* can be used to estimate the maximum capacity of the system measured by the number of units passing a counter. This channel represents a device to be passed by an order at the beginning or the end of its production process, such as dispatching or release of orders, mounting and dismounting of jobs on pallets, or quality controls.

2.1 Open queueing networks

The application of open queueing networks has been suggested by JACKSON [1957,1965]. In order to model a production system by an open queueing network, we assume:

(1) The queueing network has M interacting servers (machines) m=1,...,M; service times of each server are independent, exponentially distributed random variables with rate β_m.

(2) There are K sources of customers (k=1,...,K), representing different types of jobs to be processed; each type of jobs is characterized by a special path through the network. Interarrival times of customers from source k are independent, exponentially distributed random variables with rate ^{k}.

(3) Service times at different servers and interarrival times from different sources are mutually independent.

(4) Every customer has to pass successively all servers on his path through the network in the corresponding sequence.

(5) Servers have unlimited waiting space.

An open queueing network describing a production unit is portrayed in Fig. 1.
Due to the Markovian properties of the system, the following well-known lemmas hold:

Lemma 1:

(1) Given two mutually independent Poisson processes N_1 and N_2 with rate μ_1 and μ_2, respectively. Then, the combined process N_1+N_2 is a Poisson processes with rate $\mu_1+\mu_2$.

(2) Given a Poisson process N with rate . If some events are selected independently with probability p, then the selected events form a Poisson process with rate μp, the remaining events a Poisson process with rate $\mu(1-p)$.

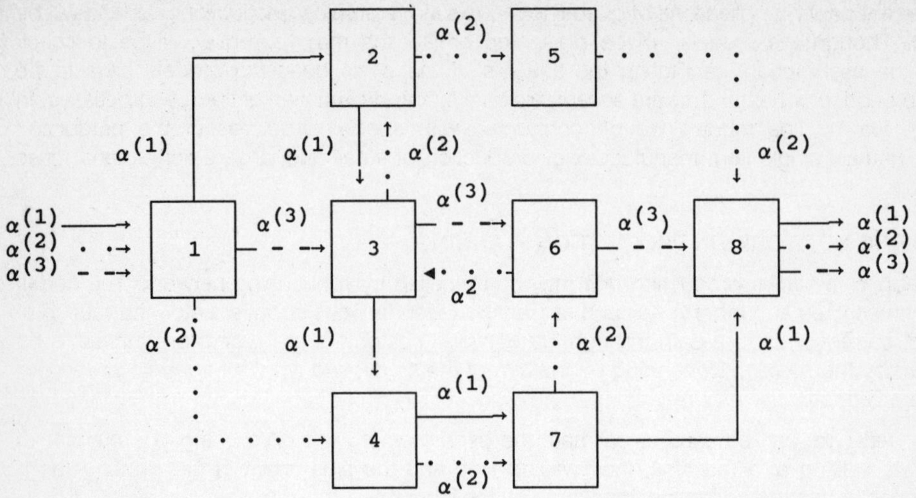

Fig. 1
Open Queueing Network

Lemma 2:

If the interarrival times of a single server and the service rates are independent exponentially distributed random variables with rates and ß, respectively, then the stationary output process, that is the sequence of completions of service, is a Poisson process with rate . Furthermore, at every epoch t, the state of the system and the sequence of completions prior to t are mutually independent.

From these lemmas the following separation theorem for queueing networks with Poisson input and exponentially distributed service times can be derived:

Separation Theorem:

If interarrival times of customers from different sources are independent Poisson processes, and service times of all servers are mutually independent, exponentially distributed random variables, then the system can be separated into independent single server systems with exponentially distributed interarrival times and service times.

If (m) is the set of sources using server m, then the arrival rate at server m is given by:

$$\alpha_m = \sum_{k \in \Gamma(m)} \alpha^{(k)}$$

The exponential distribution of interarrival times at each server follows from the lemmata stated above: According to assumption (2), input to the first server is exponentially distributed; due to lemma 1 its input rate $_1$ is equal to the sum of the rates ^{k} of all sources. The service rate of server 1 is exponentially distributed, hence, lemma 2 guarantees that the output of the first server is exponentially distributed with rate . Furthermore, it follows from lemma 1 that the output stream can be redivided into separate Poisson streams, according to the different sources. If there are no circular flows in the network, customers can be regrouped with respect to the next server on their path through the network into compound streams; due to lemma 1, the stream can be described by a Poisson process, the rate of the compound process is equal to the sum of the rates of its components. Hence, interarrival times at the successors of the first server are exponentially distributed; as service times of these servers are exponentially distributed, it follows from lemma 2, that their output is exponentially distributed, too. The same argument holds for the other servers. Furthermore, it can also be applied in the case of cyclic flows.

Due to the separation theorem, in the case of Poisson input and exponentially distributed service times, characteristics of the performance of a queueing network can be derived immediately from well-known results for (M/M/1) queueing systems. In particular, closed expressions for the distribution of the number of items waiting and the waiting times or the times spent at a server, as well as the first moments of these distributions are available. From the time spent at server D_m and the set of servers (m) to be visited by a customer of a special type, the time spent in the system can be derived:

$$D = \sum_{m \in \Gamma(m)} D_m$$

In the context of hierarchical systems of production planning, these results may be applied to support decision making on the superior level by supplying the following information on the performance of the subordinate level:

(1) For targets set by aggregate production planning - for example an aggregated rate of products to be processed and probable shares of different types of products - information about capacity utilization and bottle-necks can be forecasted. If the compound input rate $_m$ at a machine m is greater than its service rate $ß_m$, then its capacity is insufficient to fulfil the targets; hence, either additional capacity has to be provided, or targets have to be reduced. The same is true, if expected waiting times of orders at a special machine is too high.

(2) As all characteristics are available in closed form, and can easily be calculated, sensitivity analysis gives information about the effects of changes in the total amount of orders to be processed and in the mix of various types.

(3) Furthermore, limits for targets, e.g. the maximum input rates , can be calculated, which guarantee that

(a) the expected value of a critical measure remains below a given bound;
(b) no realizations of a critical measure surpasses a bound with a given probability.

In particular, these results are useful in interactive approaches to coordinate aggregate production planning and operative control.

2.2 Closed queueing networks

In a closed queueing network, a finite number N of customers cycles permanently between a finite number of servers and never leaves the system. At a given epoch, each customer is in one of the following states:

(1) The customer is served by one of the servers.
(2) The customer is waiting to be served by a server.

A special case of a queueing network is portrayed in figure 2: A customer leaves the server m=0 at random in order receive service by servers on a path chosen by chance; after completion of service at server M=7, the customer returns to server m=0 and waits to be released again.
At the first glance, a closed queueing network seems to be unsuited as a model of the flow of production in a manufacturing system, as orders should not circulate infinitely from one machine to the next; instead, they leave the system after completion of processing on the last machine. Using the following argument, the model may, however, be applied to analyse throughput, waiting times, and completion times in a manufacturing system with heavy load, and to get insight into its overall capacity (cf. SOLBERG [1977], TEMPELMEIER [1987]): In order to avoid excessive queues and buffers in the workshop, only a limited number N of jobs is admitted to the

system; whenever a job is completed and leaves the system, the next job in an infinite queue is released to replace immediately the leaving job. The process is controlled by a limited number N of items giving access to the manufacturing system, such as transport pallets, licenses, or KANBAN cards etc. Whenever a job is completed, the item is released and transferred to the next in an infinite queue of jobs. Hence, not an infinite number of jobs, but a limited number of pallets or licenses circulate in a closed network model of a manufacturing system.

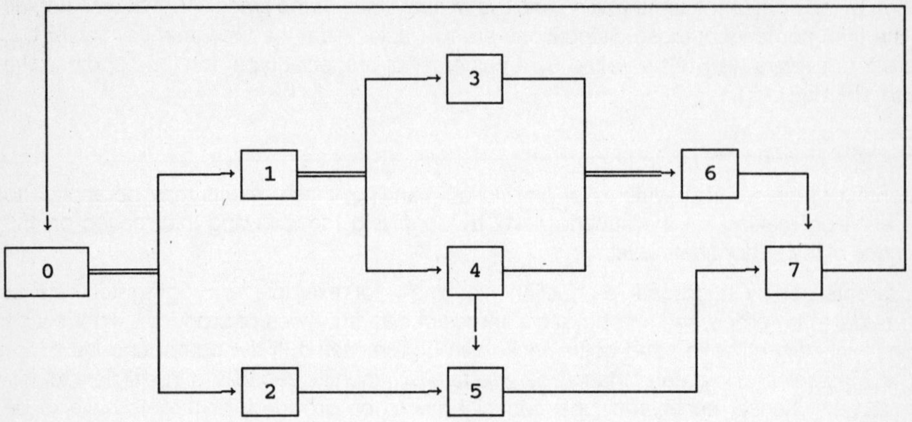

Fig. 2
Closed Queueing Network

In order to analyse the behaviour of a closed queueing network, the following assumptions have to be introduced:

(1) The network has $M+1$ servers $m=0,1,...,M$; the server $m=0$ may be considered as a dispatcher.

(2) Every job released for processing is accompanied by one of a limited number of licenses circulating in the system.

(3) Whenever a job is completed, it releases its license by returning it to the dispatcher.

(4) When a license returns to the dispatcher, the next in the line of jobs to be processed is released.

(5) When machine i has completed its processing, the job will be transferred to machine j with probability p_{ij}, and has to wait for handling at this machine. The probabilities p_{ij} are independent of the state of the system and processing times.

(6) After handling at machine $m=M$, the job is completed.

(7) If a machine is occupied, an arriving job has to wait in a queue with sufficient waiting room.

(8) Processing times at a machine m are independent, exponentially distributed random variables with rate β_m.

In contrast to open queueing models, there are no closed expressions for state probabilities and operating characteristics in a closed queueing model. Assuming exponentially distributed processing times, state probabilities can, however, be expressed in a product form, which can be solved numerically. (cf. GORDON/NEWELL [1967], BUZEN [1973], DENNING/BUZEN [1978], BRUELL/BALBO [1980]).

In particular, the following characteristics can be calculated (cf. KISTNER/SWITALSKI-STEVEN [1990a, S. 96]):

(1) The average length of the queue and the mean waiting time at machine m.
(2) The throughput of the systems, that is the average number of licenses passing the dispatcher per unit of time.
(3) The throughput of machine m and the utilization of its capacity.
(4) The probability that the number of jobs waiting at machine m exceeds a critical value N_m^*.

Using these characteristics, aggregate production planning may derive ex ante information about the maximum capacity of a manufacturing system and about bottle-necks to be considered.

2.3 Queueing Networks and the Structure of the Production Process

We have now to discuss the problem, how the queueing models described above can be applied in a hierarchical production planning system. In particular, aggregate production planning requires information about the net capacity of a manufacturing unit, which is, in turn dependent on targets set by aggregate planning. In order to derive this information, models of the behaviour of a manufacturing unit have to be considered, calculating certain measures of its performance as functions of targets set by aggregate planning.

In order to model a manufacturing unit as a queueing network, the following situation is considered:

(1) A manufacturing unit has to comply with targets set by aggregate planning. These targets are defined as aggregate quantities, for example, the average number of jobs to be processed per unit of time.

(2) The exact number and the mix of these jobs is still unknown at the time the aggregate production plan has to be formulated. Based on past experience, it may be assumed, however, that K different types of jobs k=1,..., K have to be performed. Their relative share of the total number of jobs to be processed is q_k; hence, the mean number of jobs of type k is estimated by $_k = \; q_k$.

(3) As the mix of jobs and their routing is unknown, no information is available about the exact load of machines m=1,...,M used by the manufacturing unit. However, routing of all types of jobs is known, as well as estimated processing times $E\{B_m\}$ on each machine.

(4) With respect to intervals between jobs to be released for processing, we consider two extreme cases:

(a) Jobs arrive in a Poisson stream with rate , or they are released with exponentially distributed intervals. According to lemma 1 and assumption (2), the release of jobs of type k can be described by a Poisson process with rate $_k$ (Open system).

(b) The number of jobs available is large enough to secure that, whenever a job is finished, a new one can be released for processing (Closed system).

We now have to examine whether these assumptions describe the situation of aggregate production planning appropriately. In particular, we have to define the concept of a "job" as a set of units to be processed jointly, and how stochastic fluctuations of processing times of jobs can be explained. Furthermore, we have to justify the approximation of processing times - and in the case of the open system - of interarrival times by an exponential distribution. Although more general queueing networks have been considered (cf. DISNEY/KIESSLER [1987]), existing results are not suitable for production planning models, as neither closed solutions nor efficient algorithms to calculate operating characteristics of the model are available. In order to do so, we have to distinguish between the following structures of the manufacturing process (cf. KISTNER/SWITALSKI [1988]):

(1) Single item manufacturing
(2) Small batch production
(3) Large batch production
(4) Mass production.

2.3.1 Single Item Manufacturing

Single item manufacturing is characterized by individual routing and different processing times for each job; furthermore, jobs to be processed are identical with orders placed by customers.

Nevertheless, it will be possible to identify classes of jobs with similar routing, and to consider types of jobs instead of individual jobs. Aggregate planning sets targets for these types of jobs; on the tactical level, it will usually be impossible to disaggregate these targets, and provide detailed production plans for single jobs: At the time of tactical production planning, informations about the set of orders and their processing characteristics are still incomplete; furthermore, scheduling on the tactical level is inoperable, as resulting planning models would be too complex.

From past experience, it will be possible, to forecast the average mix of job types and to estimate mean value of processing times of machines. These processing times will, however, be subject to considerable fluctuations because of the following reasons:

(1) The mix of job types is subject to stochastic variations.

(2) Job types are defined according to their routing; individual jobs assigned to a special type may differ considerably with respect to processing times.

(3) Further fluctuations of processing times are due to oscillating characteristics of the production process, such as breakdowns, and variations in the quality of material, or in the performance of manpower.

Usually, jobs are created by orders placed by different customers. Hence, it will be reasonable to assume, that interarrival times and processing times, as well as the routing of different orders will be mutually independent.

Consequently, due to the heterogeneity of orders, or jobs, respectively, the critical assumptions of queueing networks - independence and exponential distribution of interarrival and processing times - cannot be falsified ex ante. As more detailed informations about these data are not available, exponential distribution may be considered as a reasonable first approximation.

The choice of the appropriate queueing network model is dependent on the stock of jobs available at the time of tactical planning. If there is a sufficient stock of jobs to employ the manufacturing unit during the entire planning period, and if due times of many jobs exceed this period of time, it will always be possible to release another job for processing as soon as a job is completed. In this case, the production process may be described by a closed queueing network. Furthermore, the throughput - measured by the mean number of jobs passing the station 0 - is an indicator of the maximum capacity of the system.

If the number of jobs at hand is small, and further orders are expected to be placed during the planning period, and if most of these jobs have to be completed during this period of time, the behaviour of the system may be modelled by an open queueing network. Furthermore, this model may be applied to study the effects of variations in the workload and the mix of job types (measured by the probabilities q_k).

2.3.2 Batch Production

In batch production, several units (or orders) of the same product, which have to be processed on the same machines without interrupting set-up times and without set-up costs, will be combined to lots.

In hierarchical production planning, neither lot-sizes nor disaggregated production quantities, but only aggregated targets for product types are determined. Lot sizing as well as scheduling of lots will be accomplished by operative control. Aggregate production planning sets a target for the average load   of the manufacturing unit; from past experience, the mean share q_k of type k and the mean lot-size y_k can be estimated. From this, the mean number of lots of type k can be calculated:

$$\alpha^{(k)} = \frac{\alpha \cdot q_k}{y_k}$$

If orders of different sizes are placed by customers at random intervals, in the case of an open queueing network, the intervals between releases of lots of different types of products may be approximated by independent exponentially distributed random variables with rate $^{(k)}$. In the case of a closed queueing network, throughput is measured in numbers of lots, in order to get it in product units, the throughput has to be divided by the mean overall lot-size y.

According to the transport of units from one machine to the next in its routing, two types of production structures have to be considered:

(1) Closed production:

A complete lot will be delivered simultaneously to a machine; units will be processed by this machine, wait in an inventory for transport, and the entire lot is transported to the next machine, as soon as the last unit is completed. Processing of a batch can be started only if the entire lot is present at the machine. A given batch can be processed by only one machine at a time.

This type of transportation is typical for small batch manufacturing.

(2) Open production:

As soon as processing of a product (or small sub-lots) is completed at a machine, it is transported to the next in its routing; and waits there, until it can be processed at this machine. At a given instant, several units of a given batch may be processed by different machines.

This type of transportation is typical for large or medium sized batches.

As interpretation of jobs has to be different in both cases, we first consider the case of small batches, which is quite similar to the case of single item manufacturing; afterwards the case of large batches is analysed.

2.3.2.1 Small Batches

In small batch production, a lot in the network is at each period of time either processed by one of the machines, or it is waiting for service by a certain machine. As long as it is occupied by a lot, no other unit can be processed by this machine. Hence, a lot may be treated as a job following its path through the network in analogy to the case of single item manufacturing.

Furthermore, a large number of products and types with different paths through the network, with various lot-sizes, and with varying processing times have to be considered. Like in single item manufacturing, aggregate production planning, however, has no information about the actual product mix, lot sizes and processing times; only data about the average lot-sizes, mean processing times of the types, and their expected share of the total number of orders are known.

Following the argument of section 2.3.1, we can conclude, that the distribution of processing times of the lots may be approximated by an exponential distribution. The argument is strengthened by the following facts:

(1) A large number of heterogeneous products with different processing time has to be considered.

(2) Different lot sizes for different products; due to oscillating demand, sizes will be varying even for batches of the same product.

(3) Random variations of processing rate of the machines, due to the reasons stated above.

As actual product mix and lot sizes are unknown, aggregate production planning has to estimate the mean processing times $E\{B_k\} = 1/\beta_k$ for each machine from past experiences.

By the same argument used above, the assumption may be justified that the intervals between releases of successive batches of different products and their processing times are mutually independent.

Hence, for aggregate production planning, the assumptions of network queueing models - exponential distribution of processing times of all machines and exponentially distributed interarrival times in the case of an open network, as well as independence of these random variables - may be considered as reasonable approximations, if small batches of products instead of customer orders are considered.

2.3.2 Large Batches

In contrast to small batch production, in the case of large batches it is not possible to identify whole batches as jobs to be processed. As open production is predominant in this structure, a batch is not confined to one machine at a time; parts of the batch will be waiting or being processed at several machines. Due to set-up times and set-up costs caused by a switch from one batch to another, it will not be reasonable to interrupt processing of a batch and to handle another one, before all items are processed. Usually, there will be some universal machines which will be used by most types of products; hence, in large batch production, there will be only one batch in process at each period of time. Under these conditions, it will be possible, to neglect the actual flow of products through the network, and to treat the entire manufacturing unit as one single, aggregated server and batches of different products as jobs to be processed by the server. Service times of the aggregated server are equal to the completion times of batches, that is the time elapsing from the set-up of the system for a batch and the admission of the first item to the first machine in its routing up to the completion of the last item on the last machine.

If we consider that the product mix is sufficiently heterogeneous, that lot-sizes and processing times as well as set-up times at the start job are differing from product to product, and that the production rate of the machines is fluctuating, too, then it may be reasonable to assume that service times of the aggregated channel may be approximated by exponentially distributed random variables. If batches combine a large number of orders placed by different customers, the assumption that the time intervals between the release of successive batches are independent, exponentially distributed random variables may be justified.

Hence, in the case of large batch production, aggregate production planning may apply a single channel model with exponentially distributed interarrival times and exponentially distributed service times to describe the overall behaviour of the entire manufacturing unit. Furthermore, as closed solutions for single channel systems with Poisson input and independent, generally distributed service times are available, the assumption of exponentially distributed overall processing times may be dropped, if sufficient information about the distribution of the overall processing time, depending on the expected product mix are available.

2.3.3 Mass Production

In contrast to the cases of single item manufacturing and batch production, neither queueing networks nor traditional single server models of queueing theory may be used to estimate the capacities of production units disposable in aggregate production planning. This may be attributed to the following special features of mass production:

(1) In this case, an entire manufacturing unit is dedicated to each product. All items have to be processed by the same machines in the same sequence; instead of a queueing network, a series of servers has to be considered.

(2) There is just one more or less homogeneous product that has to be processed; hence it will not be justified to assume exponentially distributed processing times; in fact, they will slightly oscillate around a typical value. Production planning will try to secure a steady flow of items to be released for production, hence exponentially distributed interarrival times will be not typical for mass production. Under these conditions, there exists no closed solution for a series of servers, which may be used to describe the behaviour of the manufacturing unit.

(3) In mass production, there is neither a competition of orders for machine capacity nor the necessity of sequencing; at most, there will be a lot-sizing problem, if the production unit is not fully employed. Hence, the problem of coordination of tactical planning and operative control is not so important as it is in single item manufacturing and in batch production.

3. SEMI-CONTINUOUS QUEUEING MODELS

In the cases of single item manufacturing and small batch production, queueing networks are suitable to model the performance of manufacturing units embedded into a hierarchical production planning system. In the case of large batches, however, only the overall performance of the system, but not the characteristics of single machines can be studied using a single channel queueing model. Finally, traditional queueing models are not suited to describe the behaviour of manufacturing units with mass production, since it will not be possible to justify the assumption of exponentially distributed processing times of machines.

The cases of mass production and large batch production may be modelled, however, using semi-continuous queueing models.They will be appropriate to study the behaviour of manufacturing units, in particular, the overall capacity, capacity utilization of single machines, and bottle-necks.

A semi-continuous queueing system with servers in series or networks of servers is characterized by the following features:

(1) M servers - machines in the context of production planning - in series or a network of servers - process a continuous flow of orders with a constant rate.

(2) The rate of each server is subject to stochastic changes. The time intervals between two successive changes of the rate are independent, exponentially distributed random variables.

(3) Temporary differences between processing rates of successive servers are partially levelled out by finite buffers at each server. However, from time to time a server has to interrupt processing; it will either be blocked, because the buffer at the next server is full, or it will be starved, as its predecessor works at a slower rate and the buffer is empty.

Appropriate regeneration points can be identified; hence, the system may be analysed using Markov renewal techniques as described by SRINIVASAN [1974], SRINIVASAN/SUBRAMANIAN [1980].

A first model of this type is due to WIJNGAARD [1979], who considers a tandem systems with two states of the servers (on/off); a similar approach has been suggested by YAO/BUZACOTT [1986]. Recently, DE KOSTER [1989] discussed in detail a two-stage line with several processing rates, and generalized the model to the cases of n-stage lines and general layouts. As in closed queueing network models, the main results are the average throughput of the system, and the fractions of time, a server is starving, blocked or processing with a given rate.

This type of semi-continuous queueing network models is quite suitable for the analysis of manufacturing units for large batch or mass production. In both cases, oscillations in the production rate will be small compared with downtimes of the system due to breakdowns or lack of orders, and thus may be neglected. In both cases, breakdowns of a machine result in a production rate of zero, lack of orders may be modelled by setting the processing rate of the first machine equal to zero. Furthermore, it will be possible to consider large batch production with different types of goods by attributing individual processing rates to each type. In this case, a change of types results in switching from one rate to another.

From the point of view of mass production, the most critical assumptions, independence and exponential distribution of times between two successive changes of the processing rate of a machine, may be justified by the fact, that breakdowns occur randomly, and that orders for a single product, or different types of products are placed by different customers; furthermore lot sizes will vary randomly, and will result in processing times which may be approximated by exponentially distributed random variates

4. CONCLUSIONS

If servers are interpreted as machines or working stations, queueing networks describe well the structure of manufacturing units. There are, however, only a few applications of this type of models in traditional production planning, as the assumption of independent, exponentially distributed processing times and the goals of production planning, to smooth the flow of goods in process, and to achieve a well-balanced use of capacities have been considered to be in contradiction.

However, modern approaches of production planning, stressing hierarchical structures of the planning process, open new fields of application of queueing models. In order to set targets for manufacturing units, aggregate planning on a superordinate level relies on information about the behaviour of the manufacturing units, in particular about net capacities, which are in turn dependent on the targets to be set.

This information may be derived from a queueing network model of the manufacturing system. Single item manufacturing as well as small batch production may be described by traditional queueing network models; semi-continuous queueing networks are more suitable to model manufacturing systems for mass or large batch production. In both cases, the assumptions of independence and exponential distribution of intervals between critical events, can be justified by the fact, that orders are placed independently by a large number of customers, and that actual lot-sizing, routing, and scheduling of the manufacturing systems are unknown at the time of aggregate production planning.

REFERENCES

Bailey NTJ (1954) On Queuing Processes with Bulk Service. J. Roy. Stat.Soc. Ser. B 16:80-87

Barrer DY (1957) Queuing with Impatient Customers and Indifferent Clerks, OR 5: 644-649

Bruell StC, Balbo G (1980) Computational Algorithms for Closed Queueing Networks. Elsevier North Holland, New York

Burke PJ (1956) The Output of a Queuing System. OR 4:699-704

Buzacott, JA, Shantikumar JG (1985) On Approximate Models of Dynamic Job Shops. MS 31:870-887

Buzen JP (1973) Computational Algorithms for Closed Queueing Networks with Exponential Servers. CACM 16:527-531

Cobham A (1954) Priority Assignment in Waiting Line Problems. OR 2:70-76

de Koster MBM (1989) Capacity Orientated Analysis and Design of Production Systems. Springer, Berlin-Heidelberg

Denning PJ, Buzen JP (1978) The Operational Analysis of Queueing Networks. Computing Surveys 10:225-261

Disney RI, Kiessler, PC (1987) Traffic Processes in Queueing Networks, John Hopkins Univ. Press, Baltimore

Ferschl F (1961) Warteschlangen mit gruppiertem Input. Ufo 5:185-196

Ferschl F (1964) Zufallsabhängige Wirtschaftsprozesse. Grundlagen und Theorie der Wartesysteme. Physica, Würzburg Wien

Gaver DP (1959) Imbedded Markov Chain Analysis of Waiting Line Processes in Continuous Time. Ann.Math.Stat 30:698-720

Gaver DP (1962) A Waiting Line With Interrupted Services, Including Priorities. J.Roy.Stat.Soc.Ser. B24:73-90

Gnedenko BW, König D (1983/84) Handbuch der Bedienungstheorie, 2 Bde. Akademie-Verlag, Berlin (Ost)

Gordon WJ, Newell GF (1967) Closed Queuing Systems with Exponential Servers. OR 15: 254-265

Hax AC, Meal HC (1975) Hierarchical Integration of Production Planning and Scheduling. In: Geisler MA (Hrsg) Logistics, TIMS Studies in the Management Sciences. North Holland, Amsterdam: 53-69

Hunt GC (1956) Sequential Arrays of Waiting Lines. OR 4:674-683

Jackson JR (1957) Networks of Waiting Lines. OR 5:518-521

Jackson JR (1965) Jobshop-like Queueing Systems MS 10:131-142

Jaiswal, NK (1968) Priority Queues. Academic Press, New York

Keilson J, Kooharian A (1960), On Time Dependent Queuing Processes. Ann.Math.Stat. 31:104-112

Kendall DG (1951) Some Problems in the Theory of Queues. J.Roy. Stat.Soc. Ser. B 13:151-173

Kendall DG (1953) Stochastic Processes Occurring in the Theory of Queues and Their Analysis by the Method of Imbedded Markov Chain. Ann.Math.Stat. 24:338-354

Kistner K-P (1974) Wartesysteme mit Betriebsstörungen. Westdeutscher Verlag, Opladen

Kistner K-P (1987) Warteschlangentheorie. In: Gal T (Hrsg) Grundlagen des Operations Research, Bd. 3. Springer, Berlin Heidelberg New York: 253-289

Kistner K-P, Switalski M (1988) Produktionstypen und die Struktur des Produktionsprozesses. WISU 17 332:337

Kistner K-P, Steven/Switalski M (1990a) Warteschlangen-Netzwerke in der hierarchischen Produktionsplanung. OR Spektrum 12:89-101

Kistner K-P, Steven M (1990b) Warteschlangennetzwerke in hierarchischen Planungssystemen. Komplexe Systeme und Qualität, Sonderheft 55 der Wissenschaftlichen Zeitschrift der Hochschule für Verkehrswesen "Friedrich List" Dresden: 108-122

Lindley DV (1952) The Theory of Queues with a Single Server. Proc. Cambridge Phil.Soc 48:277-289

Little JDC (1961) A Proof of the Queueing Formula L = W. OR 9: 383-387

Morse P (1959) Queues, Inventories, and Maintenance. J. Wiley, New York

Palm C (1947) Arbetskraftnes Ferdelning Vid Baljaning av Automaskener Industidningen. Norden 75:75-80 und 119-123; Engl. Transl. The Distribution of Repairmen in Servicing Automatic Machines. J. Ind. Engin. 9 (1958):28-42.

Solberg JJ (1977) A Mathematical Model of Computing Manufacturing Systems. Proceedings of the Fourth International Conference on Production Research, Tokio:1265-1275

Srinivasan SK (1974) Stochastic Point Processes and Their Applications. Griffin, London

Srinivasan SK, Subramanian R (1980) Probabilistic Analysis of Redundant Systems. Springer, Berlin-Heidelberg

Suri R,Hildebrant RR (1984) Modelling Flexible Manufacturing Systems using Mean-Value Analysis. J of Manufacturing 3:27-38

Switalski M, Kistner K-P (1988) Produktionstypen und die Struktur des Produktionsprozesses. WISU 17:332-337

Tempelmeier H (1988) Kapazitätsplanung für flexible Fertigungssysteme. ZfB 58:963-980

Thiruvengadam K (1963) Queuing with Breakdowns. OR 11:62-71

White H und Christie LS (1958) Queuing with Preemptive Priorities or with Breakdown. OR 6:79-95

Wijngaard J (1979) The Effect of Interstage Buffer Storage on the Output of Two Unreliable Production Units in Series, with Different Production Rates, AIIE Transactions 11:1979

Yao DD, Buzacott JA (1986) Models of Flexible Manufacturing Systems with Limited Local Buffer, Intern J Prod Res 24:107-118

RELIABILITY ANALYSIS OF A COMPLEX SYSTEM USING BOOLEAN FUNCTION TECHNIQUE

YADAVALLI V.S.SARMA
DEPARTMENT OF STATISTICS, NATIONAL UNIVERSITY OF LESOTHO, LESOTHO

HOWARD P. HINES
ACTUARIAL DEPARTMENT, MUTUAL LIFE, JAMAICA

ABSTRACT

This paper deals with reliability analysis of a complex system consisting of two subsystems, connected in parallel. Each subsystem consists of a generator, one main switchboard and a given number of auxiliary switchboards. Two models of this system are studied. The reliability of the system with its imposed conditions is then calculated using the Boolean function technique and combinatorial methods. Some special cases are studied and a numerical example is provided.

1. INTRODUCTION

Today there exists a large number of systems whose performance are evaluated solely on the basis of experience but not with the aid of reliability calculations. This is usually the case when the system under consideration consists of a large number of units.

Fratta and Montanari (1973) and Gupta and Sharma (1986) have successfully applied the Boolean function technique to the reliability analysis of various complex systems which may be taken to represent power plants.

The present work is a reliability analysis of the complex system shown in figure 1.The system consists of two subsystems: subsystem 1 and subsystem 2, connected in parallel. Each subsystem consists of a generator, one main switchboard and a given number of auxiliary switchboards. Any two auxiliary switchboards belonging to the same subsystem will be taken to be identical with respect to reliability considerations. These components are connected by cables which are assumed to be one hundred percent reliable.

Two models of the system described will be studied. In the case of the first model the system requires only one route to be functional for the overall system to be functional, whereas in the case of the second, a given number of functional routes in each of the subsystems are required to be functional. The reliability of the system with its imposed conditions is then calculated using the Boolean function technique and combinatorial methods. In the special case of Weibull and exponential distributions the mean time to system failure is also computed. Numerical examples are also provided.

2 MODEL 1

2.1 Assumptions.

1. The state of each component and of the whole system is either good (operating) or bad (failed).
2. The states of all components are statistically independent.
3. The reliabilities of all components are known beforehand.
4. There is no standby or switched redundancy.
5. The pdf's of the failure times for all components are all arbitrary.
6. There is no repair facility.
7. The system fails if and only if (i) both generators fail or (ii) at least one main switchboard or the auxiliary switchboards in all the routes have failed. Otherwise the system is functional.
8. All cables are one hundred percent reliable.
9. Any two auxiliary switchboards belonging to the same subsystem have the same reliability.

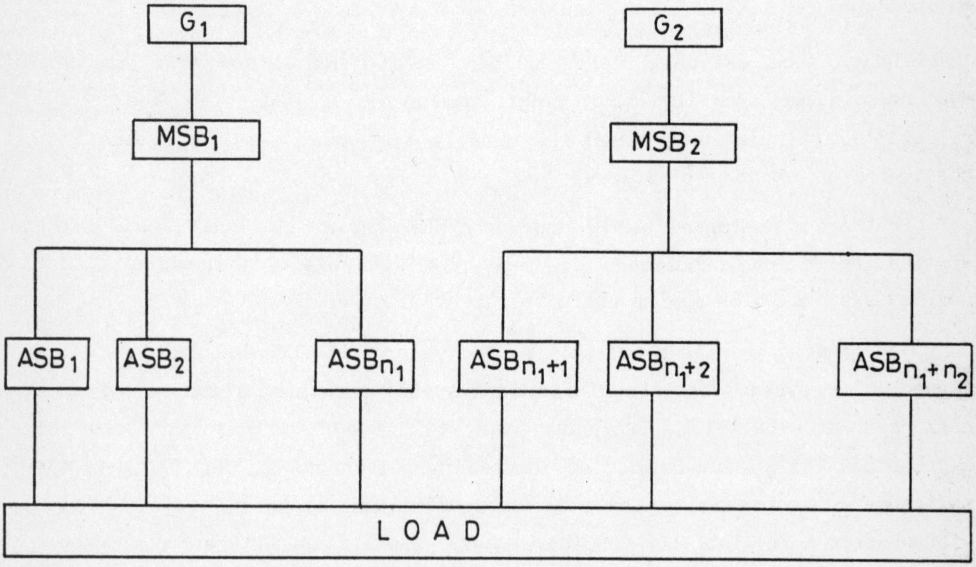

G_1 , G_2 , : generators 1 and 2.

MSB_1 , MSB_2 : main-switchboards 1 and 2.

ASB_1 , ASB_2 , ... $ASB_{n_1+n_2}$:auxiliary switchboards.

FIGURE 1

2.2 Notation

1. $X_1(t), X_3(t)$: States of generators G_1 and G_2 at time $t>0$ with reliability functions $R_1(t)$ and $R_3(t)$ respectively.
2. $X_2(t), X_4(t)$: States of MSB_1 and MSB_2 at time $t>0$ with reliability functions $R_2(t)$ and $R_4(t)$.
3. $X_5(t), X_6(t), \ldots, X_{n_1+n_2+4}(t)$: States of $ASB_1, ASB_2, \ldots, ASB_{n_1+n_2}$ at time $t>0$.

$R_5(t)$ is the reliability function of each of $ASB_1, ASB_2, \ldots, ASB_{n_1}$

$R_6(t)$ is the reliability function of each of ASB_{n_1+1} $ASB_{n_1+2} \ldots$ $ASB_{n_1+n_2}$

4. $X_k(t) = \begin{cases} 0 \text{ for component in failed state} \\ 1 \text{ for component in operable state} \end{cases}$
5. $Pr(f=1)$: The probability of the successful operation of the system where f is the Boolean function describing the system behaviour.
6. $$\begin{vmatrix} X_{11}(.) & X_{12}(.) & \ldots X_{1n}(.) \\ X_{21}(.) & X_{22}(.) & \ldots X_{2n}(.) \\ \vdots & \vdots & \vdots \\ X_{m1}(.) & X_{m2}(.) & \ldots X_{mn}(.) \end{vmatrix} = \bigvee_{i=1}^{m} \bigwedge_{j=1}^{n} X_{ij}(.)$$

2.3 Reliability Analysis

Using the Boolean function technique, the conditions for the successful operation of the complex system in terms of a logical matrix are expressed as follows

$$f(X_1, X_2, \ldots, X_{n_1+n_2+4}) = \begin{vmatrix} X_1 & X_2 & X_5 \\ X_1 & X_2 & X_6 \\ \vdots & \vdots & \vdots \\ X_1 & X_2 & X_{n_1+4} \\ X_3 & X_4 & X_{n_1+5} \\ \vdots & \vdots & \vdots \\ X_3 & X_4 & X_{n_1+n_2+4} \end{vmatrix}$$

where

$$X_i = X_i(t), \quad i = 1, 2, \ldots, n_1 + n_2 + 4.$$

Let

$$\Pi_1 = X_1 X_2 X_5, \qquad \Pi_2 = X_1 X_2 X_6$$
$$\vdots$$

$$\vdots$$

$$H_{n_1} = X_1 X_2 X_{n_1+4}; \qquad H_{n_1+1} = X_3 X_4 X_{n_1+5} \tag{1}$$

$$\vdots$$

$$H_{n_1+n_2} = X_3 X_4 X_{n_1+n_2+4}$$

then

$$f(X_1, X_2, \ldots, X_{n_1+n_2+4}) = \begin{vmatrix} H_1 \\ H_2 \\ \vdots \\ H_{n_1+n_2} \end{vmatrix} \tag{2}$$

Let the probability of successful operation of the system up to time t, i.e. the reliability of the power supply, be denoted by $R_s(t)$, then

$$R_s(t) = \Pr[f=1] \tag{3}$$

By using the inclusion-exclusion formula, i.e.

$$\Pr\left\{ \bigvee_{i=1}^{d} H_i \right\} = \sum_i \Pr(H_i) - \sum_i \sum_j \Pr(H_i \wedge H_j) + \ldots +$$

$$(-1)^{d-1} \Pr(H_1 \wedge H_2 \wedge \ldots \wedge H_d), \text{ for some } d \in Z^+ \tag{4}$$

and considering the cases

(i) $n_1 = n_2 = 1$, (ii) $n_1 = n_2 = n > 1$, (iii) $n_2 > n_1 \geq 1$, (iv) $n_1 > n_2 \geq 1$

we obtain the following :

CASE (i) : $n_1 = n_2 = 1$

$$R_s(t) = R_1(t)R_2(t)R_5(t) + R_3(t)R_4(t)R_6(t) - R_1(t)R_2(t)R_3(t)R_4(t)R_5(t)R_6(t)$$

CASE (ii) : $n_1 = n_2 = n$

$$R_s(t) = \Pr\left\{ \bigvee_{i=1}^{2n} H_i = 1 \right\}$$

$$= \sum_{0<k<n+1} (-1)^{k-1} \begin{bmatrix} n \\ k \end{bmatrix} R_1(t)R_2(t)R_5^k(t) + \sum_{0<k<n+1} (-1)^{k-1} \begin{bmatrix} n \\ k \end{bmatrix} R_3(t)R_4(t)R_6^k(t)$$

$$+ \sum_{0<k<n+2} \sum_{0<s<r} (-1)^{k-1} \begin{bmatrix} n \\ s \end{bmatrix} \begin{bmatrix} n \\ k-s \end{bmatrix} R_1(t)\, R_2(t)\, R_3(t)\, R_4(t)\, R_5^s(t)\, R_6^{k-s}(t)$$

$$+ \sum_{n+1<k<2n+1} \; \sum_{k-(n+1)<s<n+1} (-1)^{k-1} \begin{bmatrix} n \\ s \end{bmatrix} \begin{bmatrix} n \\ k-s \end{bmatrix}$$

$$\times \; R_1(t)\, R_2(t)\, R_3(t)\, R_4(t)\, R_5^s(t)\, R_6^{k-s}(t) \qquad (5)$$

CASE (iii) : $n_2 > n_1 \geq 1$

$$R_s(t) = \Pr \left\{ \bigvee_{i=1}^{n_1+n_2} H_i = 1 \right\}$$

$$= \sum_{0<k<n_1+1} (-1)^{k-1} \begin{bmatrix} n_1 \\ k \end{bmatrix} R_1(t)\, R_2(t)\, R_5^k(t)$$

$$+ \sum_{0<k<n_2+1} (-1)^{k-1} \begin{bmatrix} n_2 \\ k \end{bmatrix} R_3(t)\, R_4(t)\, R_6^k(t)$$

$$+ \sum_{1<k<n_1+2} \; \sum_{0<s<n_1+1} (-1)^{k-1} \begin{bmatrix} n_1 \\ s \end{bmatrix} \begin{bmatrix} n_2 \\ k-s \end{bmatrix}$$

$$\times \; R_1(t)\, R_2(t)\, R_3(t)\, R_4(t)\, R_5^s(t)\, R_6^{k-s}(t)$$

$$+ \sum_{n_2+1<k<n_1+n_2+1} \; \sum_{k-(n_2+1)<s<n_1+1} (-1)^{k-1} \begin{bmatrix} n_1 \\ s \end{bmatrix} \begin{bmatrix} n_2 \\ k-s \end{bmatrix}$$

$$\times \; R_1(t)\, R_2(t)\, R_3(t)\, R_4(t)\, R_5^s(t)\, R_6^{k-s}(t) \qquad (6)$$

CASE (iv): $n_1 > n_2 \geq 1$

$$R_s(t) = \Pr \left\{ \bigvee_{i=1}^{n_1+n_2} H_i = 1 \right\}$$

$$= \sum_{0<k<n_1+1} (-1)^{k-1} \begin{bmatrix} n_1 \\ k \end{bmatrix} R_1(t)\, R_2(t)\, R_5^k(t)$$

$$+ \sum_{0<k<n_2+1} (-1)^{k-1} \begin{bmatrix} n_2 \\ k \end{bmatrix} R_3(t)\, R_4(t)\, R_6^k(t)$$

$$+ \sum_{1<k<n_2+2} \; \sum_{0<s<n_2+1} (-1)^{k-1} \begin{bmatrix} n_2 \\ s \end{bmatrix} \begin{bmatrix} n_1 \\ k-s \end{bmatrix}$$

$$\times \; R_1(t)\, R_2(t)\, R_3(t)\, R_4(t)\, R_6^s(t)\, R_6^{k-s}(t)$$

$$+ \sum_{n_2+1<k<n_1+2} \sum_{0<s<_2+1} (-1)^{k-1} \begin{bmatrix} n_2 \\ s \end{bmatrix} \begin{bmatrix} n_1 \\ k-s \end{bmatrix}$$

$$\times \; R_1(t)\, R_2(t)\, R_3(t)\, R_4(t)\, R_6^s(t)\, R_5^{k-s}(t)$$

$$+ \sum_{n_1+1<k<n_1+n_2+1} \sum_{k-(n_1+1)<s<n_2+1} (-1)^{k-1} \begin{bmatrix} n_2 \\ s \end{bmatrix} \begin{bmatrix} n_1 \\ k-s \end{bmatrix}$$

$$\times \; R_1(t)\, R_2(t)\, R_3(t)\, R_4(t)\, R_5^s(t)\, R_6^{k-s}(t) \qquad (7)$$

3. MODEL 2

3.1 Introduction

In the previous section the case which required only one successful route from a generator to the critical consumer was considered.

In most of the literature so far this has also been the assumption. However, in practice, situations arise in which the power supplied via one route is inadequate and therefore more than one successful routes is required for overall system reliability.

3.2 Assumptions

The assumptions for model 2 are identical with those of model 1 except the assumption 7 which now reads as follows:

7'. The system operates successfully if and only if at least r_1 routes of subsystem 1 and r_2 routes of subsystem 2 are functional.

(The notation for this section is identical with that of the previous sections).

3.3 Reliability Analysis

Using the Boolean function technique the conditions for capability for the successful operation of the complex system may be represented as :

$$f(X_1, X_2, X_{n_1+n_2+4}) = \bigvee_{j=r_1}^{n_1} \bigvee_{i=r_2}^{n_2} \left[(H_{s_1} \wedge H_{s_2} \wedge \wedge H_{s_j} \wedge (H_{l_1} \wedge H_{l_2} \wedge \ldots \wedge H_{l_i}) \right] \qquad (8)$$

where

$$S_k \in Z^+; \quad 1 \le S_k \le n_1; \quad l_k \in Z^+; \quad n_1+1 \le l_k \le n_1+n_2, \quad \text{for all } k \in Z^+ ;$$

$S_\alpha \ne S_\beta$, $l_\alpha \ne l_\beta$ for $\alpha \ne \beta$ and the H's are given by (1)

We evaluate $R_s(t)$ defined by (3), by considering the following possibilities:

(i) $r_1, r_2 > 0$, (ii) $r_1 = 0$ and $r_2 > 0$, (iii) $r_1 > 0$ and $r_2 = 0$

CASE (i): $r_1, r_2 > 0$

$$R_s(t) = \Pr\left\{ \bigvee_{j=r_1}^{n_1} \bigvee_{i=r_2}^{n_2} \left[(H_{s_1} \wedge H_{s_2} \wedge \ldots \wedge H_{s_j}) \wedge (H_{l_1} \wedge H_{l_2} \wedge \ldots \wedge H_{l_i}) \right] = 1 \right\}$$

$$= \sum_{r_1-1<m_1<n_1+1} \; \sum_{r_2-1<m_2<n_2+1} \left\{ \left[\binom{n_1}{m_1} R_1(t)\, R_2(t)\, R_5^{m_1}(t)\, (1-R_5(t))^{n_1-m_1} \right] \right.$$

$$\left. \times \left[\binom{n_2}{m_2} R_3(t)\, R_4(t)\, R_6^{m_2}(t)\, (1-R_6(t)))^{n_2-m_2} \right] \right\} \qquad (9)$$

CASE (ii) : $r_2 > 0$ and $r_1 = 0$

$$R_s(t) = \sum_{0<m_1<n_1+1} \; \sum_{r_2-1<m_2<n_2+1} \left\{ \left[\binom{n_1}{m_1} R_1(t)\, R_2(t)\, R_5^{m_1}(t)\, (1-R_5(t))^{n_1-m_1} \right] \right.$$

$$\left. \times \left[\binom{n_2}{m_2} R_3(t) R_4(t) R_6^{m_2}(t) (1-R_6(t)))^{n_2-m_2} \right] \right\}$$

$$+ \sum_{r_2-1<k<n_2+1} \left\{ \left[\binom{n_2}{k} R_3(t)\, R_4(t)\, R_6^{m_1}(t)\, (1-R_6(t))^{n_2-k} \right] \right.$$

$$\left. \times \left[1 - R_1(t)\, R_2(t)\, (1-(1-R_5(t))^{n_1} \right] \right\} \qquad (10)$$

CASE (iii) : $r_1 > 0$ and $r_2 = 0$

$$R_s(t) = \sum_{r_1-1<m_1<n_1+1} \; \sum_{0<m_2<n_2+1} \left\{ \left[\binom{n_1}{m_1} R_1(t) R_2(t)\, R_5^{m_1}(t)(1-R_5(t))^{n_1-m_1} \right] \right.$$

$$\left. \times \left[\binom{n_2}{m_2} R_3(t)\, R_4(t)\, R_6^{m_2}(t)(1-R_6(t))^{n_2-m_2} \right] \right\}$$

$$+ \sum_{r_1-1<k<n_1+1} \left\{ \left[\binom{n_1}{k} R_1(t) R_2(t) R_5^{k}(t)(1-Rs(t))^{n_1-k} \right] \right.$$

$$\left. \times \left[1 - R_3(t) R_4(t)(1-(1-R_6(t))^{n_2} \right] \right\} \qquad (11)$$

4. NUMERICAL ILLUSTRATION FOR MODEL 1

The mean time to the system failure, MTSF, is calculated for the special case $n_1 = n_2 = 3$, when the failure times follows (i) an exponential distribution and (ii) a Weibull distribution with $\alpha_i = 2$ for $i=1,2,\dots,6$. The results obtained for the various parametric values are shown in Tables 1 and 2. As is to be expected, as a particular parametric value decreases, the MTSF increases in both the cases.

TABLE 1

$\lambda_1=1/100$, $\lambda_3=1/100$, $\lambda_4=1/50$, $\lambda_5=1/10$, $\lambda_6=1/10$

λ_2	MTSF for exponential times ($\alpha = 1$)	MTSF for Weibull times ($\alpha_i = 2$)
1/10	14.79544	13.04709
1/20	16.28040	14.07742
1/30	17.15512	14.49795
1/40	17.71554	14.67952
1/50	18.10255	14.77093
1/60	18.38513	14.82273
1/70	18.60027	14.85471
1/80	18.76943	14.87579
1/90	18.90589	14.89038
1/100	19.01825	14.90087

TABLE 2

$\lambda_1=1/100$, $\lambda_2=1/10$, $\lambda_3=1/100$, $\lambda_4=1/50$, $\lambda_5=1/10$

λ_6	MTSF for exponential times ($\alpha = 1$)	MTSF for Weibull times ($\alpha_i = 2$)
1/10	14.79544	13.04709
1/20	21.25276	22.75618
1/30	25.19369	29.72797
1/40	27.70361	34.02816
1/50	29.37281	36.50908
1/60	30.52836	37.90048
1/70	31.35465	38.67738
1/80	31.96174	39.11558
1/90	32.41813	39.36720
1/100	32.76807	39.51497

REFERENCES

1. Fratta. L. and Montanari, U.G., A Boolean Algebra method for computing the terminal reliability in a communication network, IEEE Trans.Circuit Theory, 1973, Vol. C-20, pp 203-211.

2. Gupta. R.P. and Sharma. R.K., Reliability and MTTF analysis of power plant consisting of three generators by Boolean function technique, Microelectronics and Reliability, 1986, Vol.26, pp 641-645.

THE SECOND MOMENT OF THE MARKOVIAN REWARD PROCESS

S. Subba Rao
Department of Information Systems and Operations Management
The University of Toledo
Toledo, OH 43606

ABSTRACT

In this note an m-state aperiodic irreducible Markov Chain is considered with an associated reward matrix. Expressions for the second moment of S_n, the accumulated reward in n transitions, have been obtained using elementary methods. These provide an alternative computational method compared to methods using the eigen value structure. Asmyptotic forms are also considered.

INTRODUCTION

Consider an m-state, aperiodic Markov Chain with a transition probability matrix $\underline{\underline{P}} = \{p_{ij}\}$. With every transition i j, there is an associated reward r_{ij}, and let $\underline{\underline{R}}\{\pi_{ij}\}$ be the reward matrix. The accumulated reward in n transitions given an initial state i is of great interest in applications in inventory systems, repair and maintenance, and waiting line problems. The reward process has been studied by Howard (1971, Chapter 13), where results for the expected value of the accumulated reward in n transitions have been obtained. The asymptotic forms have also been studied. However, the second moment of the accumulated reward process has not been obtained in a computationally convenient form. The availability of the second moment would be very useful in many of the applications.

MOMENTS OF S_n

Let $S_n = S_n^{(i)}$ represent the accumulated reward in n transitions given an initial state i. $S_n^{(i)}$ can also be interpreted as the reward that will be accumulated in the next n transitions given the present state i. In this note we obtain an expression for the second moment of $S_n^{(i)}$ using elementary arguments. It is hoped that the present result is computationally convenient compared to a result of Phatarfod (1965) based on eigen value structures.

If Z_r (r = 1, 2,...) is the reward at transition r, the accumulated reward at transition n is

$$S_n^{(i)} = S_{n-1}^{(i)} + Z_n = \left\{ \sum_{r=1}^{n} Z_r \mid X_0 = i \right\}$$

where X_0 is the initial state of the Markov Chain. To simplify notation, the superscript (i) will be omitted from now on except wherever necessary. Below are given three theorems relating to $E(S_n)$ and $E(S_n^2)$, the first and second moments of the reward process.

Theorem 1

Let $V_i(n) = E(S_n)$ represent the expected accumulated reward in the next n transitions, given the current state i. Then the vector $\underline{V}(n) = [V_1(n), V_2(n), \ldots V_m(n)]^T$ is given by

$$\underline{\underline{V}}(n) = \left[\underline{\underline{I}} + \sum_{k=1}^{n-1} \underline{\underline{P}}^k \right] \underline{Q} + \underline{\underline{P}}^n \underline{V}(0) \qquad (1)$$

where

$$\underline{Q} = \left[Q_1, Q_2, \ldots Q_m\right]^T,$$

$$Q_i = \sum_j p_{ij} r_{ij} = E(Z_1 \mid X_0 = i)$$

Asymptotically

$$\underline{V}(n) \sim n\underline{\underline{\pi}}\, \underline{Q} + \underline{v} \qquad (2)$$

where

$$\underline{\pi} = \begin{bmatrix} \pi_1 & \pi_2 & \ldots & \pi_m \\ \pi_1 & \pi_2 & \ldots & \pi_m \\ \pi_1 & \pi_2 & \ldots & \pi_m \end{bmatrix},$$

$\{\pi_i\}$ being the equilibrium probabilities of the Markov Chain,

$$\underline{v} = (v_2, v_2, \ldots v_m)^T,$$

$$v_i = \sum_j \sum_{n=0}^{\infty} \left[p_{ij}^{(n)} - \pi_j\right] Q_j.$$

It is to be noted that $p_{ij}^{(n)}$ is the n step transition probability of the Markov Chain. Further, in obtaining (2), the assumption $\underline{V}(0)=\underline{0}$ has been made.

Proof. The proof of this theorem is well known, based on a recursive relation. The reader is referred to Bhat (1972, p. 187-189).

$E(S_n^2)$

The main results of this note concern the second moment of S_n. These are detailed below .

We observe that S_n can be written as

$$S_n = Z_1 + \sum_{r=2}^{n} Z_r \tag{3}$$

Let the first transition take the Markov Chain from state i to state j. Then $Z_1 = r_{ij}$ with probability p_{ij}. Also, we note that $\sum_{r=2}^{n} Z_r = S_{n-1}(j)$ with probability p_{ij}. Squaring (3), taking expectations, and denoting $E(S_n^2)$ by $V_i^{(2)}(n)$, we have

$$V_i^{(2)}(n) = E(Z_1^2) + \sum_j p_{ij} V_j^{(2)}(n-1) + 2E\left\{Z_1 . S_{n-1}^{(j)}\right\} \tag{4}$$

Now

$$E(Z_1^2) = \sum_j p_{ij} r_{ij}^2 = Q_i^{(2)}, \quad \text{(say)} \tag{5}$$

$$E\left\{ Z_1 S_{n-1}^{(j)} \right\} = E(Z_1) \sum_j p_{ij} E\left\{ S_{n-1}^{(j)} \,\middle|\, Z_1 \right\} = Q_i \sum_j p_{ij} V_j(n-1) \tag{6}$$

Let

$$\underline{Q}^{(2)} = \left[Q_1^{(2)}, Q_2^{(2)}, \ldots, Q_m^{(2)} \right]^T,$$

$$\underline{V}^{(2)}(n) = \left[V_1^{(2)}(n), V_2^{(2)}(n), \ldots, V_m^{(2)}(n) \right]^T,$$

and

$$(\underline{\underline{A}})_d = \text{diag}\,(a_1, a_2, \ldots a_m).$$

Then, writing (4) in matrix form, we have the following result.

$$\underline{V}^{(2)}(n) = \underline{Q}^{(2)} + \underline{\underline{P}}\, \underline{V}^{(2)}(n-1) + 2(\underline{\underline{P}}\, \underline{V}(n-1)_d)\, \underline{Q} \tag{7}$$

Without loss of generality we can assume $\underline{V}(0) = \underline{0}$. Substituting for $\underline{V}(n-1)$ for (1) in (7) above and using $\underline{V}(0) = \underline{0}$, we get

$$\underline{V}^{(2)}(n) = \underline{Q}^{(2)} + \underline{\underline{P}}\, \underline{V}^{(2)}(n-1) + 2 \sum_{k=1}^{n-1} \left[\underline{\underline{P}}^k \underline{Q}\right]_d \underline{Q} \tag{8}$$

Putting n = 1, 2, ... successively in (8) and letting $\underline{V}^{(2)}(0) = \underline{0}$, we have the following :

Theorem 2

The vector $\underline{V}^{(2)}(n)$ of the second moments of S_n is given by

$$\underline{V}^{(2)}(n) = \left[\underline{\underline{I}} + \sum_{k=1}^{n-1} \underline{\underline{P}}^k\right] \underline{Q}^{(2)} + 2 \sum_{k=1}^{n-1} \underline{\underline{P}}^{n-k-1} \sum_{l=1}^{k} \left[\underline{\underline{P}}^k \underline{Q}\right]_d \underline{Q} \tag{9}$$

The above theorem provides a computational method for the second moment of S_n. For large n, an asymptotic form and a lower bound for the second moment is developed below. This is similar in structure to (2), the asymptotic form for the first moment.

For the ergodic Markov Chain considered here, the n-step transition probabilities $p_{ij}^{(n)}$ are such that

$$p_{ij}^{(n)} = \pi_j + e_{ij}^{(n)}$$

where π_j are the equilibrium probabilities of the chain, and $|e_{ij}^{(n)}|<cr^n$, (c>0, 0<r<1). (See Theorem 5.2.2. of Bhat (1972).) The n-step transition probability matrix can be written as

$$\underline{\underline{P}}^n = \underline{\underline{\pi}} + \underline{\underline{\eta}}^{(n)}, \tag{10}$$

where $\underline{\eta}^{(n)} = \{e_{ij}^{(n)}\}$, the matrix with elements $e_{ij}^{(n)}$. Following the same line of argument as in Bhat (1972), p. 188-189, we can show that

$$\left[\underline{\underline{I}} + \sum_{=1}^{n} \underline{\underline{P}}^k\right] \underline{Q}^{(2)} = n\underline{\underline{\pi}}\, \underline{Q}^{(2)} + \underline{v}^{(2)} \tag{11}$$

where

$$\underline{v}^{(2)} = \left[v_1^{(2)},\ v_2^{(2)},\ \ldots\ v_m^{(2)}\right]^T,$$

$$v_i^{(2)} = \sum_j \sum_{n=0}^{\infty} e_{ij}^{(n)}\, Q_j^{(2)}.$$

If we retain terms involving n, then

$$\left[\underline{\underline{I}} + \sum_{k=1}^{n} \underline{\underline{P}}^{k} \right] \underline{Q}^{(2)} \sim n\underline{\underline{\pi}}\, \underline{Q}^{(2)}$$

$$= n \sum_{j} \pi_1 Q_i^{(2)} \cdot \underline{1} \qquad (12)$$

where $\underline{1} = (1, 1, \ldots 1)^T$.

Next let us consider the second term on the right side of (9). Writing $\underline{\underline{P}}^n = \underline{\underline{\pi}} + \underline{\underline{\eta}}^{(n)}$ as before, we have

$$\sum_{k=1}^{n-1} \underline{\underline{P}}^{n-k-1} \sum_{l=1}^{k} \left(\underline{\underline{P}}^{l}\, \underline{Q} \right)_d = \sum_{k=1}^{n-1} \left(\underline{\underline{\pi}} + \underline{\underline{\eta}}^{(n-k-1)} \right) \cdot \sum_{l=1}^{k} \left(\left[\underline{\underline{\pi}} + \underline{\underline{\eta}}^{(1)} \right] \underline{Q} \right)_d$$

$$= \sum_{k=1}^{n-1} \left(\underline{\underline{\pi}} + \underline{\underline{\eta}}^{(n-k-1)} \right) \sum_{l=1}^{k} \left[(\underline{\underline{\pi}}\underline{Q})_d + (\underline{\underline{\eta}}^{(1)}\underline{Q})_d \right]$$

$$= \sum_{k=1}^{n-1} k \left(\underline{\underline{\pi}} + \underline{\underline{\eta}}^{(n-k-1)} \right) (\underline{\underline{\pi}}\, \underline{Q})_d + \sum_{k=1}^{n-1} \left(\underline{\underline{\pi}} + \underline{\underline{\eta}}^{(n-k-1)} \right) \sum_{l=1}^{k} \left[(\underline{\underline{\pi}}\, \underline{Q})_d + (\underline{\underline{\eta}}^{(1)} \underline{Q})_d \right]$$

In the above expression only the first term will have coefficients involving n. Noting that $(\underline{\underline{\pi}}\, \underline{Q})_d = \left(\sum_i \pi_i Q_i \right) \underline{\underline{I}}$, the first term can be approximated by

$$\frac{n(n-1)}{2} (\underline{\underline{\pi}} + \underline{\underline{U}}) \cdot \sum_i \pi_i Q_i \qquad (13)$$

where $\underline{\underline{U}}$ is the Unit Matrix. It is to be noted that the elements of $\underline{\underline{\eta}}$ are replaced by their maximum values in arriving at this approximation. Thus the second term on the right side of (9) is approximated by

$$n(n-1) \left(\sum_i \pi_i Q_i \right) (\underline{\underline{\pi}} + \underline{\underline{U}})\underline{Q} = n(n-1) \left[\left(\sum_i \pi_i Q_i \right)^2 + \sum_i Q_i \right] \cdot \underline{1} \qquad (14)$$

Now, combining (12) and (14) we can state the following result of Theorem 3.

Theorem 3

For large n, the vector $\underline{V}^{(2)}(n)$ of second moments of S_n is

$$\underline{V}^{(2)}(n) \sim n \sum_i \pi_i Q_i^{(2)}\, \underline{1} + n(n-1) \left[\left(\sum_i \pi_i Q_i \right)^2 + \sum_i Q_i \right] \underline{1}$$

The asymptotic variance can be computed using the above result and the result of theorem 1. A lower bound for $\underline{V}^{(2)}(n)$ can be obtained as follows.

If $r_{ij} \geq 0$, (this is not a serious restriction), then $Q_i \geq 0$. It can be easily verified that $\left[\underline{\underline{P}}^k \underline{Q} \right]_d \geq \underline{\underline{0}}$ since $\underline{\underline{P}}^k \geq \underline{\underline{0}}$. Hence the second term on the right side of (9) is non-negative. It then follows from (9) that

$$\underline{V}^{(2)}(n) \geq \left[\underline{\underline{I}} + \sum_{k=1}^{n-1} \underline{\underline{P}}^k \right] \underline{Q}^{(2)} \tag{15}$$

which provides a lower bound.

Remark 1. We have made certain assumptions like $V(0)_i = 0$ These can be removed but the results will be complicated.

Remark 2. The expression for the second moment of S_n can be given in terms of the eigen values of the matrix generating function $\underline{\underline{P}}(t) = \left\{ p_{ij} \; e^{tr_{ij}} \right\}$. In a different context, Phatarfod (1965) has obtained such an expression. If $\lambda_i(t)$, (i= 1, 2, ... m) are the m distinct eigen values of $\underline{\underline{P}}(t)$, then he shows that

$$E(S_n^2) = n\, \lambda_1''(0) + n(n-1)\left[\lambda_1'(0)\right]^2 + 2\left[\frac{d}{dt}\, \underline{t}'_1(t)\right]_{t=0} (\underline{\underline{P}}^n - \underline{\underline{I}}) \left[\frac{d}{dt}\, \underline{s}_1(t)\right]_{t=0}$$

where $\underline{t}'_1(t)$ and $\underline{s}_1(t)$ are the row and column vectors respectively corresponding to $\lambda_1(t)$. It can now be seen that the result of theorem 2 is an alternative method for the computation of the second moment of S_n.

REFERENCES

Bhat, U.N., (1972) Elements of Applied Stochastic Processes. John Wiley and Sons, Inc., New York.

Howard, R.A., (1971) Dynamic Probabilistic Systems. Vol.II: Semi Markov and Decision Processes. John Wiley and Sons, Inc., New York.

Phatarfod, R.M., (1965) Sequential Analysis of Dependent Observations-I, Biometrika, 52, 157-165.

Correlation Functions in Reliability Theory

R. Subramanian
Indian Institute of Technology
Madras 600 036, India

N. Ravichandran
Indian Institute of Management
Ahmedabad 380 015, India

ABSTRACT

The multivariate point process induced by the stochastic behaviour of a two-unit warm standby redundant repairable system is studied. Expressions for the product densities of the events corresponding to the entry into each of the states and the interval reliability are obtained. The reliability and availability are deduced as special cases.

1. INTRODUCTION

Two-unit standby redundant repairable systems have been studied in the past extensively [Osaki and Nakagawa (1976), Ashok Kumar and Agarwal (1980), Srinivasan and Subramanian (1980), Birolini (1985), Yearout *et. al.* (1986), Ravichandran (1990)]. In most of the attempts the central quantity of interest is the availability and/or the reliability of the system. However, these problems often give rise to some interesting stochastic processes which are important by themselves and have not been studied so far except in the stationary case [Srinivasan and Subramanian (1977)]. These processes are essentially non-Markov and do not necessarily fall under the semi-Markov type. In this contribution we study the multivariate stochastic point process induced by the given reliability problem. The layout of this paper is as follows. In section 2 the model is described and section 3 deals with interval reliability [Barlow and Proschan (1965)]. Section 4 is concerned with the correlation structure of the events induced by the reliability problem and obtains expressions for the first and second moments of the number of events of the various types.

2. ASSUMPTIONS AND NOTATION

1. The system consists of two units, which are identical and statistically independent. Either unit performs the system operation satisfactorily.
2. At t = 0, a new unit is switched online and another unit is kept as a warm standby. This initial condition will be denoted by E.
3. The lifetime of a unit while online is a random variable with pdf $f_1(.)$.
4. A unit while in standby state has a constant failure rate λ ; for convenience its pdf is denoted by $f_2(.)$.
5. The repair time of a unit is a random variable with pdf g(.).
6. Switch is perfect and switchover is instantaneous.
7. Each unit is 'new' after repair.

* denotes the convolution operation

$$C(t) = \int_0^t c(u)\, du , \quad c(.) \text{ any function}$$

$\bar{c}(t) = 1-c(t)$

$c^{(n)}(t)$ = n fold convolution of c(t) with itself over (0,t)

$\delta(.)$ = Dirac delta function

3. SOME PRELIMINARY RESULTS

Let Z(t) denote the state of the system at time t, representing the number of failed units at time t. Then $\{Z(t), t\geq 0\}$ is a discrete valued continuous time parameter stochastic process with state space $S = \{0,1,2\}$. The points of discontinuity of the process $\{Z(t)\}$ are the epochs corresponding to a failure or repair completion of the units. The function Z(.) has a negative jump when a repair completion takes place and a positive jump when a failure occurs, the magnitude of the jump being always unity. Thus we identify the following events in this study:

E_i : Event that the process $\{Z(t)\}$ enters state i, (i=0,1,2).

Our main aim is to study the above point events. We note that the event E which is given to have occurred at t = 0 cannot occur in the sequel. Further not all the events E_i are regenerative. The events E_0 and E_2 are always non-regenerative. The event E_1 is regenerative if it occurs in the following ways :

i. Entry into state 1 from state 2

ii. Entry into state 1 from state 0, due to the failure of the online unit.

It is not regenerative if it corresponds to the entry into state 1 from state 0, due to the failure of the standby unit. A regenerative E_1 event will be denoted by $\bar{E}_1$.

Let $\{t_i\}$ be the epochs at which the system enters the various states. Then it is clear that the $\{t_i\}$ form a stochastic process. To study this process it is convenient to use the following random variables.

X_t - random variable denoting the age of the online unit at time t, if a unit happens to be operating online at t.

Y_t - random variable denoting the elapsed repair time of a unit at time t, if a unit happens to be undergoing repair at time t.

To start with we study the stochastic behaviour of the standby unit during a failure-free operation period of the online unit. In this interval the standby unit alternates between the operable (s) and repair (r) states successively. If S(t) denotes the state of the standby at any time t, then $\{S(t), t\geq 0\}$ is an alternating renewal process. This renewal process can be characterised by the following functions.

$$P_{is}(t) = \Pr \{ S(t) = s \mid S(-0) \neq S(0) = i \}$$

$$P_{ir}(t) = \Pr \{ S(t) = r \mid S(-0) \neq S(0) = i \}$$

$$P_{ir}(t,y) = \lim_{\Delta\to 0} \frac{1}{\Delta} \Pr \{S(t) = r;\ y < Y_t \leq y + \Delta \mid S(-0) \neq S(0) = i\},\quad i = r,s.$$

We have by simple probability arguments

$$P_{ss}(t) = \bar{F}_2(t) + \gamma(t) * \bar{F}_2(t) \tag{1}$$

$$P_{rs}(t) = g(t) * P_{ss}(t) \tag{2}$$

$$P_{sr}(t,y) = f_2(t-y)\, \bar{G}(y) + [f_2(t-y) * \gamma(t-y)]\bar{G}(y) \tag{3}$$

$$P_{rr}(t,y) = \bar{G}(y)\, \delta(t-y) + \gamma(t-y)\, \bar{G}(y) \tag{4}$$

$$P_{rr}(t) = \int_0^t P_{rr}(t,y)\, dy = \bar{G}(t) + \gamma(t) * \bar{G}(t) \tag{5}$$

$$P_{sr}(t) = \int_0^t P_{sr}(t,y)\, dy = f_2(t) * P_{rr}(t) \tag{6}$$

where

$$\gamma(t) = \sum_{n=1}^{\infty} (f_2 * g)^{(n)}(t) \tag{7}$$

Sometimes we also need these state probabilities given that the standby unit is under repair initially and the elapsed repair time is y.

$$P_{ri}(t|y) = \Pr \{ S(t) = i \mid S(0) = r\ ;\ Y_0 = y \},\qquad i = s,r$$

The expressions for $P_{rs}(t|y)$ and $P_{rr}(t|y)$ are given below.

$$P_{rr}(t|y) = \frac{\bar{G}(t+y)}{\bar{G}(y)} + \frac{g(t+y)}{\bar{G}(y)} * P_{sr}(t) \tag{8}$$

$$P_{rs}(t|y) = \int_0^t \frac{g(v+y)}{\bar{G}(y)} P_{ss}(t-v)\, dv \tag{9}$$

In addition we will also need the entry probabilities into the states s and r:

$$\pi_{ij}(t) = \Pr\{\, S(t) = j \neq S(t-0) \mid S(-0) \neq S(0) = i \,\}, \quad i,j = s,r$$

$$\pi_{rj}(t|y) = \Pr\{\, S(t) = j \neq S(t-0) \mid S(0) = r\, ;\, Y_0 = y \,\},\ j = s,r.$$

These probabilities are given by

$$\pi_{ss}(t) = \pi_{rr}(t) = \gamma(t) \tag{10}$$

$$\pi_{rs}(t) = g(t) + g(t) * \gamma(t) \tag{11}$$

$$\pi_{sr}(t) = f_2(t) + f_2(t) * \gamma(t) \tag{12}$$

$$\pi_{rs}(t|y) = \frac{g(t+y)}{\bar{G}(y)} + \frac{g(t+y)}{\bar{G}(y)} * \pi_{ss}(t) \tag{13}$$

$$\pi_{rr}(t|y) = \frac{g(t+y)}{\bar{G}(y)} * \pi_{sr}(t) \tag{14}$$

We also require in our analysis the following functions which characterise the time interval between the E event and the next $\bar{E}_1$ event and the time interval between two successive $\bar{E}_1$ events. Let $N_1(t)$ denote the number of $\bar{E}_1$ events in $(0,t)$ and $f_{ij}(t)$ be defined as follows :

$$f_{01}(t) = \lim_{\Delta\to 0} \frac{1}{\Delta} \Pr\{N_1(t+\Delta) - N_1(t) = 1;\ N_1(t) = 0 \mid E \text{ at } t = 0\}$$

$$f_{11}(t) = \lim_{\Delta\to 0} \frac{1}{\Delta} \Pr\{N_1(t+\Delta) - N_1(t) = 1;\ N_1(t) = 0 \mid \bar{E}_1 \text{ at } t = 0\}$$

We have

$$f_{01}(t) = f_1(t)P_{ss}(t) + \int_0^t f_1(u)\, du \int_0^t P_{rs}(u,x) \frac{g(t-u+x)}{\bar{G}(x)}\, dx \tag{15}$$

$$f_{11}(t) = f_1(t)P_{rs}(t) + \int_0^t f_1(u)\, du \int_0^t P_{rr}(u,x) \frac{g(t-u+x)}{\bar{G}(x)}\, dx \tag{16}$$

where $P_{ij}(t)$ and $P_{ij}(u,x)$ are given by (1)-(4).

The expression for $f_{01}(t)$ is obtained by the following considerations. For the first $\bar{E}_1$ event to occur in $(t,t+\Delta)$ given that E has occurred at t=0, it is necessary that the unit switched online at t=0 should fail before t. At the epoch of failure of the online unit, the standby unit can be either operable or under repair. In the former case $\bar{E}_1$ occurs and in the latter case $\bar{E}_1$ occurs at the epoch of repair completion of the standby. The expression for $f_{11}(t)$ is obtained in a similar fashion.

We note that the epochs $\{t_i\}$ corresponding to the occurrences of $\bar{E}_1$ events form a renewal process with pdf $f_{11}(.)$. The corresponding renewal density is given by

$$\alpha(t) = \sum_{n=1}^{\infty} f_{11}^{(n)}(t) \tag{17}$$

The function $f_{11}(.)$ represents the pdf of the random variable denoting the time interval between two successive $\bar{E}_1$ events, with a possible system failure in between. However in our analysis we also require the pdf of the random variable representing the time interval between two successive $\bar{E}_1$ events, there being no system failure in between. Denoting this pdf by $\tilde{f}_{11}(.)$ we have

$$\tilde{f}_{11}(t) = f_1(t)\, P_{rs}(t) \tag{18}$$

The corresponding renewal density is given by

$$\beta(t) = \sum_{n=1}^{\infty} \tilde{f}_{11}^{(n)}(t) \tag{19}$$

Suppose that initially the system enters state 0 and x denotes the age of the online unit at that epoch, then the pdf of the time to the next $\bar{E}_1$ is given by

$$f_{01}(t|x) = \frac{f_1(x+t)}{\bar{F}_1(x)} P_{ss}(t) + \int_0^t \frac{f_1(z+x)}{\bar{F}_1(x)} \int_0^z P_{sr}(z,u) \frac{g(t-z+u)}{\bar{G}(u)}\, du \tag{20}$$

The renewal density of the $\bar{E}_1$ events given E at t=0 is

$$\psi(t) = f_{01}(t) + f_{01}(t) * \alpha(t) \tag{21}$$

If the initial condition is not E but entry into state 0, the age of the online unit at that epoch being x then the corresponding renewal density is given by

$$\psi(t|x) = f_{01}(t|x) + f_{01}(t|x) * \alpha(t) \tag{22}$$

4. OPERATING CHARACTERISTICS OF THE SYSTEM

We next proceed to obtain the interval reliability $R(t,\tau)$, an important measure of performance of the system. This is defined as the probability that the system is available at time t and is up in (t, t+τ). It is clear that the reliability R(τ) and availability A(t) of the system are obtained by setting t = 0 and τ = 0 respectively in R(t, τ). Even though this is a very useful measure it has not received the attention it deserves. There are only very few investigations which deal with interval reliability. [Subramanian and Ravichandran (1979, 1981a, 1981b) and Franken and Streller (1980)].

To obtain an expression for R(t, τ) we proceed as follows. Since the system is available at t, it should be found in one of the states 0 or 1 and for the future description of the system we need the age of the online unit, when the system is in state 0 and in addition the elapsed repair time when it is in state 1. Thus we are led to define the following functions:

$$B_0(t, x) = \lim_{\Delta \to 0} \frac{1}{\Delta} \Pr\{ Z(t) = 0;\ x < X_t \le x + \Delta \mid E \text{ at } t=0 \}$$

$$B_1(t, x, y) = \lim_{\Delta, \Delta' \to 0} \frac{1}{\Delta\Delta'} \Pr\{Z(t) = 1;\ x < X_t \le x + \Delta;\ y < Y_t \le y + \Delta' \mid E \text{ at } t=0\},\ y \le x$$

$$B_2(t,y) = \lim_{\Delta \to 0} \frac{1}{\Delta} \Pr\{ Z(t) = 2;\ y < Y_t \le y+\Delta \mid E \text{ at } t=0 \}$$

$$C_2(t,y) = \lim_{\Delta, \Delta' \to 0} \frac{1}{\Delta\Delta'} \Pr\{ E_2 \text{ occurs in } (t, t+\Delta);\ y < Y_t \le y+\Delta' \mid E \text{ at } t=0\}$$

By probabilistic arguments we get

$$B_0(t, x) = \bar{F}_1(x)\, \delta(t-x)\, P_{rs}(x) + \psi(t-x)\, \bar{F}_1(x)\, P_{rs}(x) \tag{23}$$

$$B_1(t,x,y) = \bar{F}_1(x)\, \delta(t-x)\, P_{Sr}(t,y) + \psi(t-x)\, \bar{F}_1(x)\, P_{rr}(x, y) \tag{24}$$

$$C_2(t,y) = f_1(t)\, P_{sr}(t,y) + \int_0^{t-y} \psi(u)\, f_1(t-u)\, P_{rr}(t-u,y)\, du \tag{25}$$

$$B_2(t,y) = C_2(t,y)\, \frac{\bar{G}(t-y)}{\bar{G}(y)} \tag{26}$$

To compute the interval reliability of the system R(t,τ) we make use of the following additional functions which represent the reliability of the system in (t, t+τ) under certain specified initial conditions.

$$R_0(t, \tau \mid x) = \Pr\{\text{System up in } (t, t+\tau) \mid Z(t) = 0;\ X_t = x\}$$

$$R_1(t, \tau \mid x, y) = \Pr\{\text{System up in } (t, t+\tau) \mid Z(t) = 1;\ X_t=x;\ Y_t=y\}$$

Using these functions and by considering the following mutually exclusive and exhaustive cases that

(i) the system is found in state 0 at time t or

(ii) the system is found in state 1 at time t,

we get the interval reliability $R(t, \tau)$ as

$$R(t,\tau) = \int_0^t R_0(t,\tau|x)\, B_0(t,x)\, dx + \int_0^t dx \int_0^t R_1(t,\tau|x,y)\, B_1(t,x,y)\, dy \tag{27}$$

where

$$R_0(t,\tau|x) = \frac{\bar{F}_1(x+\tau)}{\bar{F}_1(x)} + \int_0^\tau \frac{f_1(u+x)}{\bar{F}_1(x)} P_{ss}(u)\, R_1(t-u)\, du \tag{28}$$

$$R_1(t,\tau|x,y) = \frac{\bar{F}_1(x+\tau)}{\bar{F}_1(x)} + \int_0^\tau \frac{f_1(u+x)}{\bar{F}_1(x)} P_{rs}(u|y)\, R_1(\tau-u)\, du \tag{29}$$

and

$$R_1(t) = \bar{F}_1(t) + \beta(t) * \bar{F}_1(t) \tag{30}$$

On simplification equation (27) becomes

$$R(t,\tau) = F_1(t+\tau) + \int_0^t \psi(t-v)\, R_1(\tau+v)\, dv + \int_t^{t+\tau} f_1(u)\, R_1(t+\tau-u)\, du \tag{31}$$

where $\psi(t)$ is given by (21).

The stationary value of $R(t,\tau)$ as $t \longrightarrow \infty$, known as the limiting interval reliability and denoted by $R(\tau)$, is given by

$$R(\tau) = \frac{1}{\mu} \int_0^\infty R_1(\tau+v)\, dv \tag{32}$$

where

$$\mu = \int_0^\infty t\psi(t)\, dt.$$

We can obtain the availability $A(t)$ of the system conditional upon the E event at $t = 0$ in terms of $B_0(t, x)$ and $B_1(t,x,y)$:

$$A(t) = \int_0^t B_0(t,x)\, dx + \int_0^t dx \int_0^x B_1(t,x,y)\, dy \tag{33}$$

On simplification we get

$$A(t) = \bar{F}_1(t) + \psi(t) * \bar{F}_1(t) \tag{34}$$

As an easy consequence of the key renewal theorem we obtain the steady state availability of the system from equation (34)

$$A_\infty = \int_0^\infty \bar{F}_1(t)\, dt \Big/ \int_0^\infty t\, f_{11}(t)\, dt. \tag{35}$$

5. POINT PROCESSES GENERATED BY THE EVENTS E_i

We next study the point processes induced by the stochastic behaviour of the system. We have already observed that some of the E_i events are regenerative while others are not. In view of this, the point events are best studied by product densities [Ramakrishnan (1958), Srinivasan (1974)]. We first derive expressions for the product densities of the first two orders and use them to get the first and second moments of the number of E_i events in (0,t). The product densities in this case are defined as follows:

$$h_i(t) = \lim_{\Delta\to 0} \frac{1}{\Delta} \Pr\left\{ N_{E_i}(t+\Delta) - N_{E_i}(t) = 1 \mid E \text{ at } 0 \right\}$$

$$h_{ij}(t_1,t_2) = \lim_{\Delta,\Delta'\to 0} \frac{1}{\Delta\Delta'} \Pr\left\{ N_{E_i}(t_1+\Delta) - N_{E_i}(t_1) = 1; \right.$$

$$\left. N_{E_i}(t_2+\Delta') - N_{E_i}(t_2) = 1 \mid E \text{ at } t=0 \right\},\ t_2 > t_1.$$

To obtain these functions we introduce $C_0(t,x)$ and $C_1(t,x)$ where

$$C_0(t,x) = \lim_{\Delta,\Delta'\to 0} \frac{1}{\Delta\Delta'} \Pr\{ E_0 \text{ occurs in } (t,\ t+\Delta);\ x<X_t\le x+\Delta' \mid E \text{ at } t = 0 \}$$

$$C_1(t,x) = \lim_{\Delta,\Delta'\to 0} \frac{1}{\Delta\Delta'} \Pr\{ E_1 \text{ occurs in } (t,\ t+\Delta);\ x<X_t\le x+\Delta' \mid E \text{ at } t = 0 \}$$

The expression for $C_0(t,x)$ is obtained by considering the following mutually exclusive and exhaustive possibilities:

(i) The online unit does not fail in (0, t)

(ii) The online unit fails in (u, u+du), u < t.

$$C_0(t,x) = \bar{F}_1(x)\,\delta(t-x)\,\pi_{ss}(t) + \psi(t-x)\,\bar{F}_1(x)\,\pi_{rs}(x) \qquad (36)$$

Similarly we obtain the expression for $C_1(t,y)$:

$$C_1(t,y) = \psi(t)\,\delta(y) + \bar{F}_1(y)\,\delta(t-y)\,\pi_{sr}(y) + \psi(t-y)\,\bar{F}_1(y)\,\pi_{rr}(y) \qquad (37)$$

We next concentrate on the first order product densities. The product density corresponding to the E_0 events is given by

$$h_0(t) = \int_0^t C_0(t,x)\,dx = \bar{F}_1(t)\,\pi_{ss}(t) \int_0^t \psi(t-x)\,\bar{F}_1(x)\,\pi_{rs}(x)\,dx \qquad (38)$$

The product density corresponding to the E_1 events is given by

$$h_1(t) = \int_0^t C_1(t,x)\,dx$$

$$= \psi(t) + \bar{F}_1(t)\,\pi_{sr}(t) + \int_0^t \psi(t-x)\,\bar{F}_1(x)\,\pi_{sr}(x)\,dx \qquad (39)$$

while the product density corresponding to the E_2 events is given by

$$h_2(t) = \int_0^t C_2(t, y)\, dy$$

$$= f_1(t)\, P_{sr}(t) + \int_0^t \psi(t-u)\, f_1(u)\, P_{rr}(u)\, du \tag{40}$$

where $C_2(t,y)$ is given by the equation (25).

Next we proceed to obtain the second order product densities of the E_i events. Since the E_i events are not regenerative in general, for the future description of the process we need the age x of the online unit in the case of the events E_0 or E_1. When x=0, E_1 corresponds to $\bar{E}_1$.

The second order product density is given by

$$h_{ij}(t_1,t_2) = \int_0^{t_1} C_i(t_1,x)\, dx \left[\int_0^{t_2-t_1} C_{ij}(t_2-t_1, y|x)\, dy \right], \quad i,j=0,1,2 \tag{41}$$

where $C_i(t,x)$ are given by equations (36),(37) and (25).$C_{ij}(t,y|x)$ are defined as below:

$$C_{ij}(t,y|x) = \lim_{\Delta,\Delta'\to 0} \frac{1}{\Delta\Delta'} \Pr\{ E_j \text{ occurs in } (t, t+\Delta);$$

$$y<X_t\le y+\Delta' | E_i \text{ at } t=0;\ X_0 = x \},\ i,\ j = 0,\ 1.$$

$$C_{2j}(t,y|x) = \lim_{\Delta,\Delta'\to 0} \frac{1}{\Delta\Delta'} \Pr\{ E_j \text{ occurs in } (t, t+\Delta);$$

$$y<X_t\le y+\Delta' | E_2 \text{ at } t=0;\ Y_0 = x \},\ j = 0,\ 1.$$

$$C_{i2}(t,y|x) = \lim_{\Delta,\Delta'\to 0} \frac{1}{\Delta\Delta'} \Pr\{ E_2 \text{ occurs in } (t, t+\Delta);$$

$$y<Y_t\le y+\Delta' | E_i \text{ at } t=0;\ X_0 = x \},\ i = 0,\ 1.$$

$$C_{22}(t,y|x) = \lim_{\Delta,\Delta'\to 0} \frac{1}{\Delta\Delta'} \Pr\{ E_2 \text{ occurs in } (t, t+\Delta);$$

$$y<Y_t\le y+\Delta' | E_2 \text{ at } t=0;\ Y_0 = x \}$$

The expressions for $C_{ij}(t,y|x)$ can readily be obtained:

$$C_{00}(t,y|x) = \frac{\bar{F}_1(x+t)}{\bar{F}_1(x)}\, \delta(y-(x+t))\, \pi_{ss}(t) + \psi(t-y|x)\, \bar{F}_1(y)\, \pi_{rs}(y) \tag{42}$$

$$C_{01}(t,y|x) = \frac{\bar{F}_1(x+t)}{\bar{F}_1(x)}\, \delta(y-(x+t))\, \pi_{sr}(t) + \psi(t|x)\, \delta(y)$$

$$+ \psi(t-y|x)\, \bar{F}_1(y)\, \pi_{rr}(y) \tag{43}$$

$$C_{02}(t,y|x) = \frac{f_1(x+t)}{\bar{F}_1(x)} P_{sr}(t,y)\ \delta(y-(x+t) + \psi(t-y|x)\ f_1(y)\ \bar{G}(y) \qquad (44)$$

$$C_{10}(t,y|x) = \frac{\bar{F}_1(x+t)}{\bar{F}_1(x)} \delta(y-(x+t))\ \pi_{rs}(t) + \psi(t-y|x)\ \bar{F}_1(y)\ \pi_{rs}(y) \qquad (45)$$

$$C_{11}(t,y|x) = \frac{\bar{F}_1(x+t)}{\bar{F}_1(x)} \delta(y-(x+t))\ \pi_{rr}(t) + \psi(t-y|x)\ \bar{F}_1(y)\ \pi_{rr}(y) + \psi(t|x)\ \delta(y) \qquad (46)$$

$$C_{12}(t,y|x) = \frac{f_1(x+t)}{\bar{F}_1(x)} P_{rr}(t,y)\ \delta(y-(t+x)) + \psi(t-y|x)\ f_1(y)\ \bar{G}(y) \qquad (47)$$

$$C_{20}(t,y|x) = \int_0^{t-y} \frac{g(x+u)}{\bar{G}(x)} C_{10}(t-u,y|0)\ du \qquad (48)$$

$$C_{21}(t,y|x) = \frac{g(x+t)}{\bar{G}(x)} \delta(y) + \int_0^{t-y} \frac{g(x+u)}{\bar{G}(x)} C_{11}(t-u,y|0)\ du \qquad (49)$$

$$C_{22}(t,y|x) = \int_0^{t-y} \frac{g(x+u)}{\bar{G}(x)} C_{12}(t-u,y|0)\ du \qquad (50)$$

$\psi(t|x)$ is given by equation (22).

Thus the second order product densities are completely determined. The first and second moments of the number of events in the interval $(0,t)$ are obtained by integrating the first and second order product densities. For example, if N_d is the random variable representing the number of system downs then

$$E\,[N_d] = \int_0^t h_2(u)\ du \qquad (51)$$

and

$$E\,[N_d^2] = \int_0^t dt_2 \int_0^t h_{22}(t_1,t_2)\ dt_1 + \int_0^t h_2(u)\ du. \qquad (52)$$

6. Concluding Remarks

In this contribution we have analysed the multivariate stochastic point process induced by the given reliability problem and obtained expressions for the first and second order product densities leading to the the moments of the number of events of various types in an arbitrary time interval. We have also obtained the interval reliability.

REFERENCES

[1] Ashok Kumar and M. Agarwal (1980) A review of standby systems, IEEE Trans. Rel., R 29, 290-294

[2] R.E. Barlow and F. Proschan (1965) Mathematical Theory of Reliability, John Wiley.

[3] A. Birolini (1985) On the use of Stochastic Processes in Modeling Reliability Problems, Lecture Notes in Economics and Mathematical Systems, No. 252, Springer Verlag.

[4] P. Franken and A. Streller (1980) Reliability Analysis of Complex Repairable Systems by Means of Markov Point Processes, J. Appl. Prob., 17, 154-167.

[5] S. Osaki and T. Nakagawa (1976) Bibliography for Reliability and Availability of Stochastic Systems, IEEE Trans. Rel., R-25, 284-287.

[6] A. RamaKrishnan (1959) Probability and Stochastic Process in Handbuch der Physik, Vol. 3, Springer Verlag.

[7] N. Ravichandran (1990) Stochastic Methods in Reliability Theory, Wiley Eastern Limited.

[8] S.K. Srinivasan (1974) Stochastic Point Processes and Their Applications, Griffin.

[9] S.K. Srinivasan and R. Subramanian (1977) Availability of 2-unit Systems with one Repair Facility, J. Math. Phy. Sci., 11, 331-350.

[10] S.K. Srinivasan and R. Subramanian (1980) Probabilistic Analysis of Redundant Systems, Lecture Notes in Economics and Mathematical Systems, No. 175, Springer Verlag.

[11] R. Subramanian and N. Ravichandran (1979) Interval Reliability of a 2-unit Redundant System. IEEE Trans. on Rel., R28, 84.

[12] R. Subramanian and N. Ravichandran (1981 a) Interval Reliability of an n-unit System with single Repair Facility, IEEE Trans. Rel., R-30, 30-31.

[13] R. Subramanian and N. Ravichandran (1981 b) Interval Reliability of a Redundant System, J. Math. Phy. Sci., 15, 409-421.

[14] R.D. Yearout, P. Reddy and D.L. Grosh (1986) Standby Redundancy in Reliability - A Review, IEEE Trans. Rel., R-35, 285-292.

Vol. 307: T.K. Dijkstra (Ed.), On Model Uncertainty and its Statistical Implications. VII, 138 pages. 1988.

Vol. 308: J.R. Daduna, A. Wren (Eds.), Computer-Aided Transit Scheduling. VIII, 339 pages. 1988.

Vol. 309: G. Ricci, K. Velupillai (Eds.), Growth Cycles and Multisectoral Economics: the Goodwin Tradition. III, 126 pages. 1988.

Vol. 310: J. Kacprzyk, M. Fedrizzi (Eds.), Combining Fuzzy Imprecision with Probabilistic Uncertainty in Decision Making. IX, 399 pages. 1988.

Vol. 311: R. Färe, Fundamentals of Production Theory. IX, 163 pages. 1988.

Vol. 312: J. Krishnakumar, Estimation of Simultaneous Equation Models with Error Components Structure. X, 357 pages. 1988.

Vol. 313: W. Jammernegg, Sequential Binary Investment Decisions. VI, 156 pages. 1988.

Vol. 314: R. Tietz, W. Albers, R. Selten (Eds.), Bounded Rational Behavior in Experimental Games and Markets. VI, 368 pages. 1988.

Vol. 315: I. Orishimo, G.J.D. Hewings, P. Nijkamp (Eds), Information Technology: Social and Spatial Perspectives. Proceedings 1986. VI, 268 pages. 1988.

Vol. 316: R.L. Basmann, D.J. Slottje, K. Hayes, J.D. Johnson, D.J. Molina, The Generalized Fechner-Thurstone Direct Utility Function and Some of its Uses. VIII, 159 pages. 1988.

Vol. 317: L. Bianco, A. La Bella (Eds.), Freight Transport Planning and Logistics. Proceedings, 1987. X, 568 pages. 1988.

Vol. 318: T. Doup, Simplicial Algorithms on the Simplotope. VIII, 262 pages. 1988.

Vol. 319: D.T. Luc, Theory of Vector Optimization. VIII, 173 pages. 1989.

Vol. 320: D. van der Wijst, Financial Structure in Small Business. VII, 181 pages. 1989.

Vol. 321: M. Di Matteo, R.M. Goodwin, A. Vercelli (Eds.), Technological and Social Factors in Long Term Fluctuations. Proceedings. IX, 442 pages. 1989.

Vol. 322: T. Kollintzas (Ed.), The Rational Expectations Equilibrium Inventory Model. XI, 269 pages. 1989.

Vol. 323: M.B.M. de Koster, Capacity Oriented Analysis and Design of Production Systems. XII, 245 pages. 1989.

Vol. 324: I.M. Bomze, B.M. Pötscher, Game Theoretical Foundations of Evolutionary Stability. VI, 145 pages. 1989.

Vol. 325: P. Ferri, E. Greenberg, The Labor Market and Business Cycle Theories. X, 183 pages. 1989.

Vol. 326: Ch. Sauer, Alternative Theories of Output, Unemployment, and Inflation in Germany: 1960–1985. XIII, 206 pages. 1989.

Vol. 327: M. Tawada, Production Structure and International Trade. V, 132 pages. 1989.

Vol. 328: W. Güth, B. Kalkofen, Unique Solutions for Strategic Games. VII, 200 pages. 1989.

Vol. 329: G. Tillmann, Equity, Incentives, and Taxation. VI, 132 pages. 1989.

Vol. 330: P.M. Kort, Optimal Dynamic Investment Policies of a Value Maximizing Firm. VII, 185 pages. 1989.

Vol. 331: A. Lewandowski, A.P. Wierzbicki (Eds.), Aspiration Based Decision Support Systems. X, 400 pages. 1989.

Vol. 332: T.R. Gulledge, Jr., L.A. Litteral (Eds.), Cost Analysis Applications of Economics and Operations Research. Proceedings. VII, 422 pages. 1989.

Vol. 333: N. Dellaert, Production to Order. VII, 158 pages. 1989.

Vol. 334: H.-W. Lorenz, Nonlinear Dynamical Economics and Chaotic Motion. XI, 248 pages. 1989.

Vol. 335: A.G. Lockett, G. Islei (Eds.), Improving Decision Making in Organisations. Proceedings. IX, 606 pages. 1989.

Vol. 336: T. Puu, Nonlinear Economic Dynamics. VII, 119 pages. 1989.

Vol. 337: A. Lewandowski, I. Stanchev (Eds.), Methodology and Software for Interactive Decision Support. VIII, 309 pages. 1989.

Vol. 338: J.K. Ho, R.P. Sundarraj, DECOMP: an Implementation of Dantzig-Wolfe Decomposition for Linear Programming. VI, 206 pages.

Vol. 339: J. Terceiro Lomba, Estimation of Dynamic Econometric Models with Errors in Variables. VIII, 116 pages. 1990.

Vol. 340: T. Vasko, R. Ayres, L. Fontvieille (Eds.), Life Cycles and Long Waves. XIV, 293 pages. 1990.

Vol. 341: G.R. Uhlich, Descriptive Theories of Bargaining. IX, 165 pages. 1990.

Vol. 342: K. Okuguchi, F. Szidarovszky, The Theory of Oligopoly with Multi-Product Firms. V, 167 pages. 1990.

Vol. 343: C. Chiarella, The Elements of a Nonlinear Theory of Economic Dynamics. IX, 149 pages. 1990.

Vol. 344: K. Neumann, Stochastic Project Networks. XI, 237 pages. 1990.

Vol. 345: A. Cambini, E. Castagnoli, L. Martein, P Mazzoleni, S. Schaible (Eds.), Generalized Convexity and Fractional Programming with Economic Applications. Proceedings, 1988. VII, 361 pages. 1990.

Vol. 346: R. von Randow (Ed.), Integer Programming and Related Areas. A Classified Bibliography 1984–1987. XIII, 514 pages. 1990.

Vol. 347: D. Ríos Insua, Sensitivity Analysis in Multi-objective Decision Making. XI, 193 pages. 1990.

Vol. 348: H. Störmer, Binary Functions and their Applications. VIII, 151 pages. 1990.

Vol. 349: G.A. Pfann, Dynamic Modelling of Stochastic Demand for Manufacturing Employment. VI, 158 pages. 1990.

Vol. 350: W.-B. Zhang, Economic Dynamics. X, 232 pages. 1990.

Vol. 351: A. Lewandowski, V. Volkovich (Eds.), Multiobjective Problems of Mathematical Programming. Proceedings, 1988. VII, 315 pages. 1991.

Vol. 352: O. van Hilten, Optimal Firm Behaviour in the Context of Technological Progress and a Business Cycle. XII, 229 pages. 1991.

Vol. 353: G. Riccil (Ed.), Declslon Processes In Economics. Proceedings, 1989. III, 209 pages 1991.

Vol. 354: M. Ivaldi, A Structural Analysis of Expectation Formation. XII, 230 pages. 1991.

Vol. 355: M. Salomon. Deterministic Lotsizlng Models for Production Planning. VII, 158 pages. 1991.

Vol. 356: P. Korhonen, A. Lewandowski, J . Wallenius (Eds.), Multlple Crltena Decision Supporl. Proceedings, 1989. XII, 393 pages. 1991.

Vol. 358: P. Knottnerus, Linear Models with Correlaled Disturbances. VIII, 196 pages. 1991.

Vol. 359: E. de Jong, Exchange Rate Determination and Optımal Economlc Policy Under Various Exchange Rate Regimes. VII, 270 pages. 1991.

Vol. 360: P. Stalder, Regime Translations, Spillovers and Buffer Stocks. VI, 193 pages . 1991.

Vol. 361: C. F. Daganzo, Logistics Systems Analysis. X, 321 pages. 1991.

Vol. 362: F. Gehreis, Essays In Macroeconomics of an Open Economy. VII, 183 pages. 1991.

Vol. 363: C. Puppe, Distorted Probabilities and Choice under Risk. VIII, 100 pages . 1991

Vol. 364: B. Horvath, Are Policy Variables Exogenous? XII, 162 pages. 1991.

Vol. 365: G. A Heuer, U. Leopold-Wildburger. Balanced Silverman Games on General Discrete Sets. V, 140 pages. 1991.

Vol. 366: J. Gruber (Ed.), Econometric Decision Models. Proceedings, 1989. VIII, 636 pages. 1991.

Vol. 367: M. Grauer, D. B. Pressmar (Eds.), Parallel Computing and Mathematical Optimization. Proceedings. V, 208 pages. 1991.

Vol. 368: M. Fedrizzi, J. Kacprzyk, M. Roubens (Eds.), Interactive Fuzzy Optimization. VII, 216 pages. 1991.

Vol. 369: R. Koblo, The Visible Hand. VIII, 131 pages.1991.

Vol. 370: M. J. Beckmann, M. N. Gopalan, R. Subramanian (Eds.), Stochastic Processes and their Applications. Proceedings, 1990. XLI, 292 pages. 1991.